Automatic Control Systems in Biomedical Engineering

J. Fernández de Cañete · C. Galindo
J. Barbancho · A. Luque

Automatic Control Systems in Biomedical Engineering

An Interactive Educational Approach

Springer

J. Fernández de Cañete
System Engineering and Automation
University of Málaga
Málaga
Spain

J. Barbancho
Department of Electronic Technology
University of Seville
Sevilla
Spain

C. Galindo
System Engineering and Automation
University of Málaga
Málaga
Spain

A. Luque
Engineering Design Department
University of Seville
Sevilla
Spain

ISBN 978-3-319-75716-2 ISBN 978-3-319-75717-9 (eBook)
https://doi.org/10.1007/978-3-319-75717-9

Library of Congress Control Number: 2018932998

Printed on acid-free paper

This Springer imprint is published by Springer Nature
The registered company is Springer International Publishing AG
The registered company address is: Gewerbestrasse 11, 6330 Cham, Switzerland

To my beloved wife Inmaculada and to my children Julia, Rafael, Claudia, Fatima, Gabriel, Sara, Miguel, Abigail, Noemi and Pedro for their understanding and especially to God for His help.

J. Fernández de Cañete

To my two beloved A's, my wife Ana-Belén and my son Alonso.
I would need another 400 pages to express what they make me feel every day, but I have no space here.

C. Galindo

To my father, for showing me the love of always learning.

J. Barbancho

To my parents, Joaquín and Amalia, my guide, example and support. To Óscar, who has made it possible for us to look into each other's eyes while looking both in the same direction. To my children, Amalia and Óscar, reason of my sleeplessness and source of my hopes.

A. Luque

Preface

The textbook belongs to the Automatic Control field applied to clinical medicine, since there are a great variety of clinical applications that require the design of sophisticated control strategies to achieve acceptable performance even in the presence of uncertain environments.

Automatic Control is essentially a branch of engineering that addresses the problem of endowing systems with an autonomous capacity of interacting with the system under control, while keeping constant their main variables values, by using sensors, actuators and controller designing in a feedback loop structure. In this way, Automatic Control enables the use of online control devices during surgical operations and even in intensive care units under safe operation, with applications ranging from simple dosage prescriptions to more complex tasks such as the control of depth of anaesthesia. However, it must be remarked that the physician plays a key role in any devised feedback control loop and can be said that Automatic Control assists the physician rather than substitutes him during the decision-making process, thus freeing him from more demanding tasks which computers are not able to accomplish.

Computer-aided control system design and simulation tools impact nowadays the way practising engineers design automatic control systems. The computer is creating new possibilities for the biomedical engineer in regard to the control design, so that mathematical representations can be easily handled with the help of computers, thus enabling the specification of system performance and physical constraints during the control design.

Engineering methodology is essentially pluralistic, and there is usually more than one acceptable solution for engineering problems. In this way, the computer-aided design helps to obtain most of these solutions and then user is free to choose among them the most satisfactory one.

This book presents the fundamental principles and challenges encountered in the control of biomedical systems, providing practical solutions and suggesting alternatives. The perspective of the text is based on the system behaviour in the time domain both linear and nonlinear, continuous and discrete, helping the reader to be able to interpret the physical significance of mathematical results during control system analysis and design, focusing on biomedical engineering applications.

Interactive learning is promoted, endowing students with the ability to change parameters and conditions during the simulation and to see the effects of these changes, by using interactive MATLAB and SIMULINK software tools, also presenting realistic problems in order to analyse, design and develop automatic control systems.

The text also is complemented with MATLAB and SIMULINK exercise files solved to aid students to focus on the fundamental concepts treated throughout the book, following a new pedagogical approach distinct from the classical one whereby fundamental control concepts are introduced together with adequate software tools in order to gain insight into the biomedical engineering control problems.

The book is suitable for second- or third-year undergraduate students who will find the illustrative examples particularly useful to their studies of control system design and implementation. Lecturers in the control field will find the computer-aided design approach as an alternative to teaching the fundamental concepts of feedback analogic and digital control.

The book starts with Chap. 1, where main concepts of control systems engineering are introduced along with basic definitions regarding control system elements such as sensor, actuator and controller. Practical applications of control in biomedical engineering are also shown illustrating the particular characteristics of the systems under feedback control.

Chapter 2 introduces the concept of the model of a dynamic system, as a representation of a real or intended system, particularizing for continuous-time linear systems. In the same way, discrete-time linear systems are introduced and Laplace transformation and its properties are presented through a variety of examples, as well as Z transformation, presenting both of these techniques as cornerstone in analysis and design of dynamic linear systems in complex planes. The concepts of continuous and discrete transfer functions are also derived as means of system representation. This chapter ends by illustrating block diagram building and rules for their simplification, with illustrative MATLAB and SIMULINK examples with code included.

Chapter 3 presents the transient and steady-state response analysis both of continuous- and of discrete-time linear systems with a brief introduction to system identification. Data sampling and hold are described, and relation between S-plane and Z-plane is also illustrated. This chapter also introduces stability analysis of closed-loop systems in S- and Z-planes, and illustrative MATLAB and SIMULINK examples are included.

Frequency response analysis techniques both of open-loop and closed-loop systems are presented in Chap. 4, based on the Bode diagram both for continuous- and discrete-time systems together with relative stability analysis with MATLAB examples included.

Chapter 5 treats the root locus technique for analysis and design of control systems both of continuous- and discrete-time linear systems and presents the basics of the root locus graphical interface of MATLAB. System performance specifications are defined and also represented on root locus graphical interface with practical examples for better understanding.

Design of PID-based control systems is presented in Chap. 6 with the root locus graphical interface, both continuous and discrete PID controllers by using specific-purpose computational tools. Implementation issues are also tackled together with tuning methods for PID parameter selection. MATLAB and SIMULINK applications are included, focusing on biomedical engineering problems.

The remaining of the book consists of several examples and solved exercises for each chapter to aid the reader the understanding of the main concepts expressed throughout the pages.

Málaga, Spain J. Fernández de Cañete
Málaga, Spain C. Galindo
Sevilla, Spain J. Barbancho
Sevilla, Spain A. Luque

Contents

Introduction to Control Systems in Biomedical Engineering

1

This chapter is an introduction to control systems defined as a collection of interconnected elements whose goal is to achieve a desired response even when external disturbances are present. In this book, both continuous-time and discrete-time control systems will be dealt and its constitutive functional elements such as sensor, controller and actuator will also be described. Practical applications of biomedical control engineering techniques and advanced information technology are also included so as to show the potential both of analog control tand of computer control systems.

Introduction

There are two modes of control operation, either in open loop or in closed loop (feedback control). In open-loop control, the goal is to keep the system output $y(t)$ as close as possible to some desired value $r(t)$ by a suitable adjustment of the input $u(t)$ through the combined action of controller and actuator. However, because of the unknowns in the system model and the effects of external disturbances $w(t)$, open-loop control is not accurate (Fig. 1.1).

In closed-loop control, a measurement of the output of the system $y(t)$ is used to modify its input $u(t)$. The controller receives the measurement of the plant output through a sensor and then generates a suitable value of the plant input, hence closing the loop. In this way, system output is fed back and used to adjust the system input making the system self-regulated (Fig. 1.2).

In fact, the concept of feedback regulation is largely found in the overall physiology of the organism under external disturbances and the homeostasis term is frequently used when maintenance of constant physiological conditions is accomplished. Besides, the term Cybernetics is utilized when control theory is applied to physiology.

J. Fernández de Cañete et al., *Automatic Control Systems in Biomedical Engineering*,
https://doi.org/10.1007/978-3-319-75717-9_1

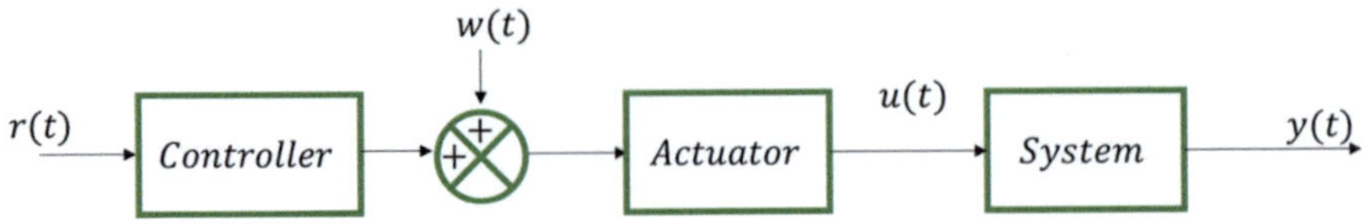

Fig. 1.1 Open-loop control configuration

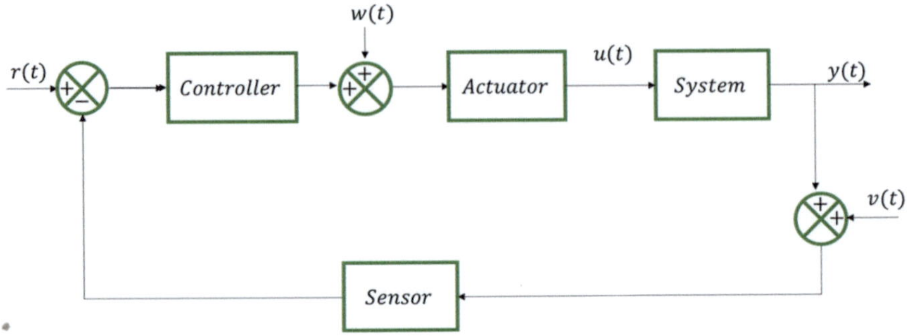

Fig. 1.2 Closed-loop control configuration

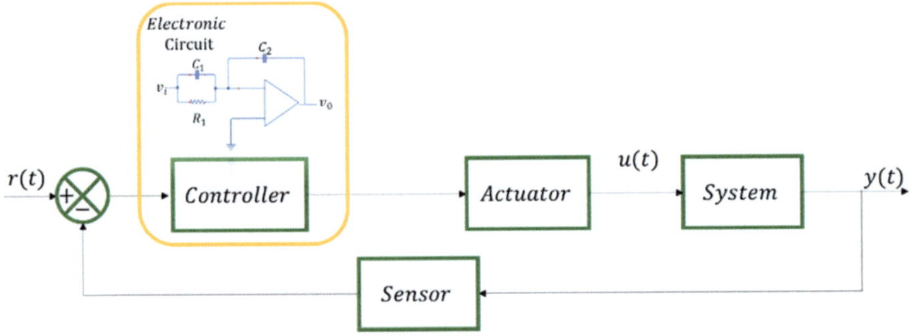

Fig. 1.3 Continuous-time control configuration

In this book, both continuous-time and discrete-time control systems will be treated. Generally, system inputs and outputs are continuous signals both in time and in amplitude. For example, consider the system to be controlled is an infusion pump. The pump input and output are defined and may change at every instant in time and can take also any value within certain range.

In continuous closed-loop control, at each time instant, the output is fed back, the error is calculated, and the controller, implemented as a physical circuit, generates a control signal (Fig. 1.3).

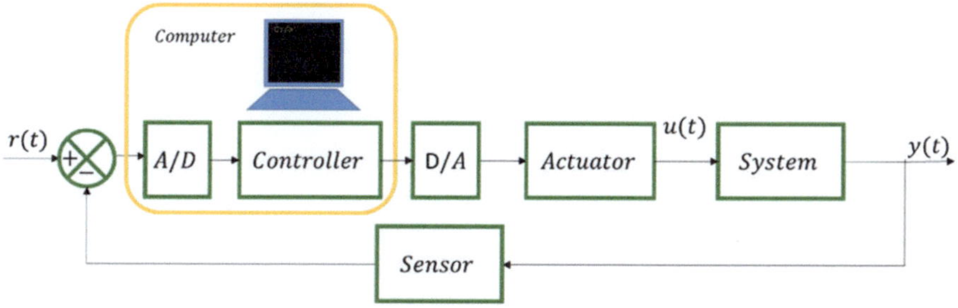

Fig. 1.4 Discrete-time control configuration

Conversely, discrete closed-loop control, digital computers and microcontrollers are used to control a continuous system, so that some form of translation is required using both digital-to-analog and analog-to-digital converters (Fig. 1.4).

There are numerous applications of control systems in biomedical engineering, from prostheses control used in the limbs of individuals with amputations or congenitally deficient extremities, drug delivery control used to regulate physiological variables, like blood pressure, blood glucose and heart rate and closed-loop control of blood sugar level by insulin infusion mimicking the endocrine pancreas among others.

Basic Control Definitions

Before discussing the structure of a control system, it is necessary to define what is meant by a system. In general, a system can be defined as a collection of interrelated elements pursuing a particular objective. For example, the reflex arc system is composed of sensory neuron, relay neuron and motor neuron with the aim of transmitting electrical signals from skin to the muscle so as to provoke contraction when skin is stimulated (see Fig. 1.5).

From a mathematical point of view, a system can be considered as a process that responds to a set of input signals or excitations to produce a corresponding set of output signals or responses.

Signals are represented as mathematical functions and can be either continuous time or discrete time, being called quantified continuous or discrete (digital) when only a finite discrete set of values are taken by the signal (Fig. 1.6).

Systems may be of a very diverse nature, and thus, classing them is needed for a better understanding and study. The following classification, regarding different system particularities, is normally applied:

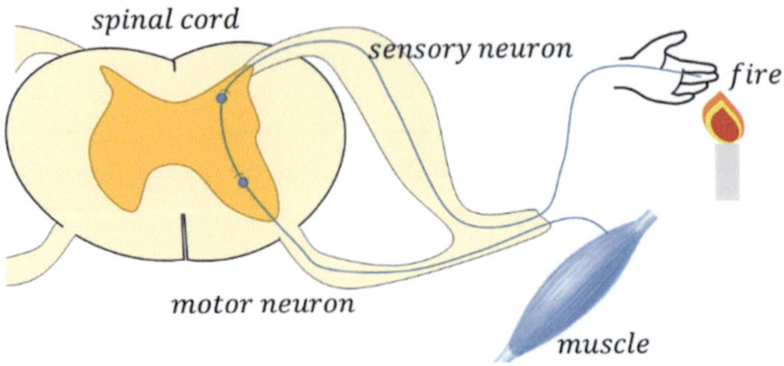

Fig. 1.5 Schematic of the nervous system

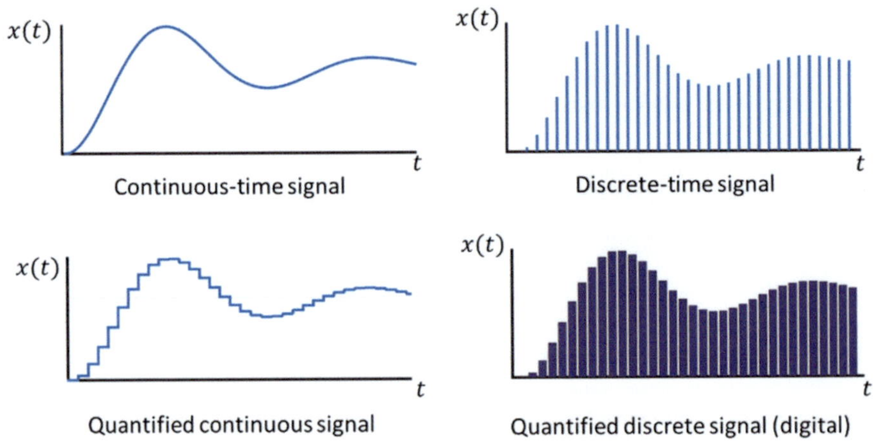

Fig. 1.6 Different types of physical signals

- **Invariant/Variant**: According to the dependence of current outputs of a system with respect to time, system can be distinguished as invariant, i.e. the system output does not depend explicitly on time, and variant systems where it exists a temporal dependence.
- **Static/Dynamics**: According to the dependence of the current outputs of a system with respect to previous ones, systems can be classed as static systems, those systems without memory, i.e. the system output only depends on the current input, and dynamic systems, that is if past outputs influence on the current one.
- **Causal/Anticipative**: According to the dependence of the current outputs of a system with respect to actual and future times, systems can be classed as causal systems, when the system output only depends on the current inputs, and anticipative systems, that is if future outputs influence on the current one.

- **Linear/Nonlinear**: Depending on the system behaviour when it is subjected to a complex input, systems can be classed as linear, when output can be described as a sum of responses to simpler inputs, and nonlinear if this condition does not hold.
- **Stochastic/Deterministic**: A system is said to be stochastic when it exhibits a certain random, unforeseeable behaviour, and thus, it may output different responses for the same input. In contrast, a system is said to be deterministic when no randomness is involved in its dynamics and therefore it always produces the same output for the same input.

Fundamental Control Concepts

In order to gain insight into control systems, it is useful to define explicitly the variables and structures involved in the problem. A block diagram of a generic control closed-loop system comprising the relationship among variables and constitutive elements is shown in Fig. 1.7.

The controlled variable is the physical quantity measured through the sensor and is the actual output of the system. The manipulated variable is the physical quantity applied to the system to modify the controlled variable and is the input of the system.

The sensor converts physical controlled variable into electrical signal, while the actuator transforms the electrical signal output of the controller into a physical manipulated variable. Actuator and sensor are influenced by unpredictable disturbances and superimposed noise which affect both to manipulated and to controlled variables, respectively.

The controller generates the electrical control signal from the deviation between measured output and desired output, this one previously transformed into the same range of measured output by applying the sensor gain as conversion factor. The feedback loop tends to reduce the discrepancy between measured and actual outputs by compensating for disturbances.

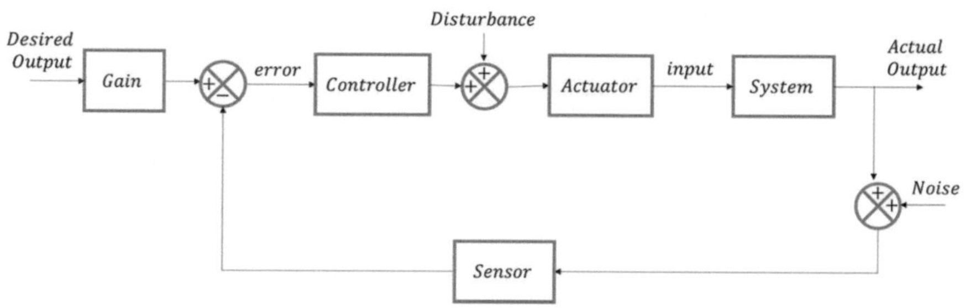

Fig. 1.7 Closed-loop control general block diagram

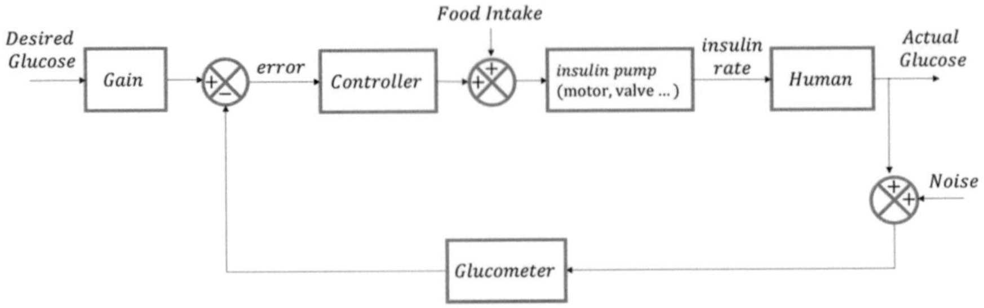

Fig. 1.8 Closed-loop control of blood glucose levels

Example 1.1
A closed-loop control system for maintaining the glycemic level in type 1 diabetes mellitus can be represented in Fig. 1.8.

The glucose–insulin system represents the system under control, with insulin injection and glycemic level as input and output, respectively. A subcutaneous glucose sensor is needed to convert glucose concentration to voltage, while an insulin pump delivery system enables the maintenance of glucose level controlled by insulin injection. An electronic circuit based on operational amplifiers is used as controller so as to minimize the glucose discrepancy between desired and measured glycemic levels, preventing the patient to undergo both hypoglycemic and hyperglycemic dangerous episodes.

Sensors and Actuators

Sensors are devices that respond to a change in a physical phenomenon (pH, temperature, force, etc.) by producing a proportional output signal of electrical nature in general. For example, a pH meter (Fig. 1.9) responds to a pH change and outputs a proportional change in voltage.

Sensors are classified as analogic or digital based on the type of output signal. Analog sensors produce continuous signals that are proportional to the sensed parameter. Digital sensors on the other hand produce digital outputs that can be directly interfaced with the digital controller.

Focusing on biomedical phenomena, we can make another classification of sensors as direct or indirect depending on the way a physiologic variable is determined, invasive or non-invasive if the sensor is inserted into the body and dynamic and static depending on the consideration of the sensor dynamics.

In the same way, sensors can be classified according to their measurement objectives in general as

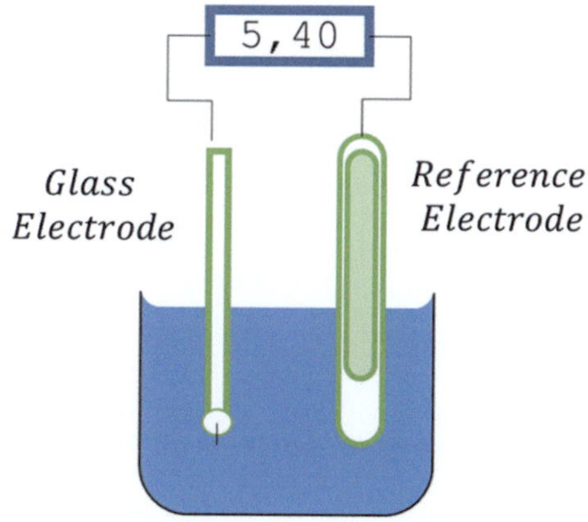

Fig. 1.9 Schema of a pH sensor

- Force, torque and pressure sensor, based on strain gage and of piezoelectric type
- Flow sensors, such as the rotameter and the turbine meters and of the electromagnetic and ultrasonic types
- Temperature sensors, being thermocouples, thermistors and resistance temperature detectors the most commonly used
- Displacement and rotational sensors, and there are both of contact-type sensors such as strain gage, LVDT, tachometer, etc., and of non-contact-type including encoders, Hall effect, resistance, inductance and capacitance among others
- Light sensors, such as phototransistors, photoresistors and photodiodes
- Chemical sensors, such as microelectrodes, surface electrodes and needle electrodes.

Nowadays, each sensor has an attached transmitter that converts, conditions and normalizes its signal to transmit at a distance along with an indicator to display analogic or digital measures (Fig. 1.10).

The output from a sensor requires to be demodulated, amplified, filtered, linearized and quantized so that the signal can be sent to the controller. In case of time-varying signals to be measured, the sensor must be calibrated in advance.

Actuators are devices that receive as inputs electrical signals from the controller and produce changes in the physical system by generating force, flow, motion, etc. (Fig. 1.11).

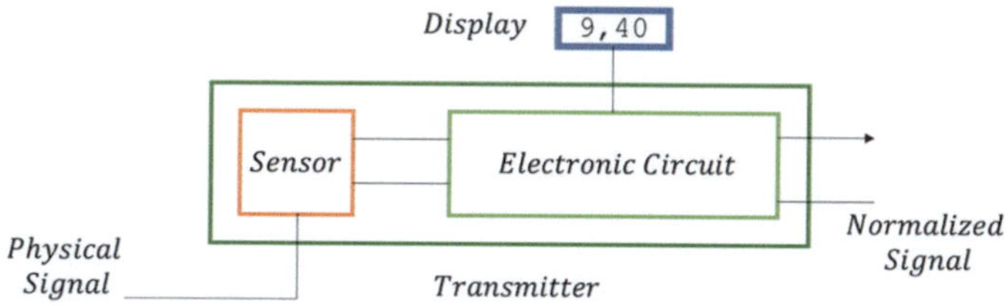

Fig. 1.10 Schematic of a sensor including transmitter and display devices

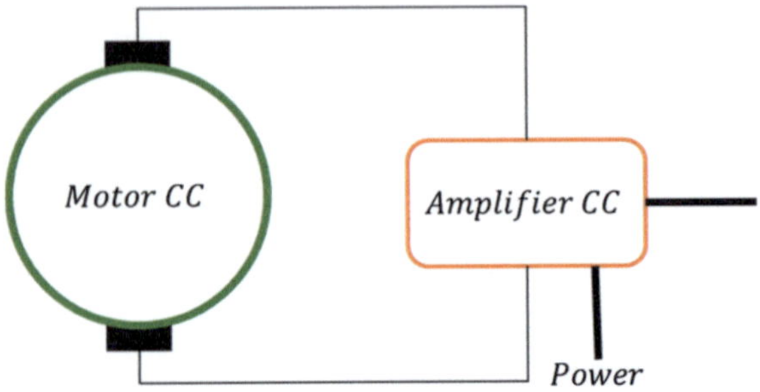

Fig. 1.11 Schematic of an actuator with power amplifier

They can be classified based on the type of energy in general as

- Electronics, such as relays, diodes, transistors and triacs.
- Electromechanical, such as AC and DC motors and stepper motors.
- Electromagnetics, mainly solenoids as constituent of valves.
- Hydraulic and pneumatics, such as pumps and pistons.

In order to select the appropriate actuator, both the power needs and the coupling mechanisms of the entire system should be considered.

Applications in Biomedical Engineering

Considerable progress is being achieved in addressing challenging problem as control of artificial organs, rehabilitation engineering, healthcare equipment and other medical systems by means of application of biomedical control engineering techniques and advanced information technology.

Rehabilitation techniques help disabled persons to improve the quality of living, like powered prostheses, functional electrical stimulation and biofeedback control.

Powered prostheses are used in the limbs of individuals with amputations or congenitally deficient extremities. A multifunction arm or leg is interfaced by means of myoelectric control (Fig. 1.12).

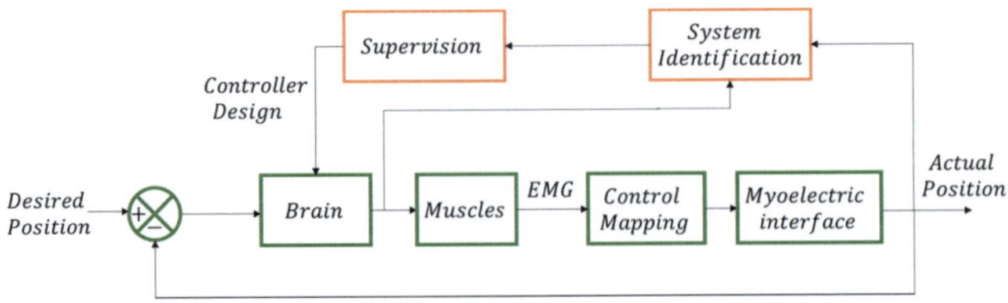

Fig. 1.12 Myoelectric control of powered prostheses

Functional electrical stimulation uses external signals to control body posture including balance, standing and walking and compensate for stroke rehabilitation, spinal cord injuries and many limb motor dysfunction (Fig. 1.13).

Biofeedback control provides tools for measuring and controlling physiological signals, like heart rate, EEG and muscle activity, acting the brain as the controller. In Fig. 1.14, it is shown a brain–computer interface (BCI)-based application.

Drugs are applied to control physiologic variables such as blood pressure, blood glucose and heart rate among others. Closed-loop control of blood glucose level is applied to compensate for the malfunction endocrine pancreas in diabetic patients. In general, a closed-loop feedback system for insulin delivery consists of a blood glucose sensor, a controller and an infusion pump (Fig. 1.8).

In the same way, automatic anaesthesia controllers are capable of taking over to provide drug administration and avoid overdosing, while anaesthesiologists are able to concentrate on critical issues that may threaten the patient's safety, though not leaving the system without the anaesthesiologist's close supervision (Fig. 1.15).

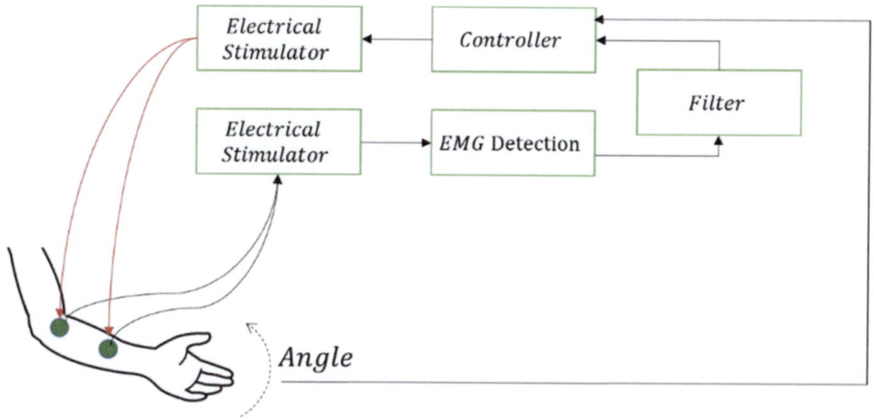

Fig. 1.13 Body control by functional electrical stimulation

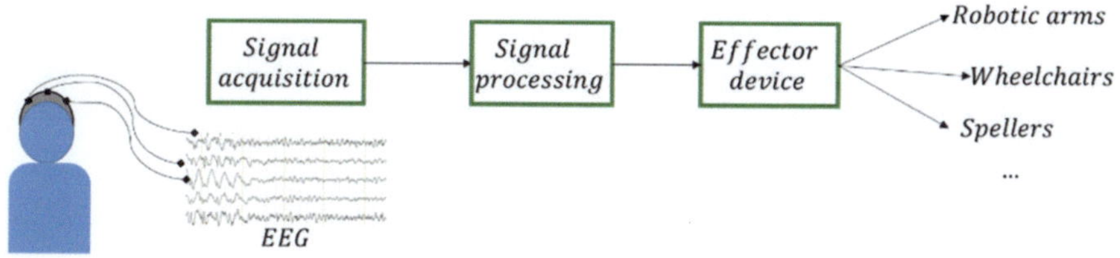

Fig. 1.14 Functional electrical stimulation through BCI

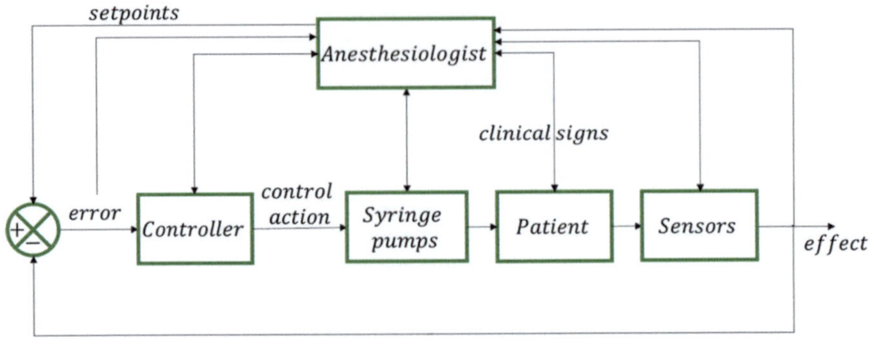

Fig. 1.15 Closed-loop control of anaesthesia with supervision

Fig. 1.16 Closed-loop control of dialysis machine

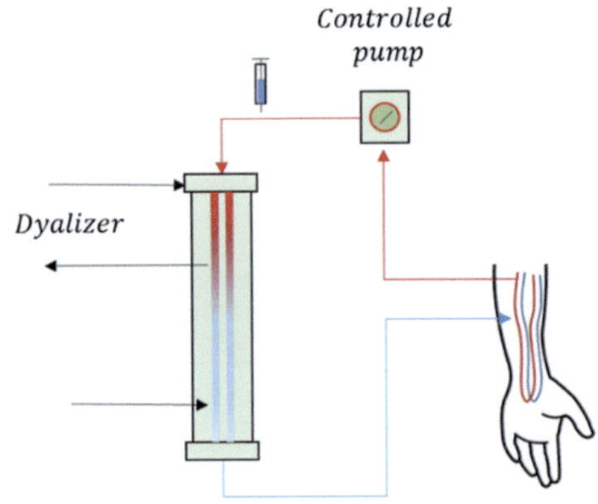

Automatic control of haemodialysis has the potential to provide a better treatment to the increasing number of end-stage renal disease patients. In order to control dose of dialysis, effective clearance must be measured and compared to prescribed clearance, so that the controller adjusts different actuators such as blood pump, dialysate pump, and treatment time to compensate for the error between output and set point (Fig. 1.16).

With regard to rehabilitation engineering, medical robotics has been influencing significantly this field at present. Robotic technologies offer enhancements to medical or surgical processes through improved precision, stability and dexterity, and robots are these days capable to improve the rehabilitation of disabled patients,

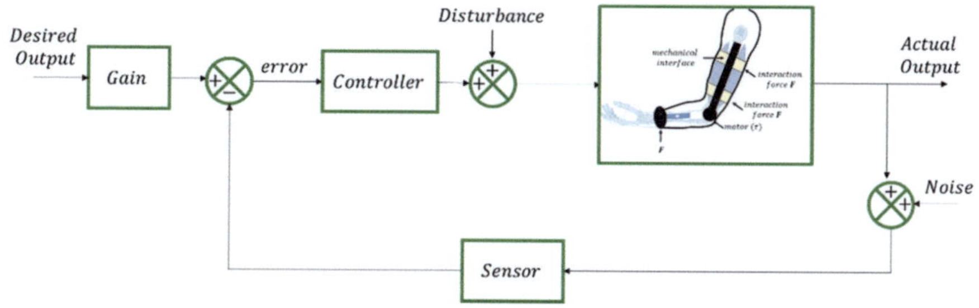

Fig. 1.17 Closed-loop control of upper limb in rehabilitation

both upper and lower limb rehabilitation and walking assistive robot among other advances (see Fig. 1.17). The design of effective control methodologies can gradually improve the performance of trajectory tracking even with disturbance rejection.

2 Mathematical Modelling Representation of Linear Systems

This chapter deals with the description of linear and time-invariant (LTI) systems in its two flavours: *continuous-time* and *discrete-time* systems. Continuous-time systems capture the inherent nature of real systems in which every magnitude has a particular value at any time. These systems can be represented through *differential equations*. Discrete-time systems are particular systems that usually come from the discretization of a continuous-time system. In a nutshell, such discretization can be seen as a decimation of the involved magnitudes at a fixed frequency. These systems can be represented as *difference equations*, being suitable for digital computers.

Mathematical tools like the Laplace and the Z transformations are used to conveniently describe continuous- and discrete-time systems, respectively. Systems' description is normally given in the form of a transfer function (in the Laplace or Z domain) which will facilitate the computation of the system response for any input signal, as well as the analysis of some interesting characteristics like stability.

This chapter also copes with the representation of complex systems using block diagrams and the techniques to reduce them into a unique transfer function. Examples in the domain of the biomedical field accompanied with the MATLAB/SIMULINK code are given.

Dynamical Systems Modelling

In almost all areas of engineering, considerable efforts are directed towards acquiring information on various aspects of system performance, what is called system analysis.

Traditionally, the system analysis was carried out by using the physical system subjected to test input signals, observing its corresponding response. Nevertheless, this is not always feasible, since either the extent to which the parameters of the physical system can be varied is very limited or there are imposed restrictions to the

J. Fernández de Cañete et al., *Automatic Control Systems in Biomedical Engineering*,
https://doi.org/10.1007/978-3-319-75717-9_2

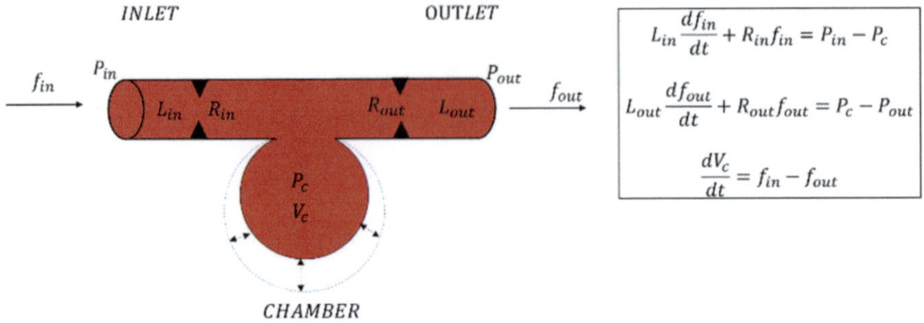

Fig. 2.1 Mathematical model of an artery with mechanical resistance, inductance and capacitance elements

application of these input tests, as it frequently happens with system analysis in human physiology.

In order to overcome these problems, a simplified representation of the physical system under analysis is developed instead, what is called system model.

System modelling is rather a difficult task, since it is necessary experience, practice and intuition to be a good system modeller. The bases for the building of mathematical models are the fundamental physical laws, such as the laws of conservation of mass, energy and momentum. Mathematical models will be obtained in the form of differential equations that describe the system dynamics. In Fig. 2.1a, simplified model on an artery tract is shown.

Continuous-Time Linear Systems

A continuous-time system is one for which the inputs, state variables and outputs are defined over some continuous range of time, and its behaviour is described by differential equations. Some of the most common input signals used to analyse the system behaviour or to design control systems are shown in Table 2.1. These signals are the impulse, step, ramp and sine.

In general, a continuous system with k inputs and j outputs (see Fig. 2.2) can be described by n independent input–output differential equations:

$$\begin{aligned}
f_1\left(y_1^{n)}, \ldots, \dot{y}_1, y_1, \ldots, y_j^{n)}, \ldots, \dot{y}_j, y_j, u_1^{m)}, \ldots, \dot{u}_1, u_1, \ldots, u_k^{m)}, \ldots, \dot{u}_k, u_k, t\right) &= 0 \\
f_2\left(y_1^{n)}, \ldots, \dot{y}_1, y_1, \ldots, y_j^{n)}, \ldots, \dot{y}_j, y_j, u_1^{m)}, \ldots, \dot{u}_1, u_1, \ldots, u_k^{m)}, \ldots, \dot{u}_k, u_k, t\right) &= 0 \\
&\vdots \\
f_j\left(y_1^{n)}, \ldots, \dot{y}_1, y_1, \ldots, y_j^{n)}, \ldots, \dot{y}_j, y_j, u_1^{m)}, \ldots, \dot{u}_1, u_1, \ldots, u_k^{m)}, \ldots, \dot{u}_k, u_k, t\right) &= 0
\end{aligned} \tag{2.1}$$

Table 2.1 Some test input signals used for system analysis

Impulse	$\delta(t)$	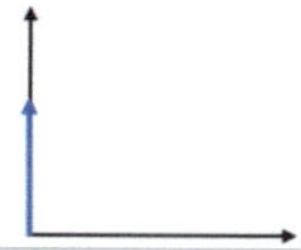
Step	$u_s(t)$	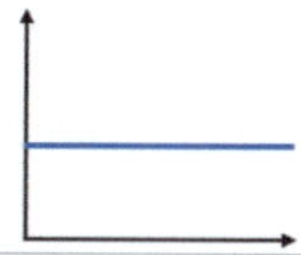
Ramp	$r(t) = tu_s(t)$	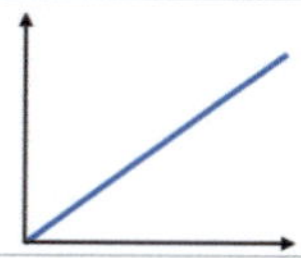
Sine	$\sin(\omega t)u_s(t)$	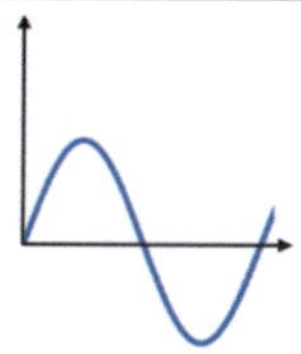

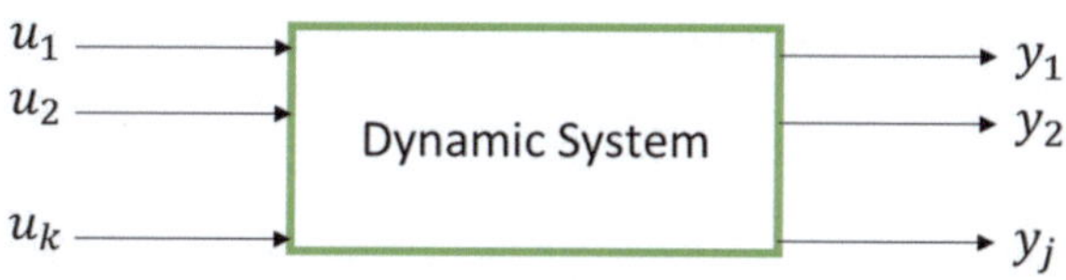

Fig. 2.2 Continuous multi-input, multi-output system

where n and m are the maximum order of the differential equations of the outputs and the inputs, respectively, with $m \leq n$ for causal behaviour, and $f_i, i = 1, \ldots, j$ nonlinear in general. The parameter n is denoted as the *system order* and plays a fundamental role for the system behaviour.

If the system is single-input and single-output, the continuous dynamic differential equation is given by Eq. (2.2).

$$f\left(y^{n)}, \ldots, \dot{y}, y, u^{m)} \ldots, \dot{u}, u, t\right) = 0 \tag{2.2}$$

where $m \leq n$ for causal behaviour, with f nonlinear in general.

Analytical methods for exact solving (2.2) given the input signals and an appropriate set of initial conditions are practical only for low-order systems. Nevertheless, the application of numerical methods will permit us to deal with high order and multi-input multi-output to obtain approximate solutions.

Let us particularize the analysis for a single-input single-output system, with f described by a linear combination of both input and output signals together with its derivatives, as in the case of (2.2), assuming also the system is stationary, and no direct dependence with time t is explicit.

Linear and Invariant Time Systems

A system is said to be linear and time-invariant when it fulfils the principle of *superposition*. The superposition property states that the system's response due to a set of inputs could be decomposed as the sum of individual responses, each one due to a given input, as graphically depicted in Fig. 2.3. When this property is not met, the system is set to be nonlinear.

A continuous-time system defined by differential equations is said to be linear if it can be expressed as a linear combination of derivatives of the output and input in the form

$$a_n y^{n)} + \cdots + a_1 \dot{y} + a_0 y = b_m u^{m)} + \cdots + b_1 \dot{u} + b_0 u \tag{2.3}$$

where the parameters $a_j, j = 1\ldots n$ and $b_k, k = 1\ldots m$ are constants or functions of time t. In the event that these parameters do not depend explicitly on the time t, the system is said to be also time-invariant, and it is not the case, the system is said to be nonlinear.

Nearly, all systems are inherently nonlinear in nature, since the most of the relationships in physics are nonlinear. Moreover, most of the linear systems are a special case of nonlinear systems in limited ranges of operation. Besides this, the nonlinear nature of the physical elements constituting the system may be an

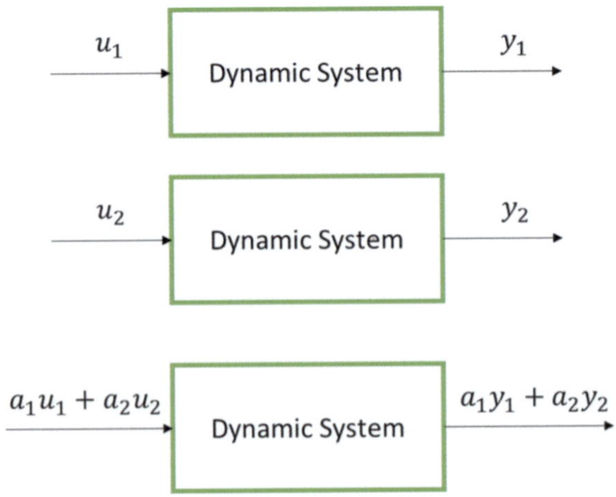

Fig. 2.3 Superposition property. A linear system yielding the output y_1 for a given input u_1, and y_2 for u_1 will produce the composed output $a_1y_1 + a_2y_2$ for the composed input $a_1u_1 + a_2u_2$

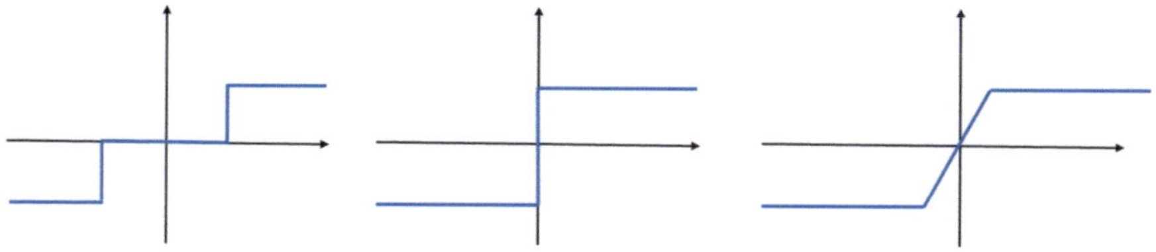

Fig. 2.4 Some examples of nonlinear relationships

essential feature of the system, causing the overall system behaviour to be nonlinear as with the dead zone, relay, saturation, etc., characteristics (see Fig. 2.4).

Systems Linearization

The objective of the linearization technique is to derive a linear approximate system whose response will agree closely with that of the original nonlinear system from which it comes.

In order to apply linearization method, we use the disturbance method, which consists in considering small variations around a nominal operating point, making into the nonlinear relationship $y = f(u)$ the following substitutions:

1. The independent variable u is replaced by $u = u_0 + \delta u$, being u_0 the operating point of u and δu the disturbance applied on u.
2. The dependent variable or curve representing the nonlinearity $y = f(u)$ is replaced by its tangent at the nominal operating point u_0 (Fig. 2.5).

Approximating $y = f(u)$ by the Taylor's expansion, we have

$$y = f(u) = f(u_0 + \delta u) = f(u_0) + \dot{f}(u_0)\delta u + \frac{1}{2!}\ddot{f}(u_0)\delta^2 u + \cdots \qquad (2.4)$$

Fig. 2.5 Linearization of $y = f(u)$ at operation point u_0

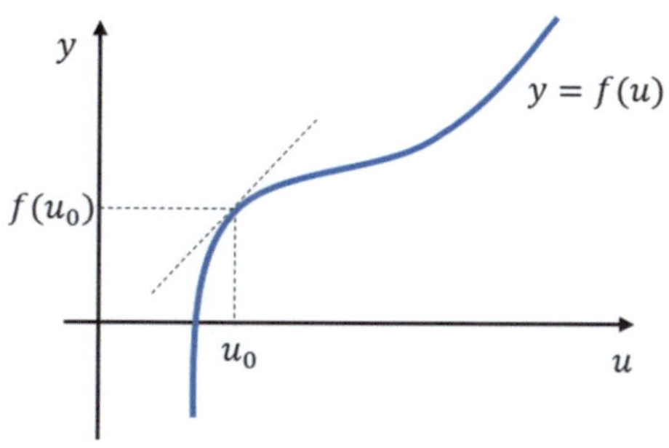

Approaching the nonlinear relation by its first derivative (tangent of the curve), it follows that

$$y = f(u) = f(u_0 + \delta u) \cong f(u_0) + \dot{f}(u_0)\delta u \tag{2.5}$$

which represents the substitution to be made on the dependent variable y whenever $u = u_0 + \delta u$.

By extending this method to a multivariable nonlinear relationship $y = f(u_1, u_2, \ldots, u_n)$ in general, we have

$$\begin{aligned} y &= f(u) = f(u_{10} + \delta u_1, \ldots, u_{n0} + \delta u_n) \\ &\cong f(u_{10}, \ldots, u_{n0}) + \frac{\partial f}{\partial u_1}\delta u_1 + \ldots + \frac{\partial f}{\partial u_n}\delta u_n \end{aligned} \tag{2.6}$$

being $\bar{u}_0 = (u_{10}, u_{20}, \ldots, u_{n0})$ the nominal operating point, where the partial derivatives of f with respect to u_i are evaluated at $\bar{u}_0$.

In case of nonlinear system relationship given by a set of differential equations, it will be applied the disturbance method in implicit form to get

$$F\left(y^{n)}, \ldots, \dot{y}, y, u^{m)}, \ldots, \dot{u}, u\right) = 0 \tag{2.7}$$

Therefore, following the same approach used in (2.6), we have

$$\begin{aligned} F\left(y^{n)}, \ldots, \dot{y}, y, u^{m)}, \ldots, \dot{u}, u\right) &= F\left(y_0^{n)}, \ldots, \dot{y}_0, y_0, u_0^{m)}, \ldots, \dot{u}_0, u_0\right) \\ &+ \frac{\partial F}{\partial y^{n)}}\delta y^{n)} + \cdots + \frac{\partial F}{\partial \dot{y}}\delta \dot{y} + \frac{\partial F}{\partial y}\delta y \\ &+ \frac{\partial F}{\partial u^{m)}}\delta u^{m)} + \cdots + \frac{\partial F}{\partial \dot{u}}\delta \dot{u} + \frac{\partial F}{\partial y}\delta u \end{aligned} \tag{2.8}$$

The nominal operating point $\bar{u}_0 = (y_0, u_0)$ around which we will make the linearization is necessarily the solution of the differential equation describing the system dynamics and is derived from Eq. (2.7), making zero the derivatives of the implicit differential equation at $u = u_0$ in

$$F\left(y_0^{n)}, \ldots, \dot{y}_0, y_0, u_0^{m)}, \ldots, \dot{u}_0, u_0\right) = 0 \tag{2.9}$$

In this way, the second term on the right-hand side of (2.8) is the differential equation linearized around the nominal operating point $\bar{u}_0 = (y_0, u_0)$, that is approximated by

$$\frac{\partial F}{\partial y^{n)}}\delta y^{n)} + \cdots + \frac{\partial F}{\partial y}\delta y + \frac{\partial F}{\partial u^{m)}}\delta u^{m)} + \cdots + \frac{\partial F}{\partial u}\delta u = 0 \tag{2.10}$$

where the partial derivatives of F w.r.t. to each independent variables of (2.7) are evaluated at $\bar{u}_0$, and whose validity is restricted to a small region $\delta\bar{u} = \bar{u} - \bar{u}_0$ around $\bar{u}_0$.

Example 2.1
A continuous flow stirred-tank reactor (see Fig. 2.6) is a system in which a chemical reaction takes place, involving an input flowrate and concentration of a given substance into the tank. The dynamics of this system is described by the set of nonlinear differential equations

$$\begin{aligned} \frac{dC}{dt} &= f_1(C, V, F_{in}, C_{in}) = \frac{F_{in}}{V}(C_{in} - C) - kC^2 \\ \frac{dV}{dt} &= f_2(C, V, F_{in}, C_{in}) = F_{in} - \beta V \end{aligned} \tag{2.11}$$

being C and V the concentration and volume inside the tank; F_{in} and C_{in} the input flowrate and its concentration; k the reaction rate constant and β the discharge coefficient.

In order to linearize the system, we firstly express the system equations in implicit form, calculating the operation point $\bar{u}_0 = (C_0, V_0, F_{in0}, C_{in0})$.

$$\begin{aligned} F_1(\dot{C}, C, \dot{V}, V, F_{in}, C_{in}) &= \frac{dC}{dt} - \frac{F_{in}}{V}(C_{in} - C) + kC^2 = 0 \\ F_2(\dot{C}, C, \dot{V}, V, F_{in}, C_{in}) &= \frac{dV}{dt} - F_{in} + \beta V = 0 \end{aligned} \tag{2.12}$$

Making zero the derivatives in (2.12), we have a system equation for solving C_0 and V_0 as a function of F_{in0} and C_{in0}:

$$\begin{aligned} -\frac{F_{in}}{V}(C_{in} - C) + kC^2 = 0 \\ \beta V = F_{in} \end{aligned} \tag{2.13}$$

Now, the partial derivatives of F_1 and F_2 with respect to the dependent variables at $\bar{u}_0 = (C_0, V_0, F_{in0}, C_{in0})$ are calculated:

$$\begin{aligned} &\frac{\partial F_1}{\partial \dot{C}} = 1 \quad \frac{\partial F_1}{\partial C} = \frac{F_{in0}}{V_0} + 2kC_0 \quad \frac{\partial F_1}{\partial C_{in}} = -\frac{F_{in0}}{V_0} \\ &\frac{\partial F_1}{\partial \dot{V}} = 0 \quad \frac{\partial F_1}{\partial V} = \frac{F_{in0}(C_{in0} - C)}{V_0^2} \quad \frac{\partial F_1}{\partial F_{in}} = -\frac{C_{in0}}{V_0} \\ &\frac{\partial F_2}{\partial \dot{C}} = 0 \quad \frac{\partial F_2}{\partial C} = 0 \quad \frac{\partial F_2}{\partial C_{in}} = 0 \\ &\frac{\partial F_2}{\partial \dot{V}} = 0 \quad \frac{\partial F_2}{\partial V} = \beta \quad \frac{\partial F_2}{\partial F_{in}} = -1 \end{aligned} \tag{2.14}$$

Fig. 2.6 Continuous stirred-tank reactor. Given an input flowrate and concentration, we are interested in the output concentration as the result of a particular chemical reaction

Thus, defining the incremental variables as $\delta C = C - C_0$, $\delta V = V - V_0$, $\delta C_{\mathrm{in}} = C_{\mathrm{in}} - C_{\mathrm{in}0}$ and $\delta F_{\mathrm{in}} = F_{\mathrm{in}} - F_{\mathrm{in}0},$, we have the linearized equation set

$$\begin{aligned} &\frac{\mathrm{d}\delta C}{\mathrm{d}t} + \left(\frac{F_{\mathrm{in}0}}{V_0} + 2kC_0\right)\delta C + \left(\frac{F_{\mathrm{in}0}(C_{\mathrm{in}0} - C_0)}{V_0^2}\right)\delta V - \left(\frac{F_{\mathrm{in}0}}{V_0}\right)\delta C_{\mathrm{in}} - \left(\frac{C_{\mathrm{in}0}}{V_0}\right)\delta F_{\mathrm{in}} = 0 \\ &\frac{\mathrm{d}\delta V}{\mathrm{d}t} + \beta\delta V - \delta F_{\mathrm{in}} = 0 \end{aligned} \tag{2.15}$$

In Fig. 2.7, it is shown the SIMULINK diagrams corresponding to the nonlinear two-tank system and its linearized approximation in Fig. 2.8. For parameters $k = 2$ and $\beta = 1$, the operation point $\bar{u}_0 = (C_0, V_0, F_{\mathrm{in}0}, C_{\mathrm{in}0}) = (0.5, 1, 1, 1)$. Thus, for a set of parameters closed to this operation point, both linear and linearized models behave similarly, while their responses diverge as long as we move from such operation point.

We can observe that as long as we modify any of the parameters from the operation point, for instance, adding δF_{in} to F_{in}, the linearized and nonlinearized responses vary. For small changes, close to the operation point, the linearized response is slightly similar to the one of the nonlinear system[1] (see Fig. 2.9a). Higher values will imply we are getting far from the operation point, and thus, the linearized model does not represent the original system, as shown in Fig. 2.9c).

[1]Note the axis scale.

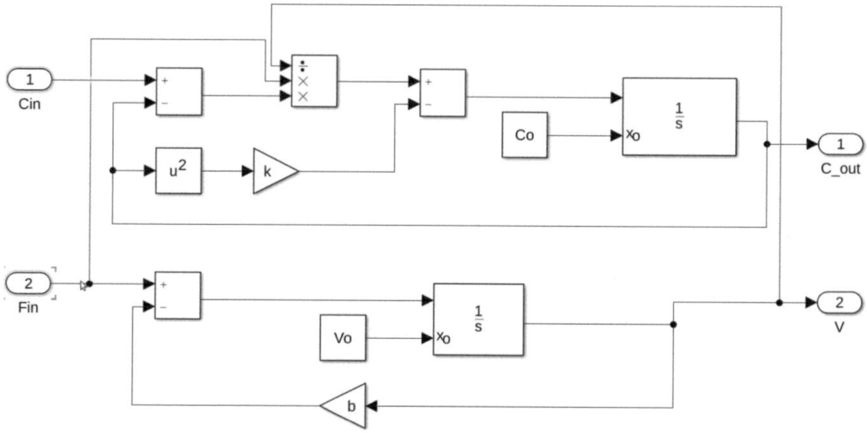

Fig. 2.7 SIMULINK diagram corresponding to the nonlinear two-tank system

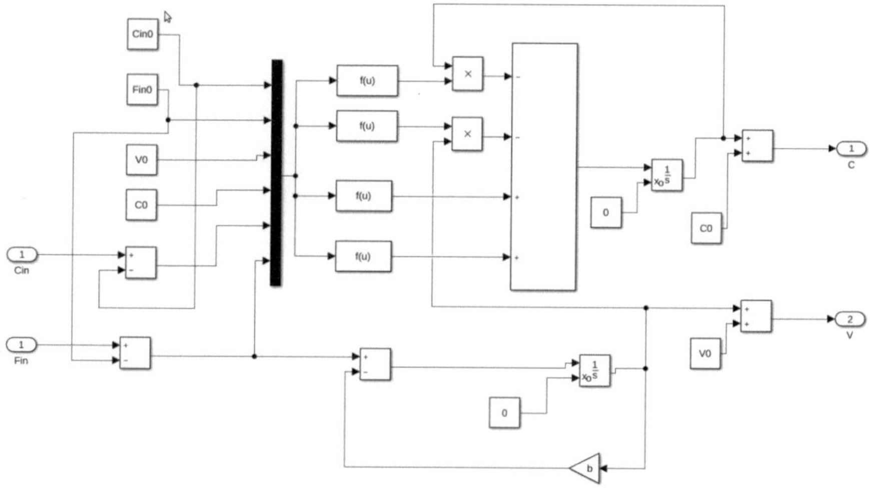

Fig. 2.8 SIMULINK diagram corresponding to the linearized two-tank system

The Laplace Transform

The technique of Laplace transform is used to solve the linear differential equations with constant coefficients expressed in the time domain which define the linear system behaviour, thus transforming then into linear algebraic equations expressed in a complex domain.

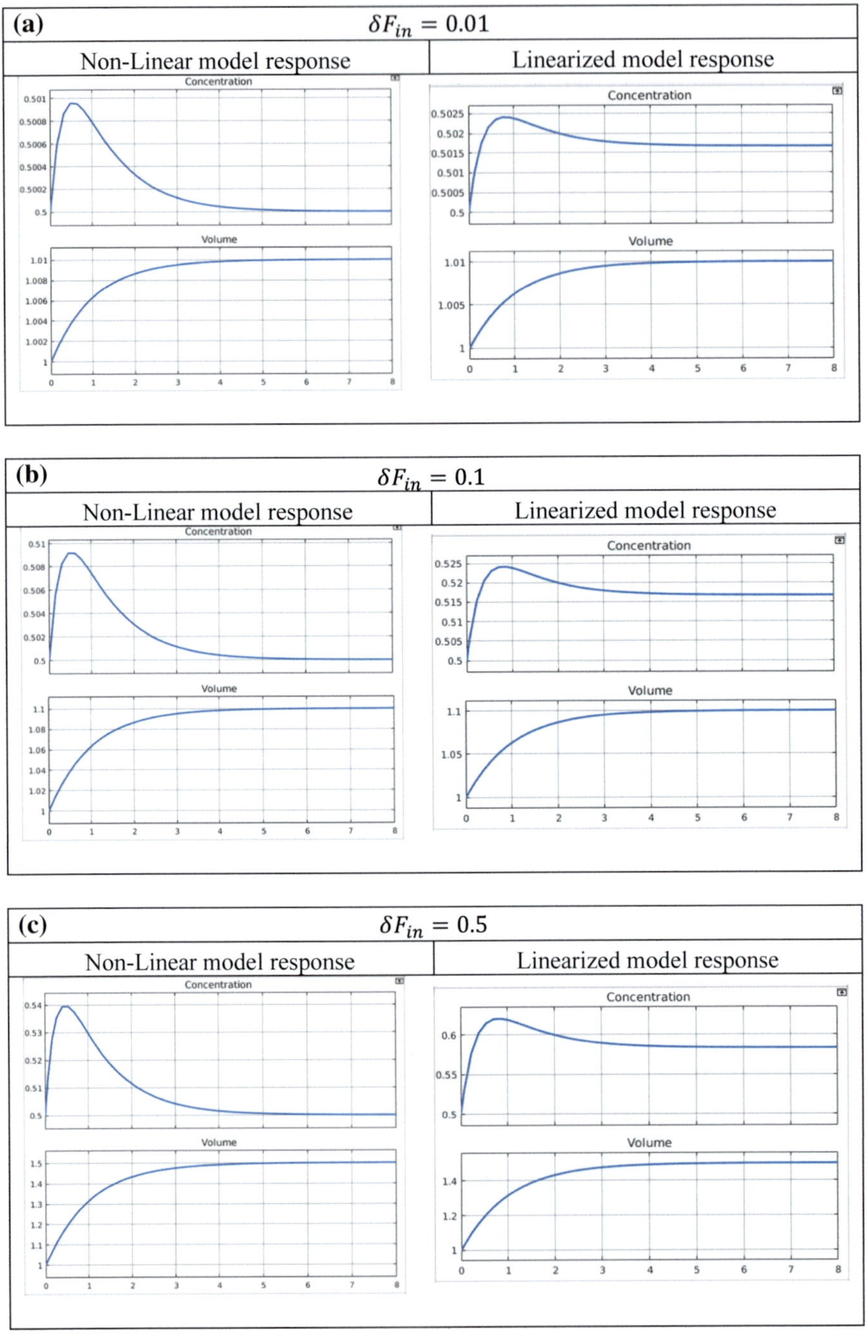

Fig. 2.9 Response to incremental changes in F_{in} corresponding both to the nonlinear (left columns) and to the linearized (right columns) two-tank system

The Laplace transform of $f(t)$ is represented symbolically as $\mathcal{L}(f(t))$ and is defined as

$$\mathcal{L}(f(t)) = F(s) = \int_0^\infty f(t)\mathrm{e}^{-st}\mathrm{d}t \tag{2.16}$$

where $s = \sigma + j\omega$ is a complex variable.

Applying Eq. (2.16), it is possible to derive a pair $f(t) \underset{\mathcal{L}}{\Rightarrow} F(s)$ for each function, so that it is possible to derive a complete table for the most representative input and output signals appearing in dynamic continuous system.

Example 2.2 Derive the Laplace transform of the impulse, step and ramp signals.

Solution

1. $\mathcal{L}(\delta(t)) = \int_0^\infty \delta(t)\mathrm{e}^{-st}\mathrm{d}t = 1$, therefore $\delta(t) \underset{\mathcal{L}}{\Rightarrow} 1$
2. $\mathcal{L}(u_s(t)) = \int_0^\infty u_s(t)\mathrm{e}^{-st}\mathrm{d}t = \int_0^\infty \mathrm{e}^{-st}\mathrm{d}t = \frac{1}{s}$, therefore $u_s(t) \underset{\mathcal{L}}{\Rightarrow} \frac{1}{s}$
3. $\mathcal{L}(r(t)) = \int_0^\infty tu_s(t)\mathrm{e}^{-st}\mathrm{d}t = \int_0^\infty t\mathrm{e}^{-st}\mathrm{d}t$, applying the integration by parts, $\mathcal{L}(r(t)) = \frac{1}{s}\int_0^\infty \mathrm{e}^{-st}\mathrm{d}t = \frac{1}{s^2}$, therefore $r(t) \underset{\mathcal{L}}{\Rightarrow} \frac{1}{s^2}$

Nevertheless, in order to find the transform pairs for complex signals and to solve for the responses of dynamic linear systems given by (2.16), we illustrate some useful properties of the Laplace transform:

(a) Linearity

$$\mathcal{L}(a_1f_1(t) + a_2f_2(t)) = a_1F_1(s) + a_1F_2(s) \tag{2.17}$$

(b) Damping

$$\mathcal{L}(\mathrm{e}^{-at}f(t)) = F(s+a) \tag{2.18}$$

(c) Displacement in time

$$\mathcal{L}(f(t-T)u_s(t-T)) = \mathrm{e}^{-sT}F(s) \tag{2.19}$$

(d) Differentiation

$$\mathcal{L}\left(f^{n)}(t)\right) = s^nF(s) - s^{n-1}f(0) - s^{n-2}\dot{f}(0) - \cdots - f^{n-1)}(0) \tag{2.20}$$

being $f(0), \dot{f}(0), \ldots, f^{n-1)}(0)$ the set of initial conditions for $f(t)$.

(e) Integration

$$\mathcal{L}\left(\int_0^t f(\tau)\mathrm{d}\tau\right) = \frac{1}{s}F(s) \tag{2.21}$$

(f) Multiplication by time

$$\mathcal{L}(t^n f(t)) = (-1)^n F^{n)}(s) \tag{2.22}$$

(g) Multiplication

$$\mathcal{L}(f_1(t)f_2(t)) \neq F_1(s)F_2(s) \tag{2.23}$$

(h) Final value

$$f(\infty) = \lim_{t\to\infty} f(t) = \lim_{s\to 0} sF(s) \tag{2.24}$$

Then, by combining the Laplace transform of simple signals with the properties above described, it is possible to build a table of Laplace transforms to derive the transform of common functions which appear frequently when the system response is demanded (see Table 2.2).

Example 2.3
The absorption of certain substances, e.g. drugs, in the body can be studied using compartmental models in which the input and output flow of the drug determines the concentration in each compartment. For instance, for an oral administration medicine we can model its concentration in the stomach (actually in the gastrointestinal system) and its absorption to the circulatory system as (see Fig. 2.10).

$$\begin{aligned}\frac{\mathrm{d}q_s}{\mathrm{d}t} &= i(t) - (k_{s1} + k_{s2})q_s \\ \frac{\mathrm{d}q_c}{\mathrm{d}t} &= k_{s2}q_s - k_c q_c\end{aligned} \tag{2.25}$$

where q_s and q_c represent the concentration of the drug within the gastrointestinal system and circulatory system, respectively, $i(t)$ is the input flow, k_{s2} is the transfer coefficient, and k_{s1} and k_c model the loss transfer coefficients.

Applying the Laplace transform properties to Eq. (2.25), we get

$$\begin{aligned}Q_s(s)s &= I(s) - (k_{s1} + k_{s2})Q_s(s) \\ Q_c(s)s &= k_{s2}Q_s(s) - k_c Q_c(s)\end{aligned} \tag{2.26}$$

Table 2.2 Laplace transform table

Function $f(t)$	Laplace transform $F(s)$
$\delta(t)$	1
$u_s(t)$	$\frac{1}{s}$
$tu_s(t)$	$\frac{1}{s^2}$
$e^{-at}u_s(t)$	$\frac{1}{s+a}$
$te^{-at}u_s(t)$	$\frac{1}{(s+a)^2}$
$\sin(\omega t)u_s(t)$	$\frac{1}{s^2+\omega^2}$
$\cos(\omega t)u_s(t)$	$\frac{s}{s^2+\omega^2}$
$t^n u_s(t)$	$\frac{n!}{s^{n+1}}$
$t^n e^{-at}u_s(t)$	$\frac{n!}{(s+a)^{n+1}}$
$\frac{1}{b-a}(e^{-at}-e^{-bt})u_s(t)$	$\frac{1}{(s+a)(s+b)}$
$\frac{1}{b-a}(be^{-bt}-ae^{-at})u_s(t)$	$\frac{s}{(s+a)(s+b)}$
$e^{-at}\sin(\omega t)u_s(t)$	$\frac{\omega}{(s+a)^2+\omega^2}$
$e^{-at}\cos(\omega t)u_s(t)$	$\frac{(s+a)}{(s+a)^2+\omega^2}$

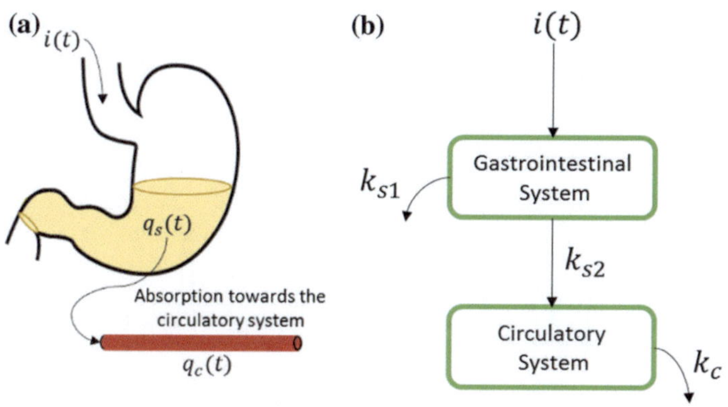

Fig. 2.10 Absorption of drugs modelled as a compartmental model. **a** Visual illustration of the system. **b** Schematic representation. Parameters k_{ij} are the transfer coefficients between the involved systems

If we are interesting in deriving the relation between the input dose $i(t)$ and its concentration into the circulatory system $q_c(t)$, we have to substitute the value of $q_s(t)$ from the first equation into the second one. In the Laplace domain, we have

$$Q_c(s) = \frac{k_{s2}}{(s+k_c)(s+k_{s1}+k_{s2})} I(s) \tag{2.27}$$

for values of $k_{s1} = 0.02\ \text{min}^{-1}$, $k_{s2} = 0.1\ \text{min}^{-1}$, $k_c = 0.05\ \text{min}^{-1}$, considering initial conditions as zero (Figs. 2.11 and 2.12)

$$Q_c(s) = \frac{0.1}{s^2 + 0.17s + 0.006} I(s) \tag{2.28}$$

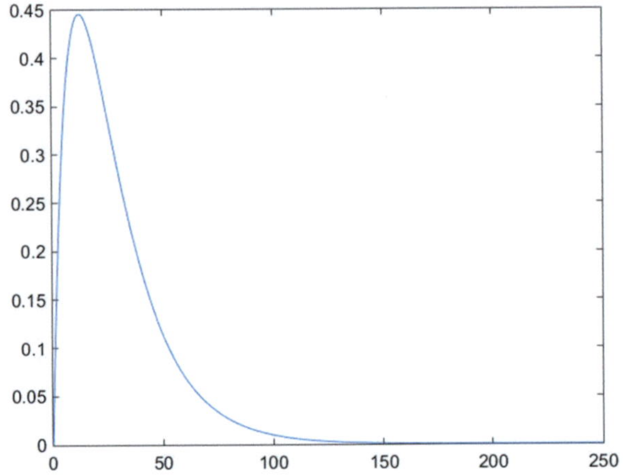

Fig. 2.11 Impulse response to the system depicted in Fig. 2.10

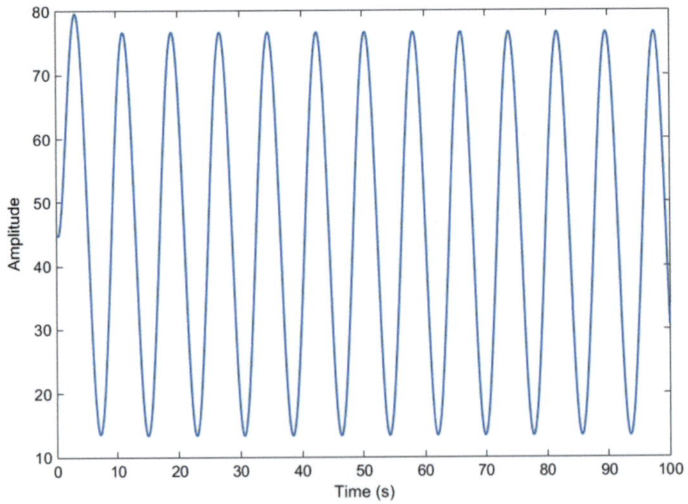

Fig. 2.12 Response to the system described by Eq. (2.44). The considered input is $u(t) = 45\ \sin\left(\frac{60}{72}t\right)$

Expression (2.28) yields the system response $Q_c(s)$ for an arbitrary input. Particularizing this for an impulse input, i.e. $I(s) = 1$, we obtain

$$Q_c(s) = \frac{0.1}{s^2 + 0.17s + 0.006} \tag{2.29}$$

Once we get the Laplace transform of the response of a continuous linear and invariant system as described, it is necessary to derive the temporal response, in this case $q_c(t)$, by applying the inverse Laplace transform, which enables us to restore the time response format.

The inverse Laplace transform of $Y(s)$ is represented symbolically as $\mathcal{L}^{-1}(Y(s))$ and is defined as the integral in the complex domain

$$\mathcal{L}^{-1}(Y(s)) = \int_{\sigma - j\omega}^{\sigma + j\omega} Y(s)e^{st}\mathrm{d}s = y(t), \quad t \geq 0 \tag{2.30}$$

where σ is a selected value to the right of all the singularities of $Y(s)$ in the s-plane.

In practice, this relation is not commonly used, as the Laplace transform $Y(s)$ is given as a proper rational function in s

$$Y(s) = \frac{N(s)}{D(s)} = \frac{n_m s^m + n_{m-1}s^{m-1} + \cdots + n_1 s + n_0}{s^n + d_{n-1}s^{n-1} + \cdots + d_1 s + d_0} \tag{2.31}$$

with $m \leq n$. This general expression can be separated as a sum of terms that can be individually transformed along with their corresponding time responses listed in Table 2.2.

The method consists of a partial fraction expansion method of the expression at hand, by firstly calculating the roots of $D(s)$, that is

$$D(s) = 0 = (s - p_1)^{r_1}(s - p_2)^{r_2} \ldots (s - p_l)^{r_l} \tag{2.32}$$

for $p_i, i = 1 \ldots l$ different roots, with multiplicity of r_l.

By factoring the rational function $Y(s)$ into its partial fraction expansion, we have in case of single, different roots

$$Y(s) = \frac{N(s)}{D(s)} = \frac{K_1}{(s + p_1)} + \frac{K_2}{(s + p_2)} + \cdots + \frac{K_l}{(s + p_l)} \tag{2.33}$$

where the constants K_j or residuals of $Y(s)$ are given by

$$K_j = \lim_{s \to -p_j} Y(s)(s + p_j) \qquad j = 1 \ldots l \tag{2.34}$$

while in case of multiple roots $Y(s)$ is decomposed as

$$
\begin{aligned}
Y(s) = \frac{N(s)}{D(s)} &= \frac{K_{11}}{(s+p_1)} + \frac{K_{12}}{(s+p_1)^2} + \cdots + \frac{K_{1r_1}}{(s+p_1)^{r_1}} + \frac{K_{21}}{(s+p_2)} \\
&+ \frac{K_{22}}{(s+p_2)^2} + \cdots + \frac{K_{2r_2}}{(s+p_2)^{r_2}} + \cdots + \frac{K_{l1}}{(s+p_l)} \\
&+ \frac{K_{l2}}{(s+p_l)^2} + \cdots + \frac{K_{lr_l}}{(s+p_l)^{r_l}}
\end{aligned} \tag{2.35}
$$

where the residuals K_{ji} are defined by

$$
K_{ji} = \frac{1}{(r_j - i)!} \lim_{s \to -p_j} \frac{\mathrm{d}^{r_j - i}\left(Y(s)(s+p_j)^{r_j}\right)}{\mathrm{d}s^{r_j - i}} \quad j = 1\ldots l, \quad i = 1, \ldots r_j \tag{2.36}
$$

Once the residuals are determined, we proceed to calculate the inverse Laplace transforms of each individual fraction by identifying the corresponding time function listed in Table 2.2 and then use the superposition relation to obtain the desired function $y(t)$.

In case of complex conjugate roots, it is necessary to group the corresponding partial fractions prior to identify the inverse transform. Assuming that the complex root is located at $p_1 = -\sigma + j\omega$ and $p_2 = -\sigma - j\omega$, the corresponding partial fractions are given by

$$
Y(s) = \frac{N(s)}{D(s)} = \cdots + \frac{K_1}{s + \sigma - j\omega} + \frac{K_2}{s + \sigma + j\omega} + \cdots \tag{2.37}
$$

Applying (2.34), we obtain the residuals

$$
\begin{aligned}
K_1 &= \lim_{s \to -\sigma + j\omega} Y(s)(s + \sigma - j\omega) \\
K_2 &= \lim_{s \to -\sigma - j\omega} Y(s)(s + \sigma + j\omega)
\end{aligned}
$$

where $K_1 = \alpha_1 + j\beta_1$ and $K_2 = \alpha_1 - j\beta_1$, that is a pair of complex conjugate residuals.

Grouping the two partial fractions, we have

$$
Y(s) = \frac{N(s)}{D(s)} = \cdots + \frac{\hat{K}_1(s+\sigma)}{(s+\sigma)^2 + \omega^2} + \frac{\hat{K}_2\omega}{(s+\sigma)^2 + \omega^2} + \cdots \tag{2.38}
$$

with $\hat{K}_1 = 2\alpha_1$ and $\hat{K}_2 = -2\beta_1$, so it is now possible to calculate the inverse Laplace transforms of each individual fraction by identifying the corresponding time function listed in Table 2.2, which correspond to an attenuated sine and cosine functions.

Example 2.4
Obtain the impulse response $q_c(t)$ corresponding to the expression (2.29).

Solution
Following the aforementioned procedure, we start by obtaining the roots of the denominator, i.e.

$$s^2 + 0.17s + 0.006 = 0 \tag{2.39}$$

by applying the MATLAB code

```
≫ den = [1 0.17 0.006];
≫ roots(den);

ans =
    -0.1200
    -0.0500
```

In this way, we can rewrite $Q_c(s)$ as the partial fraction expansion

$$Q_c(s) = \frac{0.1}{s^2 + 0.17s + 0.006} = \frac{K_1}{(s + 0.12)} + \frac{K_2}{(s + 0.05)} \tag{2.40}$$

Using (2.34), we obtain the residues

$$K_1 = \lim_{s \to -0.12} Q_c(s + 0.12) = -1.4286$$
$$K_2 = \lim_{s \to -0.05} Q_c(s + 0.05) = 1.4286$$

We can check the results obtained by applying the MATLAB code.

```
≫ num = [0.1];den = [1 0.17 0.006];
≫ [r,p] = residue(num,den)
r =
    -1.4286
     1-4286
p =
    -0.12
    -0.05
```

In this way, the partial fraction expansion becomes

$$Q_c(s) = \frac{0.1}{s^2 + 0.17s + 0.006} = \frac{-1.4286}{(s + 0.12)} + \frac{1.4286}{(s + 0.05)} \tag{2.41}$$

so that the time response can be calculated by adding the corresponding inverse transform of each fraction in (2.41), yielding the solution

$$q_c(t) = -1.4286e^{-0.12t} + 1.4286e^{-0.05t}$$

This function can be plotted in MATLAB through the code

```
>>t = 0:0.1:250;
>>y = -1.4286*exp(-0.12*t) + 1.4286*exp(-0.05*t);
>> plot(t,y)
```

Example 2.5
The circulatory system can be seen as a hydraulic system, where the heart acts as a pump and the blood vessels are the pipes. Following this analogy, we can model an aortic segment as

$$\frac{L\mathrm{d}f}{\mathrm{d}t} + Rf(t) + \frac{1}{C}\int f(t)\mathrm{d}t = p_a(t) \tag{2.42}$$

$$p_s(t) = \frac{1}{C}\int f(t)\mathrm{d}t \tag{2.43}$$

where $L, R,$ and C are the inheritance, resistance and capacitance of the aortic segment, respectively, $f(t)$ is the aortic blood flow, and $p_a(t)$ and $p_s(t)$ are the arterial and systemic pressures, respectively. If we are interested in the relation between $p_a(t)$ and $p_s(t)$, we can eliminate $f(t)$ by substituting its value from Eq. (2.42) into Eq. (2.43)

$$\frac{LC\mathrm{d}p_s}{\mathrm{d}t} + \frac{RC\mathrm{d}p_s}{\mathrm{d}t} + p_s(t) = p_a(t) \tag{2.44}$$

In order to get the response of the system, $p_s(t)$, for a given input, $p_a(t)$, we have to obtain the Laplace transformation. Applying Laplace properties, we get

$$LCs^2P_s(s) + RCsP_s(s) + P_s(s) = P_a(s) \tag{2.45}$$

So that the response turns into

$$P_s(s) = \frac{P_a(s)}{LCs^2 + RCs + 1} \tag{2.46}$$

for values of $L = 0.000825, R = 1, C = 1.46,$ and initial conditions $p_s(0) = 0, \frac{\mathrm{d}p_s}{\mathrm{d}t}(0) = 0$, Eq. (2.46) turns into

$$P_s(s) = \frac{P_a(s)}{0.0012s^2 + 1.46s + 1} \tag{2.47}$$

In case of a sinusoidal input $p_a(t) = A\ \sin(\omega t)$, its Laplace transformation is $P_a(s) = \frac{A}{s^2+\omega^2}$, and the system response for $A = 45, \omega = 60/72$ is given by

$$P_s(s) = \frac{45}{0.0012s^4 + 1.46s^3 + 1.001s^2 + 1.014s + 0.6944} \tag{2.48}$$

The roots of the denominator of Eq. (2.48) can be computed as

```
≫ den = [0.0012 1.45 1.001 1.014 0.6944];
≫ roots(den)

ans =

  1.0e + 03 *

 -1.2114  +  0.0000i
 -0.0000  +  0.0008i
 -0.0000 - 0.0008i
 -0.0007  +  0.0000i
```

Thus, we can expand the expression $P_a(s)$ into partial fractions as

$$V_O(s) = \frac{0.1}{s(0.001s^2 + 0.012s + 1)} = \frac{100}{s(s^2 + 12 + 1000)} \tag{2.49}$$

$$P_a(s) = \frac{K_1}{s + 1211.4} + \frac{K_2}{s + 0 - 0.8j} + \frac{K_3}{s + 0 + 0.8j} + \frac{K_4}{s + 0.7} \tag{2.50}$$

Obtaining the residues through the residue MATLAB command

```
≫ r = residue(45,den)

r =

-0.0000 + 0.0000i

 -13.2534 -10.8841i
 -13.2534  + 10.8841i
  26.5069  +  0.0000i
```

and thus, Eq. (2.50) becomes

$$P_a(s) = \frac{-13.2534 + 10.8841j}{s + 0 - 0.8j} + \frac{-13.2534 - 10.8841j}{s + 0 + 0.8j} + \frac{26.5069}{s + 0.7} \quad (2.51)$$

By grouping the two complex fractions as indicated in (2.38), we have

$$P_a(s) = \frac{-26.5068s}{s^2 + 0.8^2} + \frac{21.7682 * 0.8}{s^2 + 0.8^2} + \frac{26.5069}{s + 0.7} \quad (2.52)$$

so that the time response can be calculated by adding the corresponding inverse transform of each fraction in (2.52), yielding the solution

$$p_a(t) = -26.5068\cos(0.8 \cdot t) + 21.7682\sin(0.8 \cdot t) + 26.5069e^{-0.7t}$$

which can be plotted in MATLAB as

```
≫ t = 0:0.1:100;
≫       out = -2*13.2534*cos(0.8*t) + (2*10.884*0.8)*sin
(0.8*t) +
26.5069*exp(-0.7*t);
≫ plot(t,45+out)
```

Discrete-Time Linear Systems

A discrete-time system is one for which the inputs, state variables and outputs are defined over discrete instant of time, and its behaviour is described by difference equations. In case the discrete-time signal is generated by sampling a continuous-time signal with T period, the signal is called a sampled data signal. Besides, if this sampled data signal has quantized amplitude, then it is called a digital signal usually defined as a sequence of binary numbers (see Fig. 2.13).

Some of the most common input signals used to analyse the discrete-time system behaviour or to design control systems are shown in Table 2.3 and correspond to sampled counterpart versions of these signals already defined in Table 2.2.

In general, a discrete-time system with i inputs and j outputs (see Fig. 2.14) can be described by m independent input–output difference equations where a sequence of values arising from the sampling operation of input and output signals $u(kT)$ and $y(kT)$ is given by

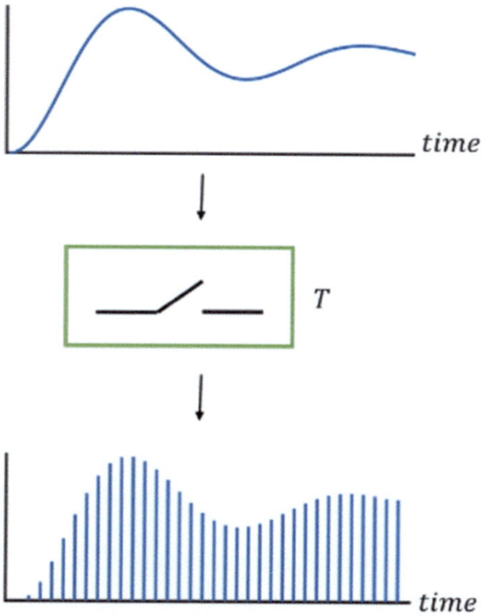

Fig. 2.13 Continuous- and discrete-time signals

Table 2.3 Some test input signals used for system analysis

Impulse	$\delta(k)$	
Step	$u_s(k)$	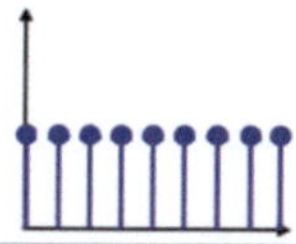
Ramp	$r(k) = ku_s(k)$	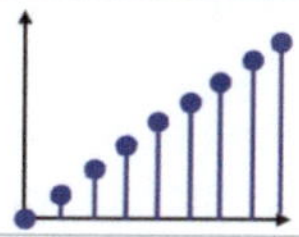
Sine	$\sin(\omega k)u_s(k)$	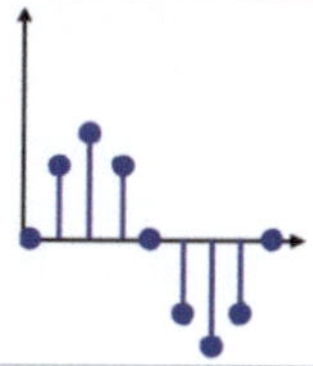

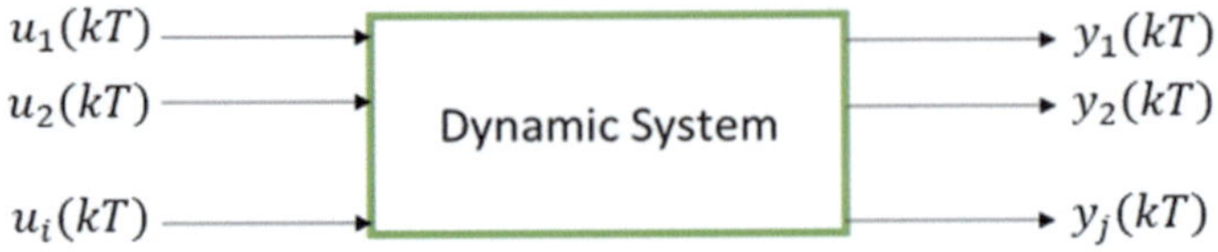

Fig. 2.14 Discrete-time multi-input, multi-output system

$$\begin{aligned}
& f_1(y_1(kT),\ldots,y_1((k-n)T),\ldots,y_j(kT),\ldots,y_j((k-n)T),\\
& \quad u_1(kT),\ldots,u_1((k-n)T),\ldots,u_i(kT),\ldots,u_i((k-n)T),kT)=0\\
& f_2(y_1(kT),\ldots,y_1((k-n)T),\ldots,y_j(kT),\ldots,y_j((k-n)T),\\
& \quad u_1(kT),\ldots,u_1((k-n)T),\ldots,u_i(kT),\ldots,u_i((k-n)T),kT=0\\
& f_m(y_1(kT),\ldots,y_1((k-n)T),\ldots,y_j(kT),\ldots,y_j((k-n)T),\\
& \quad u_1(kT),\ldots,u_1((k-n)T),\ldots,u_i(kT),\ldots,u_i((k-n)T),kT=0
\end{aligned} \tag{2.53}$$

where n is the maximum order of delayed samples of $y_j(kT)$ and $u_i(kT)$, with $f_l, l = 1,\ldots,m$ nonlinear in general. The parameter n is denoted as the *discrete-time system order* and plays a fundamental role for the system behaviour.

If the discrete-time system is single-input and single-output, the dynamic difference equation is given by

$$f(y(kT),\ldots,y((k-n)T),u(kT),\ldots,u((k-n)T),kT)=0 \tag{2.54}$$

with f nonlinear in general. The difference equation can be commonly written in a similar way by making $k-n=l$ so that

$$f(y((l+n)T),\ldots,y(lT),u((l+n)T),\ldots,u(lT),lT)=0 \tag{2.55}$$

or considering index l to represent time samples

$$f(y(l+n),\ldots,y(l),u(l+n),\ldots,u(l),l)=0 \tag{2.56}$$

Analytical methods for exact solving (2.4) given the input signal $u(l)$ and an appropriate set of initial conditions are available by using recursion in case $y(l+n)$ could be solved in terms of its delayed values.

Let us particularize the analysis for a single-input single-output system, with f described by a linear combination of both delayed samples of input and output signals, as in the case of (2.2), assuming also the system is stationary, and no direct dependence with time k is explicit.

A discrete-time system defined by difference equations is said to be linear if it can be expressed as a linear combination of delayed samples of the output and the input in the form

$$a_n y(k+n) + \cdots + a_1 y(k+1) + a_0 y(k) = b_n u(k+n) + \cdots + b_1 u(k+1) + b_0 u(k) \tag{2.57}$$

where the parameters $a_j, j = 1 \ldots n$ and $b_k, k = 1 \ldots n$ are constants or functions of time k. In the event that these parameters do not depend explicitly on the time k, the system is said to be also time-invariant, and it is not the case, the system is said to be nonlinear.

In this case, recursion enables us to solve $y(l+n)$ as a set of initial conditions $y(0), y(1), \ldots, y(n-1)$ and input signal $u(k)$ are given, so that clearing up $y(k+n)$ and assuming also $a_n \neq 0$ we have

$$y(k+n) = \frac{1}{a_n}(b_n u(k+n) + \cdots + b_1 u(k+1) + b_0 u(k) - \cdots - a_1 y(k+1) - a_0 y(k)) \tag{2.58}$$

so by substituting $k = 0, \ldots, \infty$, correspondent values of $y(n), y(n+1), \ldots, y(\infty)$ are elicited. This recursive equation can be easily implemented in a computer by updating both output and input signals $y(k)$ and $u(k)$.

The *Z* Transform

The Z transform is a mathematical tool used to solve the linear difference equations with constant coefficients expressed in the discrete-time domain which define the linear system behaviour, thus transforming then into linear algebraic equations expressed in a complex domain, just in the same way it was done for continuous-time systems throughout Laplace transform.

The Z transform of $f(kT)$ is represented symbolically as $Z(f(kT))$ and is defined as

$$Z(f(kT)) = F(z) = \sum_{k=0}^{\infty} f(kT) z^{-k} z \in \mathbb{C} \tag{2.59}$$

$$F(z) = f(0) + f(T)z^{-1} + f(2T)z^{-2} + \cdots \tag{2.60}$$

where $z = |z|e^{j\phi}$ is a complex variable, and it plays the same role as the Laplace transform. In fact, by applying Eq. (2.59) it is possible to derive a pair $f(kT) \underset{Z}{\Rightarrow} F(z)$ for each function, so that it is possible to derive a complete table for the most representative input and output signals appearing in dynamic discrete-time systems.

Example 2.6 Derive the Z transform of the impulse, step and exponential signals

Solution

1. $Z(\delta(kT)) = 1$, therefore $\delta(kT) \underset{Z}{\Rightarrow} 1$
2. $Z(u_s(kT)) = \sum_{k=0}^{\infty} 1 \cdot z^{-k} = \sum \sum_{k=0}^{\infty} 1 \cdot (z^{-1})^k = \frac{1}{1-z^{-1}} = \frac{z}{z-1}$ if $|z^{-1}| < 1$,

   therefore $u_s(kT) \underset{Z}{\Rightarrow} \frac{z}{z-1}$
3. $Z\left(e^{-akT} \cdot u_s(kT)\right) = \sum_{k=0}^{\infty} e^{-akT} \cdot z^{-k} = \sum_{k=0}^{\infty} (e^{-aT} z^{-1})^k$ if $\left|e^{-aT} z^{-1}\right| < 1$

   therefore $e^{-akT} \cdot u_s(kT) \underset{Z}{\Rightarrow} \frac{z}{z-e^{-aT}} \frac{1}{1 - e^{-aT} z^{-1}} = \frac{z}{z - e^{-aT}}$

Nevertheless, in order to find the transform pairs for complex signals and to solve for the responses of dynamic linear systems given by (2.57), we illustrate some useful properties of the Z transform:

(a) Linearity

$$Z(a_1 f_1(k) + a_2 f_2(k)) = a_1 F_1(z) + a_1 F_2(z) \tag{2.61}$$

(b) Damping

$$Z\left(c^k f(k)\right) = X\left(\frac{z}{c}\right) \quad \text{with} \quad c = e^{-aT} \tag{2.62}$$

(c) Backward time shift

$$Z(f(k-n))) = z^{-n} F(k) \tag{2.63}$$

(d) Forward time shift

$$Z(f(k+n)) = z^n \left(F(k) - \sum_{k=0}^{\infty} x(l) z^{-l} \right) \tag{2.64}$$

(e) Multiplication by time

$$Z(kf(k)) = -z \frac{dF(z)}{dz} \tag{2.65}$$

(f) Multiplication

$$Z(f_1(k) f_2(k)) \neq F_1(z) F_2(z) \tag{2.66}$$

(g) Final value

$$f(\infty) = \lim_{k\to\infty} f(k) = \lim_{z\to 1} \frac{z-1}{z} F(z) \quad (2.67)$$

Then, by combining the *Z* transform of simple signals with the properties above described, it is possible to build a table of *Z* transforms to derive the transform of common functions which appear frequently when the discrete-time system response is demanded (Table 2.4).

Example 2.7
Let us consider an automatically controlled insulin pump (see Fig. 2.15).

The digital control algorithm that is used to activate the operation of the insulin pump is given by a difference equation relating the control action $u_c(k)$ to the discrepancy between glucose set point and measured glucose $e_g(k)$. In case of a proportional-derivative digital controller, this relation is given by

Table 2.4 *Z* transform table

Function $f(kT)$	*Z* transform $F(z)$
$\delta(kT)$	1
$u_s(kT)$	$\frac{z}{z-1}$
$kTu_s(kT)$	$\frac{Tz}{(z-1)^2}$
$c^k u_s(kT) \quad (c = e^{-aT})$	$\frac{z}{z-c}$
$kTc^k u_s(kT)$	$\frac{Tcz}{(z-c)^2}$
$\sin(\omega kT)u_s(kT)$	$\frac{z\sin(\omega T)}{z^2-2z\cos(\omega T)+1}$
$\cos(\omega kT)u_s(kT)$	$\frac{z(z-cos(\omega T))}{z^2-2zcos(\omega T)+1}$
$c^k \sin(\omega kT)u_s(kT)$	$\frac{c\cdot z\sin(\omega T)}{z^2-2cz\cos(\omega T)+c^2}$
$c^k \cos(\omega kT)u_s(kT)$	$\frac{z(z-c\cdot\cos(\omega T))}{z^2-2c\cdot z\cos(\omega T)+c^2}$

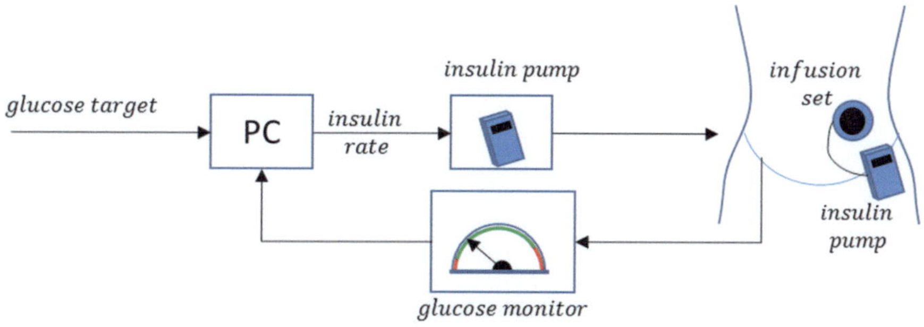

Fig. 2.15 Discrete-time controller for glycemic regulation

$$u_c(k) = K_p e_g(k) + \frac{K_d}{T}\left(e_g(k) - e_g(k-1)\right) \tag{2.68}$$

where K_p, K_d and T are parameters of a digital proportional-derivative controller and the sampling time, respectively. Applying the Z transform properties, we can get

$$U_c(z) = K_p E_g(z) + \frac{K_d}{T} E_g(z)\left(1 - z^{-1}\right) \tag{2.69}$$

A direct relation between the controller response and input discrete-time signals $u_c(k)$ and $e_g(k)$ can be obtained in the Z domain rearranging (2.69) so that

$$U_c(z) = \frac{(K_p T + K_d)z - K_d}{z} E_g(z) \tag{2.70}$$

Particularizing for values of $K_p = 5$, $K_d = 0.1$ and $T = 0.1$ s, considering initial conditions as zero, then

$$U_c(z) = \frac{0.6z - 0.1}{z} E_g(z) \tag{2.71}$$

whose step response (shown in Fig. 2.16), i.e. $E_g(z) = \frac{z}{z-1}$, is defined in z as

$$U_c(z) = \frac{0.6z^2 - 0.1z}{z^2 - z} \tag{2.72}$$

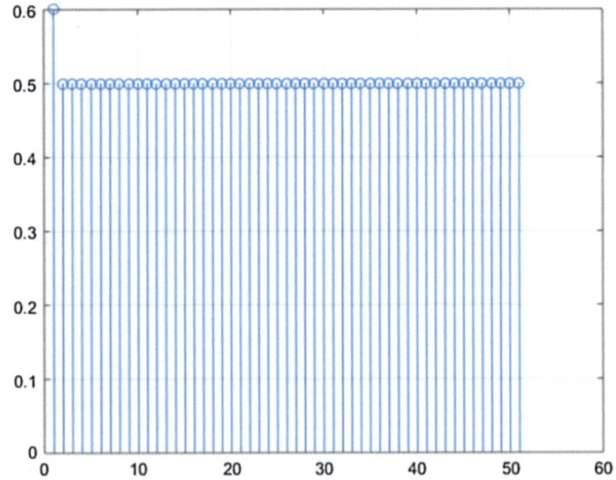

Fig. 2.16 Step response to the system depicted in Fig. 2.15

Once we get the Z transform of the response of a discrete-time linear and invariant system as described, it is necessary to derive the temporal response, in this case $u_c(k)$, by applying the inverse Z transform, which enables us to restore the discrete-time response format.

The inverse Z transform of a generic output signal $Y(z)$ is represented symbolically as $Z^{-1}(Y(z))$, and as $Y(z)$ is usually given as a proper rational function in z when a linear and invariant discrete-time system is considered, after applying an arbitrary input, $Y(z)$ is given by

$$Y(z) = \frac{N(z)}{D(z)} = \frac{n_m z^m + n_{m-1} z^{m-1} + \cdots + n_1 z + n_0}{z^n + d_{n-1} z^{n-1} + \cdots + d_1 z + d_0}$$

with $m \leq n$. This general expression can be separated as a sum of terms that can be individually transformed along with their corresponding time responses listed in Table 2.4.

The method consists of a partial fraction expansion method of the expression at hand, by firstly calculating the roots of $D(z)$ by following the same scheme already described for continuous-time systems, that is

$$D(z) = 0 = (z - q_1)^{r_1} (z - q_2)^{r_2} \ldots (z - q_l)^{r_l}$$

for $q_i, i = 1 \ldots l$ different roots, with multiplicity of r_l. Nevertheless, the partial fraction expansion of $\frac{Y(z)}{z}$ rather than $Y(z)$ will be obtained in order to get factored the rational function, so that in case of single, different roots

$$\frac{Y(z)}{z} = \frac{M(z)}{D(z)} = \frac{K_1}{(z - q_1)} + \frac{K_2}{(z - q_2)} + \cdots + \frac{K_l}{(z - q_l)}$$

where the constants K_j or residuals of $\frac{Y(z)}{z}$ are given by

$$K_j = \lim_{z \to q_j} \frac{Y(z)}{z} (z - q_j) \quad j = 1 \ldots l \tag{2.73}$$

while in case of multiple roots $\frac{Y(z)}{z}$ is decomposed as

$$\begin{aligned} \frac{Y(z)}{z} = \frac{M(z)}{D(z)} &= \frac{K_{11}}{(z - q_1)} + \frac{K_{12}}{(z - q_1)^2} + \cdots + \frac{K_{1r_1}}{(z - q_1)^{r_1}} + \frac{K_{21}}{(z - q_2)} \\ &+ \frac{K_{22}}{(z - q_2)^2} + \cdots + \frac{K_{2r_2}}{(z - q_2)^{r_2}} + \cdots + \frac{K_{l1}}{(z - q_l)} \\ &+ \frac{K_{l2}}{(z - q_l)^2} + \cdots + \frac{K_{lr_l}}{(z - q_l)^{r_l}} \end{aligned} \tag{2.74}$$

where the residuals K_{ji} are defined by

$$K_{ji} = \frac{1}{(r_j - i)!} \lim_{z \to q_j} \frac{d^{r_j - i}\left(\frac{Y(z)}{z}(z - q_j)^{r_j}\right)}{dz^{r_j - i}} \quad j = 1 \ldots l, \quad i = 1, \ldots r_j \tag{2.75}$$

Once the residuals are determined, we proceed to calculate the inverse Z transforms of each individual fraction by identifying the corresponding time function listed in Table 2.4 and then use the superposition relation to obtain the output response $y(k)$.

In case of complex conjugate roots, it is necessary to group the corresponding partial fractions prior to identify the Z inverse transform in a similar way as it was done for Laplace inverse transform.

Example 2.8 Obtain the step response $u_c(k)$ corresponding to Example 2.7.

Solution

Following the aforementioned procedure, the roots of the denominator of $U_c(z)$ are obtained, i.e.

$$z^2 - z = 0 \tag{2.76}$$

by applying the MATLAB code

```
≫ denz= [1 -1 0];
≫ roots(denz);
```

```
ans =
     0
    -1
```

In this way, we can rewrite $U_c(z)/z$ as the partial fraction expansion

$$\frac{U_c(z)}{z} = \frac{0.6z - 0.1}{z^2 - z} = \frac{K_1}{(z - 1)} + \frac{K_2}{z} \tag{2.77}$$

Using (2.73), we obtain the residues

$$K_1 = \lim_{z \to 1} \frac{U_c(z)}{z}(z - 1) = 0.5$$
$$K_2 = \lim_{z \to 0} \frac{U_c(z)}{z}(z) = 0.1$$

We can verify the results obtained by applying the MATLAB code

```
≫ num=[0.6 0.1];den=[1 -1 0];
≫ [r,p]=residue(num,den)
r =
      0.5000
      0.1000
p =
      1
      0
```

Then, the partial fraction expansion of $U_c(z)$ will be given by

$$U_c(z) = \frac{0.5z}{(z-1)} + 0.1$$

so that the discrete-time response can be calculated by adding the corresponding Z inverse transform of each fraction in (), yielding the solution

$$u_c(k) = 0.5u_s(k) + 0.1\delta(k) \tag{2.78}$$

This function can be plotted in MATLAB through the code

```
≫T=0.1 ;
≫k=0:T:5;
≫us=ones(length(k));delta=zeros(length(k));delta(1)=1;
≫uc=0.5*us+0.1*delta;
plot(k,uc,'o' )
```

Continuous and Discrete Transfer Functions

The transfer function of an invariant linear system is defined as the relation between the Laplace transform of system output $Y(s)$ and the transform of the system input $U(s)$ under the assumption of null initial conditions. The transfer function is in fact a representation of the system at hand, since it describes the response of the system for any arbitrary input.

Considering the single-input single-output continuous-time system defined by a linear and invariant differential equation in the form

$$a_n y^{n)} + \cdots + a_1\dot{y} + a_0 y = b_m u^{m)} + \cdots + b_1\dot{u} + b_0 u \tag{2.79}$$

and taking the Laplace transform of both sides, we have

$$a_n s^n Y(s) + \cdots + a_1 sY(s) + a_0 Y(s) = b_m s^m U(s) + \cdots + b_1 sU(s) + b_0 U(s) \tag{2.80}$$

by assuming the initial conditions are null. Taking the ratio between $Y(s)$ and $U(s)$, we obtain the transfer function of this system as defined by

$$G(s) = \frac{Y(s)}{U(s)} = \frac{b_m s^m + \cdots + b_1 s + b_0}{a_n s^n + \cdots + a_1 s + a_0} \tag{2.81}$$

That is, the transfer function is a property of the system itself, since it does not depend on the input, and can be represented as a block diagram, see Fig. 2.17.

In fact, the output response $Y(s)$ for an input signal $U(s)$ can be obtained by using the relation

$$Y(s) = G(s)U(s) \tag{2.82}$$

assuming there are no initial conditions.

The highest power of the denominator of $G(s)$ (characteristic equation) is called the system *order*. The roots of the characteristic equation are called the system *poles*, while the roots of the numerator are called the system *zeros*.

This form of representation corresponds to the external description of the system, which provides no information on its internal structure. Moreover, the transfer function of different systems may be the same, which is the case of analogous systems.

Example 2.9
Obtain the transfer function of the Westheimer model for the saccadic movements of the human eye, given by:

$$\frac{Jd\theta^2}{dt} + \frac{Bd\theta}{dt} + K\theta(t) = K_\tau \tau(t) \tag{2.83}$$

in which $\theta(t)$ is the output, i.e. the angular position of the eye, and $\tau(t)$ represents the input generated by the eye muscles.

The Laplace transform of (2.83) results in:

$$\left(Js^2 + Bs + K\right)\theta(s) = K_\tau \tau(s) \tag{2.84}$$

Thus, using (2.81)

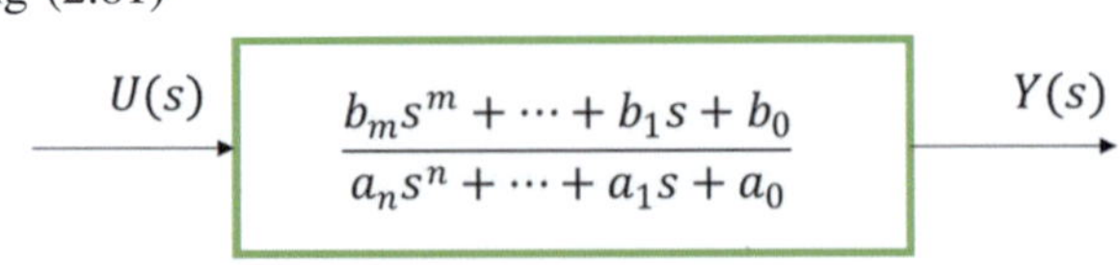

Fig. 2.17 Block diagram of the transfer function $G(s)$

$$G(s) = \frac{Y(s)}{U(s)} = \frac{\theta(s)}{\tau(s)} = \frac{K_\tau}{Js^2 + Bs + K} \tag{2.85}$$

For parameters $J = 2, B = 3, K = 1, K_\tau = 6$, the transfer function (2.85) becomes:

$$G(s) = \frac{6}{2s^2 + 3s + 1} \tag{2.86}$$

Having the transfer function $G(s)$, the output for a particular input can be easily computed by multiplying $G(s)$ times the Laplace transform of the input. Thus, for example, for the step response, we obtain $\tau(s) = \mathcal{L}(\tau(t)) = \frac{1}{s}$, so that

$$\theta(s) = \frac{6}{2s^2 + 3s + 1}\tau(s) = \frac{6}{s(2s^2 + 3s + 1)} \tag{2.87}$$

Following the procedure to obtain the inverse of the Laplace transform, the response (2.87) is expressed as:

$$\theta(s) = \frac{6}{s(2s^2 + 3s + 1)} = \frac{K_1}{s} + \frac{K_2}{(s + p_1)} + \frac{K_3}{(s + p_2)}$$

Calculating the residues and poles through the MATLAB code

```
>> num=[6];den=[2 3 1 0];
>> [r,p]=residue(num,den)

r =

     6
   -12
     6

p =

    -1
    -0.5
     0
```

and the partial fraction expansion becomes

$$\theta(s) = \frac{6}{s} + \frac{6}{(s + 1)} - \frac{12}{(s + 0.5)} \tag{2.88}$$

so that the time response can be calculated by adding the corresponding inverse transform of each fraction in (2.88), yielding the solution

$$\theta(t) = \left(6 + 6e^{-t} - 12e^{-0.5t}\right)u_s(t) \tag{2.89}$$

The step response of the system is shown in Fig. 2.18 using the MATLAB code:

```
≫ t = 0:0.1:15; y=6+6*exp(-t)-12*exp(-0.5*t);
≫ plot(t,y)
```

Alternatively, we can create the transfer function directly by:

```
≫ s=tf('s');
≫ G=6/(2*s^2+3*s+1)

G =

          6
  --------------------
  2 s^2 + 3 s + 1

Continuous-time transfer function.
```

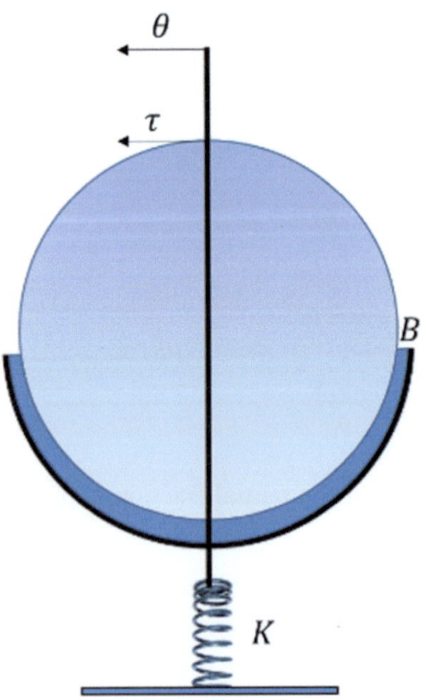

Fig. 2.18 Westheimer model of human saccadic eye movement

Fig. 2.19 Step response for the temporal function (2.89)

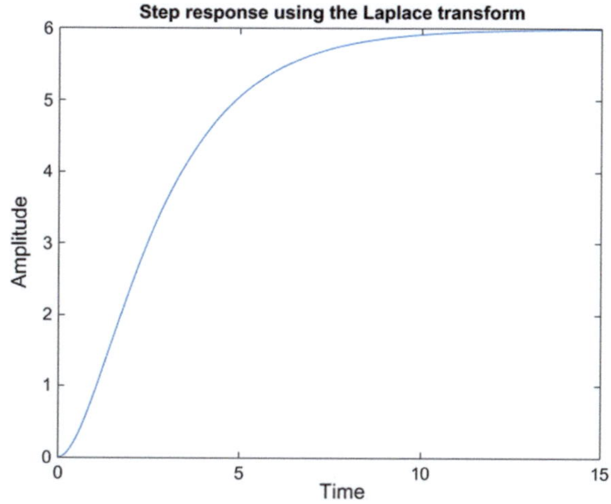

Once the transfer function is declared in MATLAB, a number of operations can be done, for instance producing the system output for some typical input signals, like step, using the command `step(G,15)` (see Fig. 2.19). MATLAB computes and displays the output of the system represented by the transfer function G for a step input during 15 s. Alternatively, for the impulse response, the command is (Fig. 2.20):

```
≫ impulse(G,15)
```

Regarding the discrete transfer function, the definition is equivalent to the one discussed in the continuous case. That is, the discrete transfer function is the relation between the Z transform of the system output and the Z transform of the input, considering null initial conditions. That is, from the difference equation:

$$\begin{aligned} a_n y(k+n) + a_{n-1} y(k+n-1) + \cdots a_1 y(k+1) + a_0 y(k) \\ = b_m r(k+m) + \cdots b_1 r(k+1) + b_o r(k) \end{aligned} \tag{2.90}$$

The discrete transfer function is defined as

$$G(z) = \frac{Y(z)}{R(z)} = \frac{b_z z^m + \cdots + b_1 z + b_0}{a_n z^n + \cdots + a_1 z + a_0} \tag{2.91}$$

Example 2.10
Consider an alternative digital controller for Example 2.7 given by the equation

$$u_e(k+2) = 0.5u_e(k+1) + 0.2u_e(k) + e_g(k+1) + 0.5e_g(k) \tag{2.92}$$

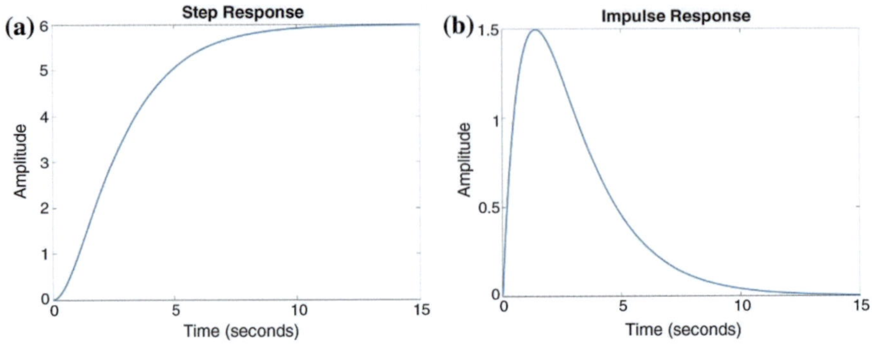

Fig. 2.20 Step (**a**) and impulse (**b**) responses of a system obtained by the step and impulse MATLAB commands

where u_e is the control action and e_g is the input to the controller, i.e. the error between the output of the glucose sensor and the reference. Obtain the discrete transfer function.

The Z transform is:

$$z^2U_e(z) - 0.5zU_e(z) - 0.2U_e(z) = zE_g(z) + 0.5E_g(z)$$

and thus, the transfer function is

$$G(z) = \frac{U_e(z)}{E_g(z)} = \frac{z + 0.5}{z^2 - 0.5z - 0.2}$$

Continuous and Discrete Matrix Transfer Function

In the case of multi-input multi-output continuous or discrete systems (Fig. 2.21), we define the transfer matrix $\bar{G}(s)(\bar{G}(z))$, whose elements $G_{ij}(s)\left(G_{ij}(z)\right)$ relate each output $Y_i(s)(Y_i(z))$ to each input $U_j(s)\left(U_j(z)\right)$, assuming zero both the remaining entries $U_k(s)(U_k(z)), k \neq j$ and the initial conditions.

For the case of continuous systems (the same definitions are valid for the discrete case), each individual transfer function element $G_{ij}(s)$ is given by the relation

$$G_{ij}(s) = \frac{Y_i(s)}{U_j(s)} \tag{2.93}$$

and assuming no initial conditions we derive each of the system output, following the expressions

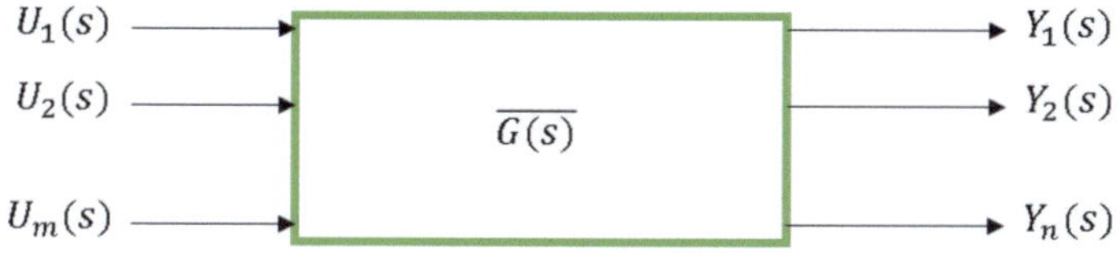

Fig. 2.21 Continuous multi-input, multi-output system

$$\begin{aligned} Y_1(s) &= G_{11}(s)U_1(s) + \cdots + G_{1m}(s)U_m(s) \\ Y_2(s) &= G_{21}(s)U_1(s) + \cdots + G_{2m}(s)U_m(s) \\ &\vdots \\ Y_n(s) &= G_{n1}(s)U_1(s) + \cdots + G_{nm}(s)U_m(s) \end{aligned} \tag{2.94}$$

which can be transformed to a vector notation

$$\bar{Y}(s) = \bar{G}(s)\bar{U}(s) \tag{2.95}$$

assuming there are no initial conditions, with $\bar{Y}(s) = [Y_1(s)\ldots Y_n(s)]$ and $\bar{U}(s) = [U_1(s)\ldots U_m(s)]$.

Block Diagrams

Complex systems may involve several, interrelated components which exhibit a particular dynamics captured by their particular equations. A convenient way to tackle the complexity of some systems is to draw a diagram that conveniently combines the transfer functions of each component. This is the idea behind the block diagrams: an intuitive manner to transform system equations (differential or difference) into a graphical diagram that captures the system dynamics and their components.

A block diagram connects blocks that represent transfer functions through lines which represent the system variables. In this way, the output of a block can be the input of others. In addition, mathematical operations like the addition/subtraction of signals and distribution points that transmit system variables through different components are used to design a block diagram. Some of these elements are shown in Fig. 2.22.

The procedure to construct a block diagram starts by transforming the equations of the system using the Laplace and connecting them following the system dynamics.

Example 2.11
Obtain the block diagram of the electromechanical hydraulic system constituted by an infusion pump operated by a DC motor (see Fig. 2.23) that follows the dynamic equations (2.96)–(2.100).

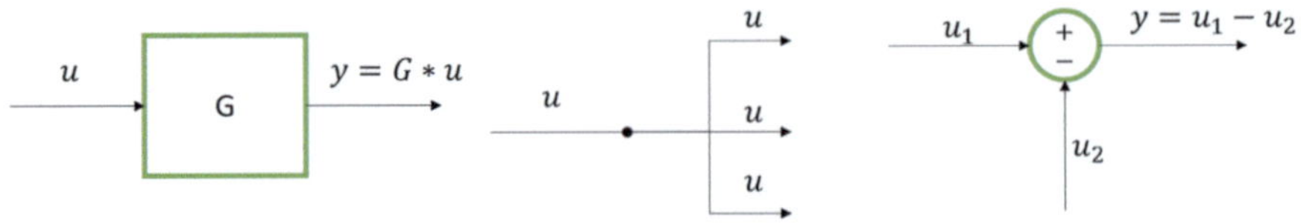

Fig. 2.22 Elements used in a block diagram. From left to right: block (transfer function), pickoff point and junction point implementing an addition/subtraction operation

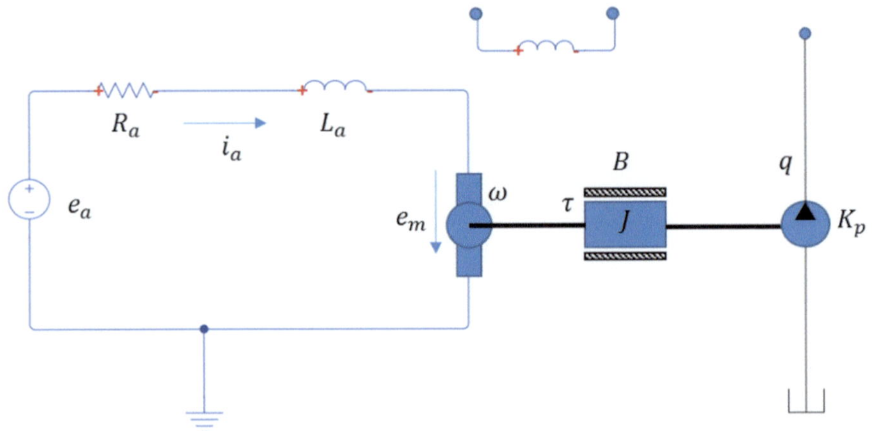

Fig. 2.23 Electromechanical hydraulic system

$$L_a \frac{\mathrm{d}i_a}{\mathrm{d}t} + R_a i_a(t) + e_m(t) = e_a(t) \tag{2.96}$$

$$e_m(t) = K_m \omega(t) \tag{2.97}$$

$$\tau(t) = K_m i_a(t) \tag{2.98}$$

$$J \frac{\mathrm{d}\omega}{\mathrm{d}t} + B\omega = \tau(t) \tag{2.99}$$

$$q(t) = K_p \omega(t) \tag{2.100}$$

by defining $\omega(t)$ and $e_a(t)$ as the system output and input, respectively. Represent the block diagram by using SIMULINK and obtain the step response for the parameter values of $L_a = 1$, $R_a = 5$, $K_m = 0.8$, $J = 1.5$, $B = 0.2$, and $K_p = 1$.

Solution
Applying the Laplace transform to the dynamic equations (2.96)–(2.100),

$$L_a s I_a(s) + R_a I_a(s) + E_m(s) = E_a(s) \tag{2.101}$$

$$E_m(s) = K_m\Omega(s) \tag{2.102}$$

$$T(s) = K_m I_a(s) \tag{2.103}$$

$$Js\Omega(s) + B\Omega(s) = T(s) \tag{2.104}$$

$$Q(s) = K_p\Omega(s) \tag{2.105}$$

Solving for $I_a(s)$ in (2.101), we get

$$I_a(s) = \frac{1}{L_a s + R_a}(E_a(s) - E_m(s)) \tag{2.106}$$

and solving for $\Omega(s)$ in (2.104), we obtain

$$\Omega(s) = \frac{1}{Js + B}T(s) \tag{2.107}$$

Based on the previous equations, we can draw the block diagram of the system with input $E_a(s)$ and output $Q(s)$ as depicted in Fig. 2.24.

In Figs. 2.25 and 2.26 are shown the corresponding block diagram developed by using SIMULINK together and its step response, respectively, for the parameter values indicated: $L_a = 1$, $R_a = 5$, $K_m = 0.8$, $J = 1.5$, $B = 0.2$ and $K_p = 1$.

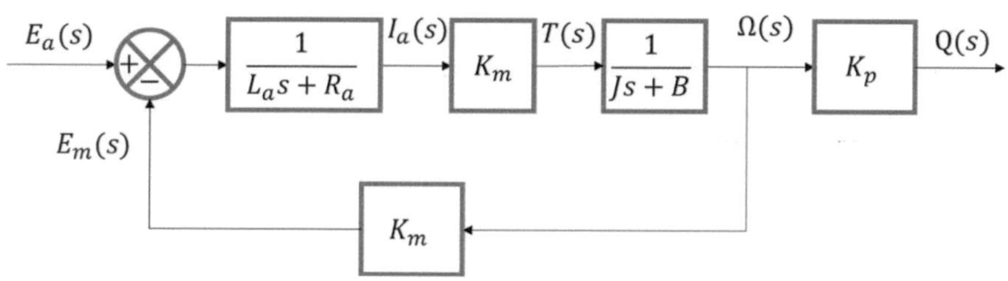

Fig. 2.24 Block diagram of the electromechanical system for Example 2.11

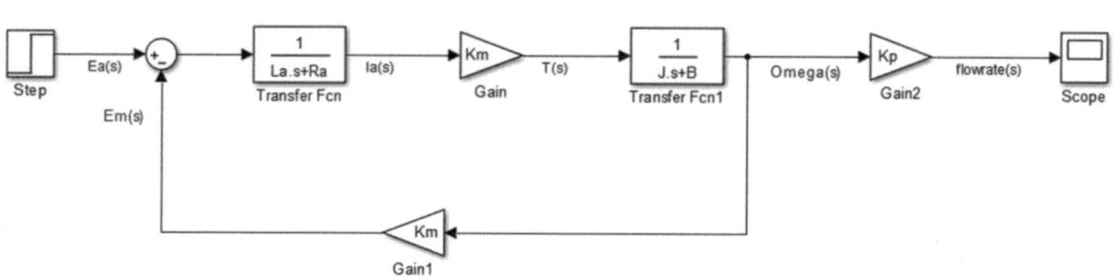

Fig. 2.25 SIMULINK diagram for Example 2.11

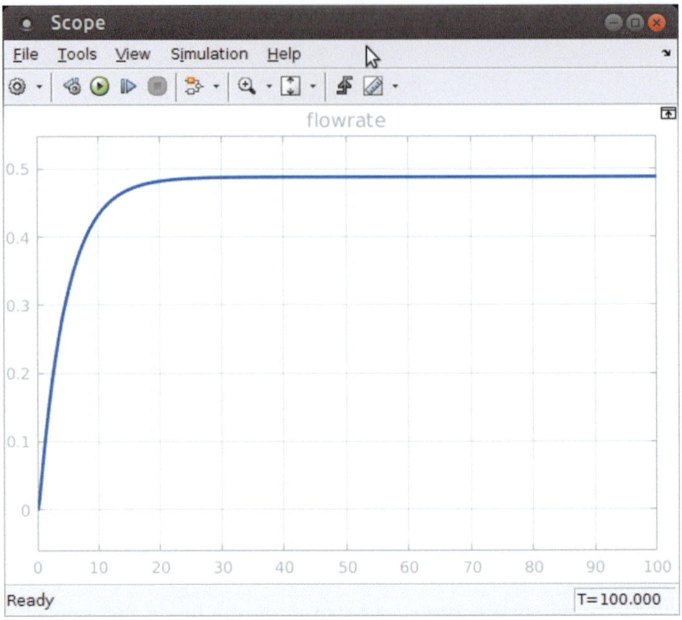

Fig. 2.26 Step response of the infusion pump obtained by the SIMULINK diagram shown in Fig. 2.25

From the block diagram of a system, we can make changes to simplify or reduce the original diagram to a single matrix transfer function or a single transfer function in case of a single-input single-output system, so that we can apply the Laplace transform technique in order to find the system time response.

For this purpose, we are going to apply some reduction rules, to transform a more complicated system into a single equivalent system. Table 2.5 shows the rules commonly used to simplify the block diagrams of dynamic systems.

By simplifying a block diagram, remember the following rules:

1. The product of transfer functions in the direction of the direct path must be the same.
2. The product of transfer functions around the loop must be the same.

Example 2.12

Reduce the block diagram of Fig. 2.27 to a single transfer function by applying the adequate reduction rules.

Solution

Firstly, we move the summing junction between G_1 and G_2 to the left and then solve for the feedback involving G_1, G_2 and H. We follow by reducing the feedback

Table 2.5 List of rules commonly used to reduce a block diagram

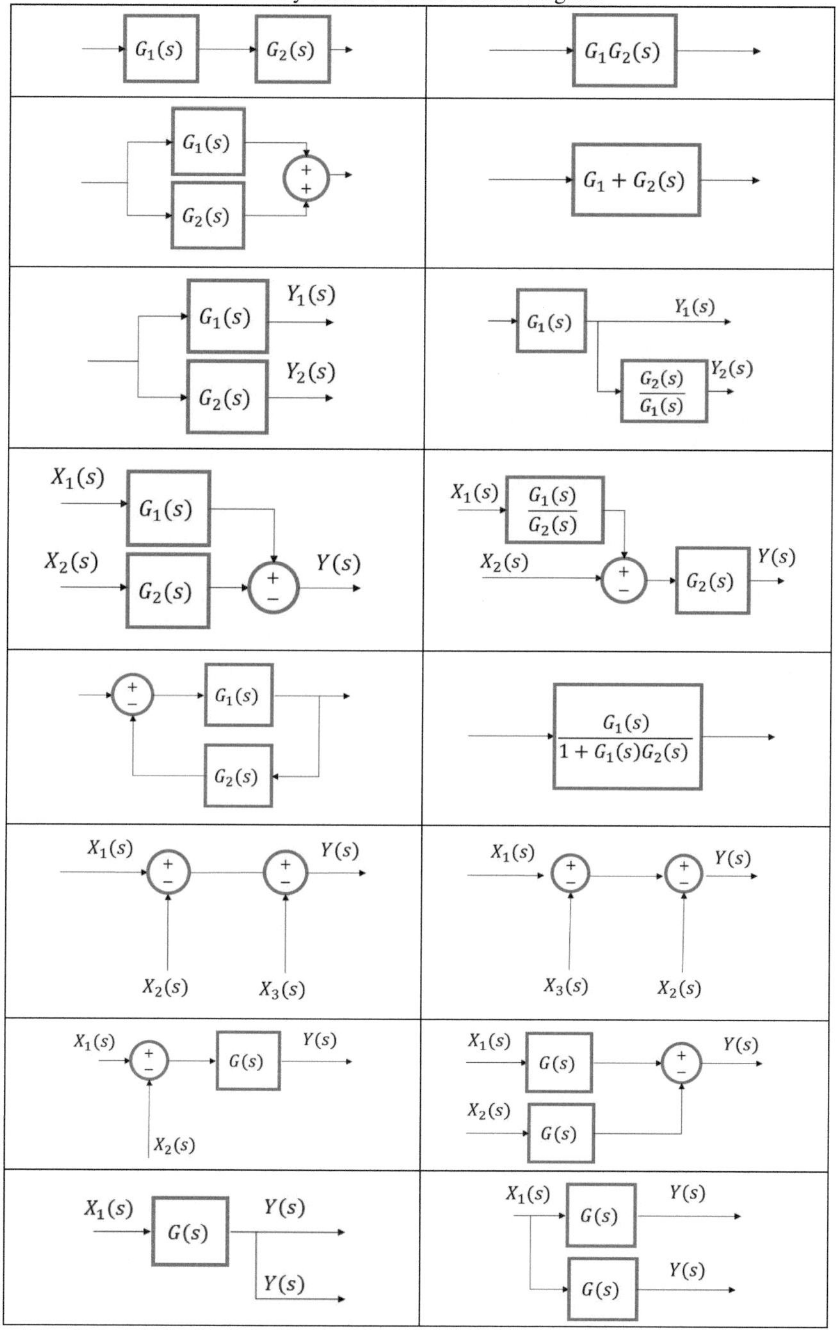

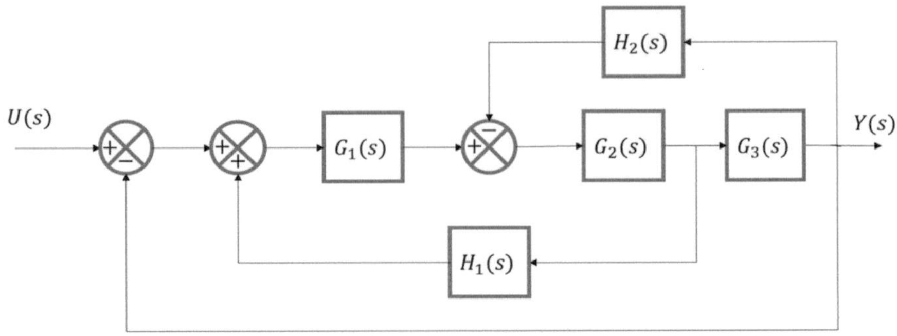

Fig. 2.27 Block diagram for Example 2.12

implying H_2/G_1 and finally solve for the unity feedback to get the single transfer function demanded $G(s) = Y(s)/U(s)$ (see Fig. 2.28).

MATLAB also includes commands that allow the reduction of block diagrams automatically to get the single transfer function representative of the whole system.

Example 2.13
Reduce the block diagram of Fig. 2.29 to a single transfer function by applying the necessary MATLAB commands for $G_1(s) = 10$, $G_2(s) = \frac{s+2}{s+10}$, $G_3(s) = \frac{1}{s+2}$, $H_1(s) = 5$, $H_2(s) = -2s$ and $H_3(s) = \frac{3}{s+2}$. Draw also the impulse response for the equivalent transfer function.

Solution
Firstly, we define each one of the individual transfer functions

```
≫ s=tf('s')
≫ G1=10;
≫ G2=(s+2)/(s+10);
≫ G3=1/(s+2);
≫ H1=5;
≫ H2=-2*s;
≫ H3=3/(s+2);
```

Then, we calculate the series product of $G_2(s)$ and $G_3(s)$

```
≫ G23=G2*G3
```

```
Transfer function:
        s + 2
------------------------
s^2 + 12 s + 20
```

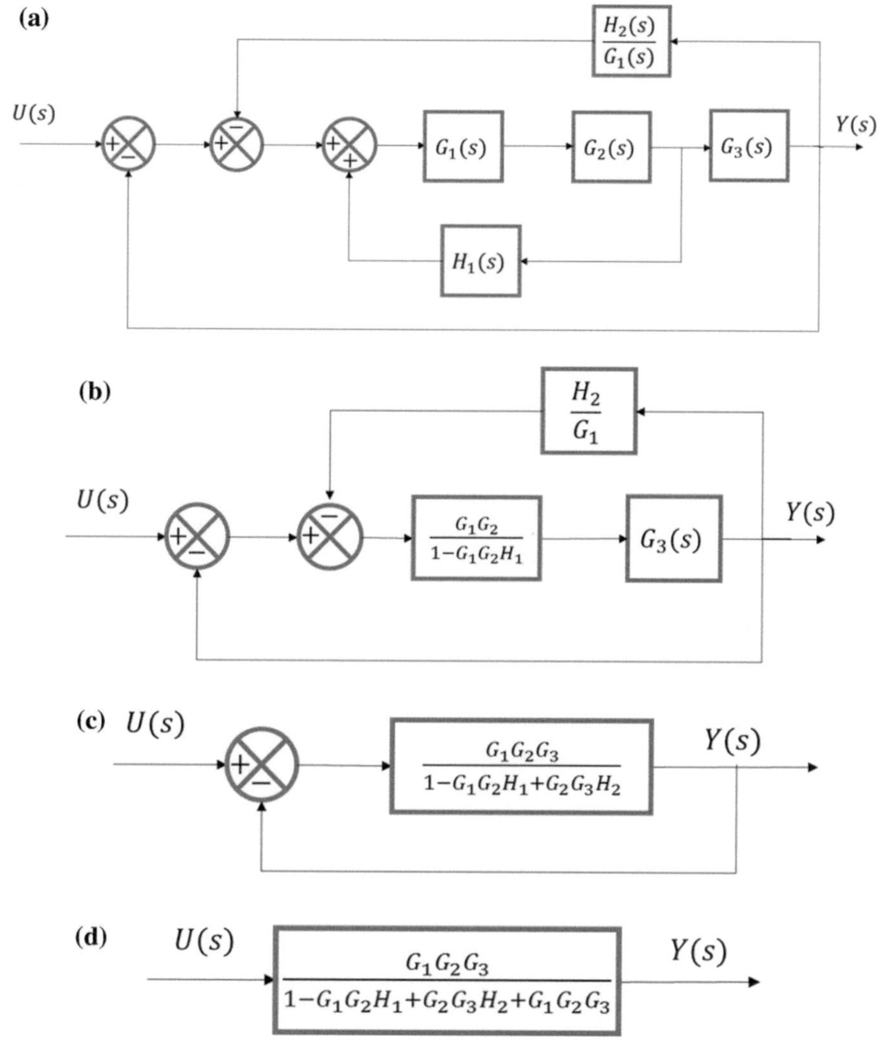

Fig. 2.28 Steps followed in reducing the block diagram for Example 2.12

We can group $H_1(s)$, $H_2(s)$ and $H_3(s)$ in a parallel way yielding

```
≫ H123=H1-H2+H3

Transfer function:
      2 s^2 + 9 s + 13
---------------
      s + 2
```

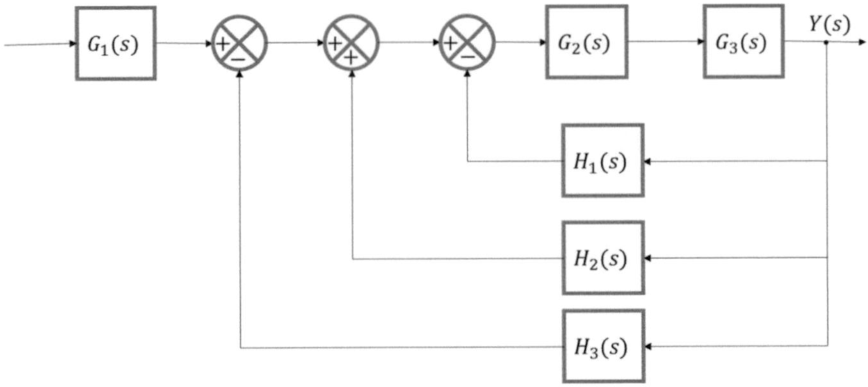

Fig. 2.29 Block diagram for Example 2.13

Below, we reduce the negative feedback configuration formed by G23 and H123

```
≫ Gfeed_G23_H123=feedback(G23,H123,-1)

Transfer function:
        s^2 + 4 s + 4
--------------------------
3 s^3 + 27 s^2 + 75 s + 66
```

Finally, we compute the series product of the last transfer function with $G_1(s)$, yielding the reduced transfer function

```
≫ G_reduced=G1*Gfeed_G23_H123

Transfer function:
     10 s^2 + 40 s + 40
-----------------------------
3 s^3 + 27 s^2 + 75 s + 66
```

Alternatively, we can rely on the MATLAB commands `connect` and `sumblk` to proceed in a different, more direct way. For that, we have to specify the input and output of each block (including summing blocks) by naming the diagram branches as shown in Fig. 2.30.

```
%%% Definition of the transfer functions
s=tf('s')
G1=tf(10,1);
G2=(s+2)/(s+10);
G3=1/(s+2);
```

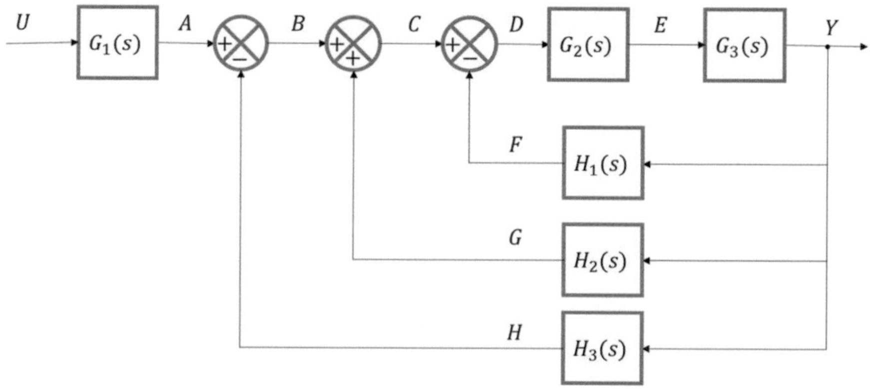

Fig. 2.30 Block diagram for Example 2.13, naming the inputs and outputs of each block

```
H1=tf(5,1);
H2=-2*s;
H3=3/(s+2);

%%% Specification of their inputs and outputs based on branch
%%% labels

G1.u = 'U';  G1.y = 'A';
G2.u = 'D';  G2.y = 'E';
G3.u = 'E';  G3.y = 'Y';

H1.u = 'Y';  H1.y = 'F';
H2.u = 'Y';  H2.y = 'G';
H3.u = 'Y';  H3.y = 'H';

%%% Specification of the inputs and outputs of the summing
%%% blocks

Sum1 = sumblk('B = A - H');
Sum2 = sumblk('C = B + G');
Sum3 = sumblk('D = C - F');

%%%% Connecting the model. The command connect returns the %%%%
State-Variable
%%%% representation
T = connect(G1,G2,G3,H1,H2,H3,Sum1,Sum2,Sum3,'U','Y');

%%% Converting the result into a transfer function based on
%%% tf
T=tf(T)
```

which produces the following, simplified, output.

```
T =

  From input "U" to output "Y":
  3.333 s^2 + 13.33 s + 13.33
  ------------------------------
  s^3 + 9 s^2 + 25 s + 22

Continuous-time transfer function.
```

Solved Problems

Problem 2.1

Figure 2.31 shows a block diagram of an ultrafiltration dialysis process. The "patient" system can be modelled by three subsystems or blocks ("trunk", "arms" and "legs"). In each of these blocks, there is a volume of liquid (plasma) defined, respectively, by $V_T(t)$, $V_A(t)$ and $V_L(t)$. There is a transfer of liquid (plasma) among blocks, according to the transfer constants k_{TA}, k_{TL}, k_A and k_L. Finally, the ultrafiltration process removes a quantity of liquid defined by $UFR(t)$ from the "trunk" block.

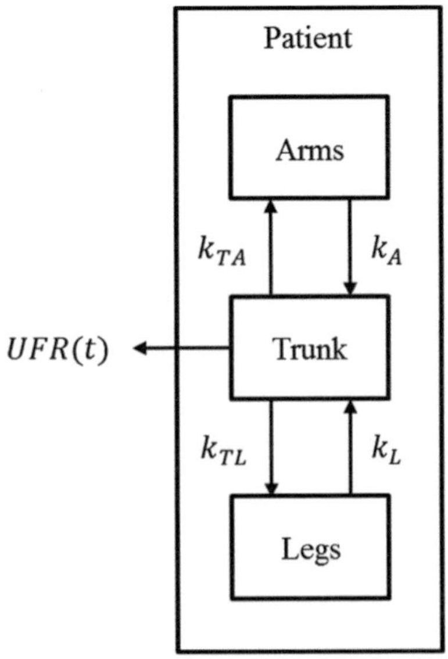

Fig. 2.31 Scheme of the ultrafiltration dialysis process

The dynamic equations defining the transfer of liquid between blocks are as follows:

$$\frac{dV_T(t)}{dt} = -(k_{TA} + k_{TL})V_T(t) + k_A V_A(t) + k_L V_L(t) - UFR(t)$$

$$\frac{dV_A(t)}{dt} = -k_A V_A(t) + k_{TA} V_T(t)$$

$$\frac{dV_L(t)}{dt} = -k_L V_L(t) + k_{TL} V_T(t)$$

(a) Obtain the block diagram in the s domain that form the "patient" system, considering $UFR(t)$ the input and $V_T(t)$ the output.
(b) Obtain the transfer function for the "patient" system.
(c) Transform the block diagram until it is in the form of a simple feedback system.
(d) Determine the transfer function, its zeros and poles and plot a map of zeros and poles, when the transfer constants have the following values:

$$k_A = 0.15; \quad k_L = 0.25; \quad k_{TA} = 0.33; \quad k_{TL} = 0.28$$

(e) The system is controlled by a feedback closed loop, where the quantity UFR of liquid to be removed is proportional to the difference between the trunk volume V_T and its desired value R (the reference trunk volume) according to the expression $UFR = K(V_T - R)$. Obtain the closed-loop transfer function when $K = 2$.
(f) Plot the closed-loop impulse and step responses.

Solution 2.1

(a) The "patient" system, considered as a whole, can be described according to Fig. 2.32.

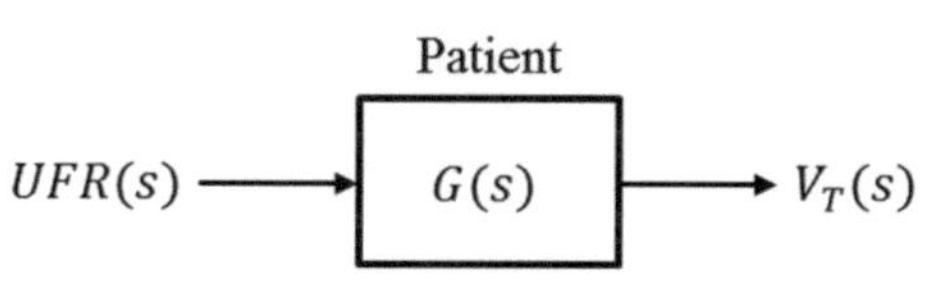

Fig. 2.32 Simplification of the patient system

In order to obtain the breakdown of its building blocks in the s domain, the Laplace transforms of the system's dynamic equations are computed,

$$\mathcal{L}\left[\frac{\mathrm{d}V_T(t)}{\mathrm{d}t}\right] = \mathcal{L}[-(k_{TA}+k_{TL})V_T(t)+k_A V_A(t)+k_L V_L(t)-UFR(t)].$$

$$\mathcal{L}\left[\frac{\mathrm{d}V_A(t)}{\mathrm{d}t}\right] = \mathcal{L}[-k_A V_A(t)+k_{TA} V_T(t)].$$
$$\mathcal{L}\left[\frac{\mathrm{d}V_L(t)}{\mathrm{d}t}\right] = \mathcal{L}[-k_L V_L(t)+k_{TL} V_T(t)].$$

Obtaining

$$sV_T(s) = -(k_{TA}+k_{TL})V_T(s)+k_A V_A(s)+k_L V_L(s)-UFR(s).$$
$$sV_A(s) = -k_A V_A(s)+k_{TA} V_T(s).$$
$$sV_L(s) = -k_L V_L(s)+k_{TL} V_T(s).$$

Grouping terms leads to

$$(s+k_{TA}+k_{TL})V_T(s) = k_A V_A(s)+k_L V_L(s)-UFR(s).$$
$$(s+k_A)V_A(s) = k_{TA} V_T(s).$$
$$(s+k_L)V_L(s) = k_{TL} V_T(s).$$

Now isolating $V_T(s)$,

$$V_T(s) = \frac{k_A}{s+k_{TA}+k_{TL}}V_A(s)+\frac{k_L}{s+k_{TA}+k_{TL}}V_L(s)-\frac{1}{s+k_{TA}+k_{TL}}UFR(s).$$
$$V_A(s) = \frac{k_{TA}}{s+k_A}V_T(s).$$
$$V_L(s) = \frac{k_{TL}}{s+k_L}V_T(s).$$

The first of these equations for $V_T(s)$ can be represented by the following block diagram shown in Fig. 2.33.

In this diagram, the values of $V_A(s)$ and $V_L(s)$ can be derived from the other two dynamic equations. The integration of these equations into the block diagram leads to the following figure.

As it can be seen in Fig. 2.34, the input $V_T(s)$ in the two new blocks is the same as that of the output for the whole system, so it can be obtained by feedback as indicated in Fig. 2.35, which represents the block diagram for the "patient" system.

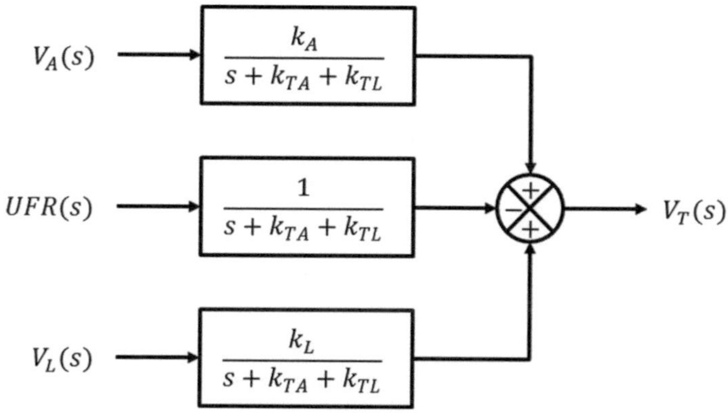

Fig. 2.33 Graphical representation of the system equations

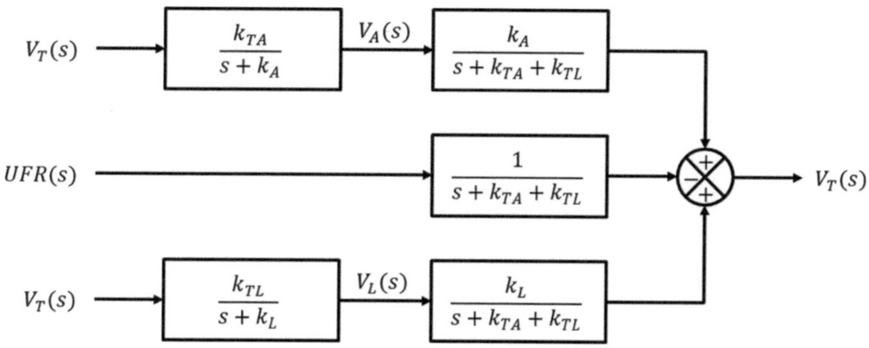

Fig. 2.34 Graphical representation of the system equations including the values of V_a and V_L

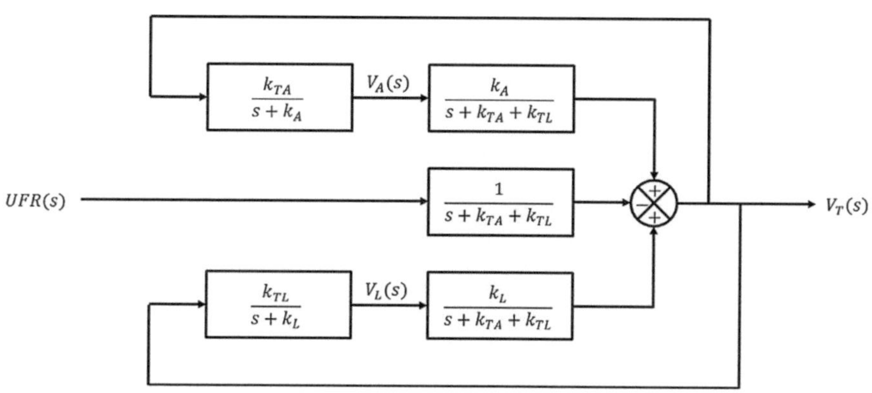

Fig. 2.35 Complete graphical representation of the system equations

(b) To obtain the transfer function $G(s)$ for the patient system, let us substitute the expression of $V_A(s)$ and $V_L(s)$ in the first of the dynamic equations, obtaining

$$V_T(s) = \frac{k_A}{s + k_{TA} + k_{TL}} \frac{k_{TA}}{s + k_A} V_T(s) + \frac{k_L}{s + k_{TA} + k_{TL}} \frac{k_{TL}}{s + k_L} V_T(s) - \frac{1}{s + k_{TA} + k_{TL}} UFR(s).$$

$$V_T(s) = \frac{k_A k_{TA}}{(s + k_{TA} + k_{TL})(s + k_A)} V_T(s) + \frac{k_L k_{TL}}{(s + k_{TA} + k_{TL})(s + k_L)} V_T(s) - \frac{1}{s + k_{TA} + k_{TL}} UFR(s).$$

$$V_T(s) \left[\frac{k_A k_{TA}}{(s + k_{TA} + k_{TL})(s + k_A)} + \frac{k_L k_{TL}}{(s + k_{TA} + k_{TL})(s + k_L)} - 1 \right] = \frac{1}{s + k_{TA} + k_{TL}} UFR(s).$$

The transfer function is, therefore,

$$G(s) = \frac{V_T(s)}{UFR(s)} = \frac{\frac{1}{s + k_{TA} + k_{TL}}}{\frac{k_A k_{TA}}{(s + k_{TA} + k_{TL})(s + k_A)} + \frac{k_L k_{TL}}{(s + k_{TA} + k_{TL})(s + k_L)} - 1}.$$

Simplifying,

$$G(s) = \frac{\frac{1}{s + k_{TA} + k_{TL}}}{\frac{k_A k_{TA}(s + k_L) + k_L k_{TL}(s + k_A) - (s + k_{TA} + k_{TL})(s + k_A)(s + k_L)}{(s + k_{TA} + k_{TL})(s + k_A)(s + k_L)}}.$$

$$G(s) = \frac{(s + k_A)(s + k_L)}{k_A k_{TA}(s + k_L) + k_L k_{TL}(s + k_A) - (s + k_{TA} + k_{TL})(s + k_A)(s + k_L)} = \frac{n(s)}{d(s)}.$$

The numerator is

$$n(s) = s^2 + s(k_A + k_L) + k_A k_L.$$

On the other hand, the denominator can be expanded according to the following expression

$$d(s) = k_A k_{TA} s + k_A k_{TA} k_L + k_L k_{TL} s$$
$$+ k_L k_{TL} k_A \cdots - \left(s^2 + k_A s + k_{TA} s + k_{TA} k_A + k_{TL} s + k_{TL} k_A\right)(s + k_L).$$
$$d(s) = k_A k_{TA} s + k_A k_{TA} k_L + k_L k_{TL} s$$
$$+ k_L k_{TL} k_A \cdots - s^3 - k_A s^2 - k_{TA} s^2 - k_{TA} k_A s - k_{TL} s^2$$
$$- k_{TL} k_A s \cdots - s^2 k_L - k_A k_L s - k_{TA} k_L s - k_{TA} k_A k_L - k_{TL} k_L s - k_{TL} k_A k_L.$$

Simplifying, grouping and sorting terms, it is obtained that

$$d(s) = -s^3 - k_A s^2 - k_{TA} s^2 - k_{TL} s^2 - k_{TL} k_A s - s^2 k_L - k_A k_L s - k_{TA} k_L s.$$
$$d(s) = -s^3 - s^2(k_A + k_L + k_{TA} + k_{TL}) - s(k_A k_L + k_A k_{TL} + k_L k_{TA}).$$

The transfer function is, therefore,

$$G(s) = -\frac{s^2 + s(k_A + k_L) + k_A k_L}{s^3 + s^2(k_A + k_L + k_{TA} + k_{TL}) + s(k_A k_L + k_A k_{TL} + k_L k_{TA})}.$$

An alternative way to obtain the transfer function is by using the following MATLAB code

```
%% Obtaining the transfer function from the system's
equations

% Defining symbolic variables
syms s VT VA VL UFR
syms kTA kTL kA kL

% Defining the system's equations
VA = kTA/(s+kA) * VT;
VL = kTL/(s+kL) * VT;
eq = VT == (kA/(s+kTA+kTL)) * VA +...
        (kL/(s+kTA+kTL)) * VL - (1/(s+kTA+kTL)) * UFR;

% Solving the system's equation, obtaining VT as a function
of UFR
VT = solve(eq,VT);

% Obtaining the system's transfer function
G = VT/UFR;
G = collect(G); %Write G as a quotient of polynomials
pretty(G);
```

The result obtained is as follows:

```
                     2
                - s  + (- kA - kL) s - kA kL
------------------------------------------------------------------
 3                          2
s  + (kA + kL + kTA + kTL) s  + (kA kL + kA kTL + kL kTA) s
```

which is the same as that previously derived.

(c) The block diagram obtained in paragraph (a) can be redrawn as shown in Fig. 2.36.

The expressions for the transfer function of each block in this graph are the following:

$$G_1(s) = \frac{1}{s + k_{TA} + k_{TL}}.$$
$$G_2(s) = \frac{k_{TA}}{s + k_A} \cdot \frac{k_A}{s + k_{TA} + k_{TL}}.$$
$$G_3(s) = \frac{k_{TL}}{s + k_L} \cdot \frac{k_L}{s + k_{TA} + k_{TL}}.$$

Blocks $G_2(s)$ and $G_3(s)$ are connected in parallel so the block diagram can be simplified as shown in Fig. 2.37a.

In order for the block diagram to have the form of a simple feedback system, it is convenient to change the signs in the addition node, as shown in Fig. 2.37b .

In this system

$$V_T(s) = -G_1(s)UFR(s) + [G_2(s) + G_3(s)]V_T(s).$$
$$[1 - G_2(s) + G_3(s)]V_T(s) = -G_1(s)UFR(s).$$

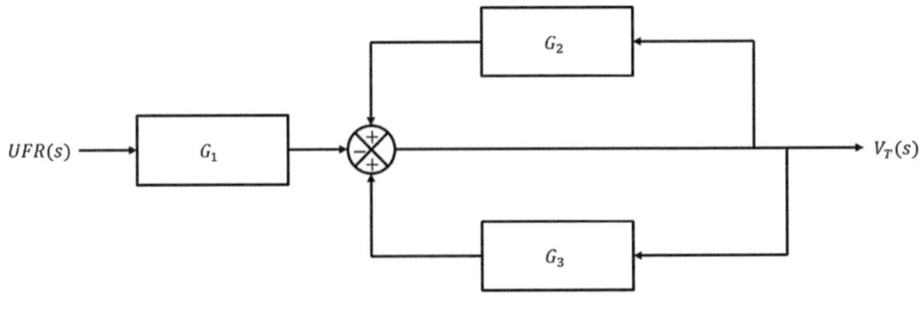

Fig. 2.36 Graphical representation of the system equations from the paragraph (a)

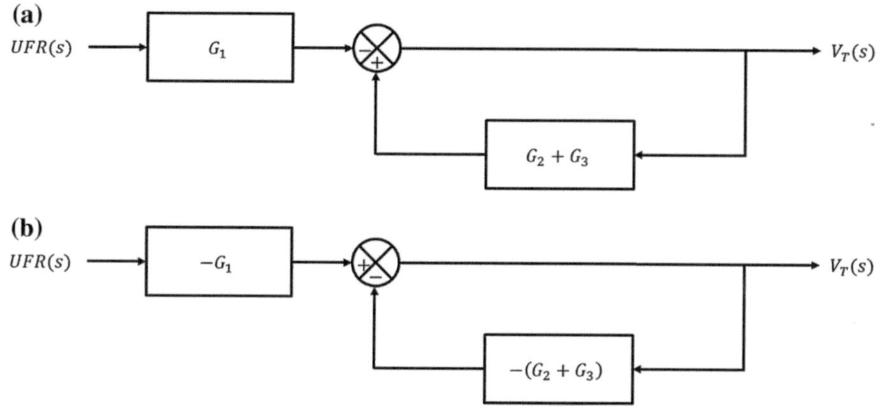

Fig. 2.37 Block diagram simplification

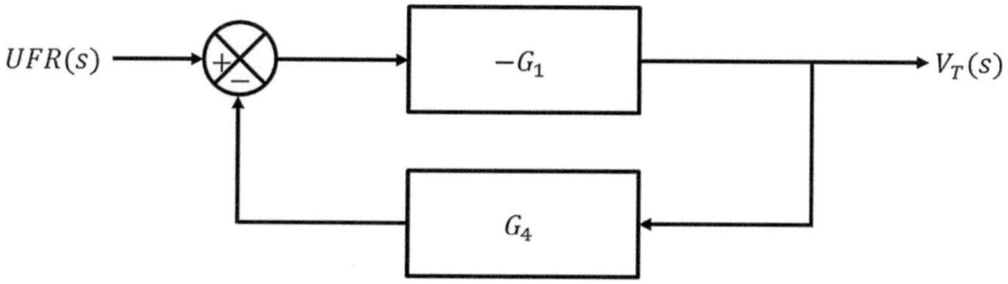

Fig. 2.38 Simplified block diagram

Therefore, the overall transfer function is

$$G(s) = \frac{V_T(s)}{UFR(s)} = \frac{-G_1(s)}{1 - G_2(s) + G_3(s)}$$

The block diagram can be transformed by inserting the first block into the feedback loop, but then having a new value $G_4(s)$, as shown in Fig. 2.38 which already has the form of a simple feedback system.

The $G(s)$ transfer function for the overall system is

$$G(s) = \frac{-G_1(s)}{1 - G_1(s) \cdot G_4(s)}.$$

Combining both expressions for $G(s)$, we get

$$\frac{-G_1(s)}{1 - G_2(s) + G_3(s)} = \frac{-G_1(s)}{1 - G_1(s) \cdot G_4(s)},$$

where we can derive that

$$G_2(s) + G_3(s) = G_1(s) \cdot G_4(s).$$

$$G_4(s) = \frac{G_2(s) + G_3(s)}{G_1(s)}.$$

To obtain the value of the transfer function, the following MATLAB code can be used:

```
%% Obtaining the transfer function from the reduced block
diagram

% Defining blocks' transfer functions
G1 = 1/(s+kTA+kTL);
G2 = kTA/(s+kA) * kA/(s+kTA+kTL);
G3 = kTL/(s+kL) * kL/(s+kTA+kTL);

% Obtaining reduced transfer function
G4 = (G2+G3) / G1;

% Obtaining the system's transfer function
G = -G1/(1-G1*G4);
G = collect(G); %Write G as a quotient of polynomials
pretty(G);
```

The result obtained is as follows:

```
                        2
                     - s   + (- kA - kL) s - kA kL
--------------------------------------------------------------
 3                                 2
s   + (kA + kL + kTA + kTL) s   + (kA kL + kA kTL + kL kTA) s
```

which is the same as that previously derived.

(d) With the values given to the constants, the transfer function is

$$G(s) = \frac{-(s^2 + s(0.15 + 0.25) + 0.15 \cdot 0.25)}{s^3 + s^2(0.15 + 0.25 + 0.33 + 0.28) + s(0.15 \cdot 0.25 + 0.15 \cdot 0.28 + 0.25 \cdot 0.33)}.$$

$$G(s) = -\frac{s^2 + 0.4s + 0.0375}{s^3 + 1.01s^2 + 0.162s}.$$

The zeros are obtained by solving the equation

$$s^2 + 0.4s + 0.0375 = 0.$$

$$s = \frac{-0.4 \pm \sqrt{0.4^2 - 4 \cdot 1 \cdot 0.0375}}{2} = \frac{-0.4 \pm 0.1}{2} = \begin{cases} s = -0.15 \\ s = -0.25 \end{cases}.$$

Likewise, the poles are computed by solving equation

$$s^3 + 1.01s^2 + 0.162s = 0.$$

$$s\left(s^2 + 1.01s + 0.162\right) = 0.$$

Therefore, a pole is placed at $s = 0$, and the other two are obtained by solving

$$s^2 + 1.01s + 0.162 = 0.$$

$$s = \frac{-1.01 \pm \sqrt{1.01^2 - 4 \cdot 1 \cdot 0.162}}{2} = \frac{-1.01 \pm 0.61}{2} = \begin{cases} s = -0.81 \\ s = -0.20 \end{cases}.$$

This paragraph can also be solved, using MATLAB, by adding the following code:

```
%% Obtaining the numeric transfer function, its zeros and its
poles

% Substituting the transfer constants by their values
(symbolic)
G = subs(G, [kA,kL,kTA,kTL], [0.15,0.25,0.33,0.28]);
%Symbolic
[numGs,denGs] = numden(G); %Obtains the num. and den. of G(s)

% Converting the symbolic G(s) to a transfer function object
numG = sym2poly(numGs); %Obtains the num. of G(s)as a numeric
vector
denG = sym2poly(denGs); %Obtains the den. of G(s)as a numeric
vector
G = tf(numG,denG); %Obtains G(s) as a transfer function
object
G = minreal(G); %Obtains the minimal realization of G(s)
display(G);
```

```
% Obtaining the zeros and poles of G(s)
zerosG = zero(G);
polesG = pole(G);
display(zerosG);
display(polesG);

% Plotting the pole-zero map of G(s)
pzmap(G);
```

The execution of this code results in the following transfer function, zeros and poles:

```
G =
     -s^2 - 0.4 s - 0.0375
   ------------------------
   s^3 + 1.01 s^2 + 0.162 s
Continuous-time transfer function.

zerosG =
    -0.2500
    -0.1500

polesG =
          0
    -0.8100
    -0.2000
```

These results are the same as that previously obtained. The code also generates the pole–zero map as shown in Fig. 2.39.

(e) Considering that $UFR = K(V_T - R)$, the closed-loop system can be described according to Fig. 2.40:

Note that the signs in the feedback do not follow the usual convention, so the closed-loop transfer function has to be carefully derived according to the following expression:

$$V_T(s) = UFR(s)G(s) = K[V_T(s) - R(s)]G(s) = KV_T(s)G(s) - KR(s)G(s).$$
$$KR(s)G(s) = KV_T(s)G(s) - V_T(s) = V_T(s)[KG(s) - 1].$$

So, the closed-loop transfer function is

$$M(s) \equiv \frac{V_T(s)}{R(s)} = \frac{KG(s)}{KG(s) - 1}.$$

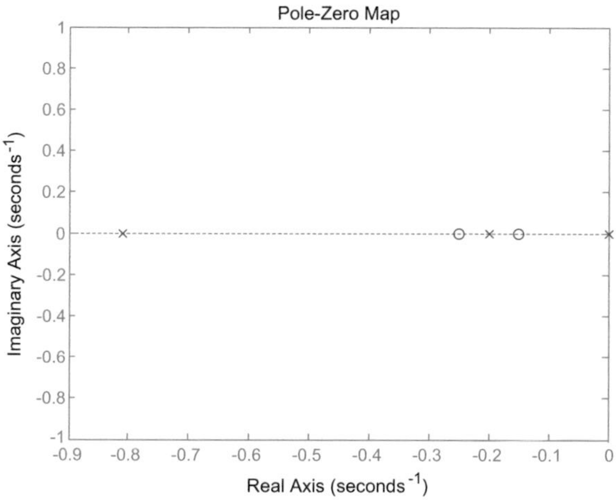

Fig. 2.39 Pole–zero map

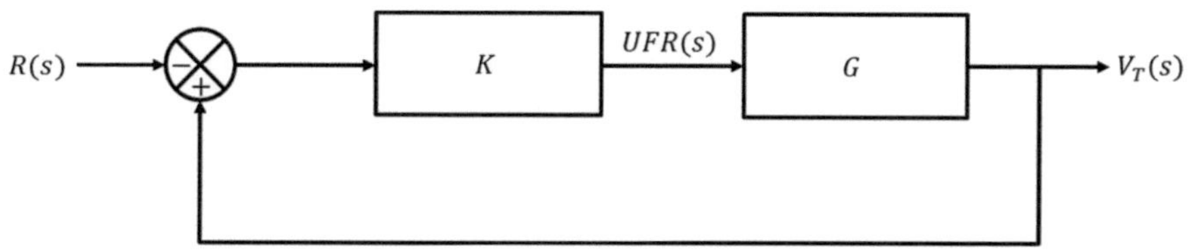

Fig. 2.40 Closed-loop system ultrafiltration dialysis process

$$M(s) = \frac{-2\frac{s^2+0.4s+0.0375}{s^3+1.01s^2+0.162s}}{-2\frac{s^2+0.4s+0.0375}{s^3+1.01s^2+0.162s}-1} = \frac{2\frac{s^2+0.4s+0.0375}{s^3+1.01s^2+0.162s}}{1+2\frac{s^2+0.4s+0.0375}{s^3+1.01s^2+0.162s}}.$$

$$M(s) = \frac{2\frac{s^2+0.4s+0.0375}{s^3+1.01s^2+0.162s}}{\frac{(s^3+1.01s^2+0.162s)+2(s^2+0.4s+0.0375)}{s^3+1.01s^2+0.162s}}.$$

$$M(s) = \frac{2s^2+0.8s+0.075}{(s^3+1.01s^2+0.162s)+(2s^2+0.8s+0.075)}.$$

$$M(s) = \frac{2s^2+0.8s+0.075}{s^3+3.01s^2+0.962s+0.075}.$$

This paragraph can also be solved by adding the following MATLAB code:

```
%% Obtaining the closed-loop transfer function

K = 2; %Proportional controller
M = (K*G) / (K*G-1); %Obtains the closed-loop transfer function
M = minreal(M); %Obtains the minimal realization of M(s)
display (M);
```

The execution of this code obtains the following result:

```
M =

     2 s^2 + 0.8 s + 0.075
  ---------------------------------
  s^3 + 3.01 s^2 + 0.962 s + 0.075
```

(f) To plot the closed-loop impulse and step responses, the following MATLAB code is added:

```
%% Obtaining the closed-loop impulse & step responses

% Obtaining the responses
[ir,t] = impulse(M); %Obtaining the impulse response
sr = step(M,t); %Obtaining the step response

% Plotting the responses
figure;
plot(t,ir,'b',t,sr,'r');
axis tight;
legend({'Impulse response','Step
response'},'Location','Best');
xlabel('Time');
ylabel('V_T(t)');
```

The result obtained is depicted in Fig. 2.41.

Problem 2.2

A certain surgical robotic arm is made up of three links, the "upper arm", the "forearm" and the "hand". The links are connected by seven joints or degrees of freedom, "shoulder" flexion and abduction; "elbow" flexion and abduction; and "wrist" flexion, abduction and pronation. Consider the joint equivalent to the flexion movement of the robotic shoulder which is defined by its angle θ with the vertical. Assume the case when the arm is fully extended and forming an angle

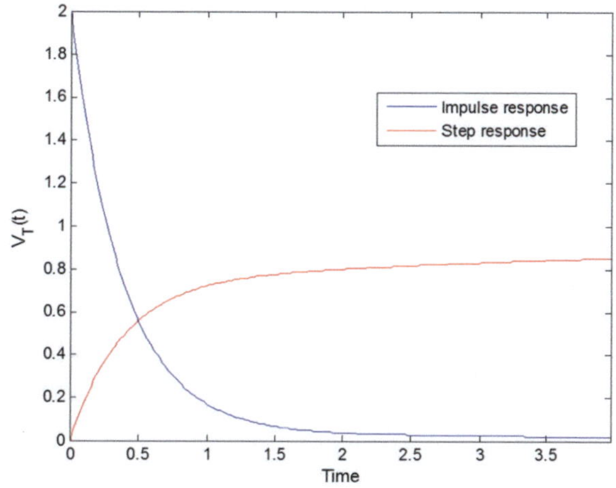

Fig. 2.41 System response for impulse and step inputs

$\theta = \theta_0$. This joint is rotated by a DC electrical motor driven by a voltage v. Although the $\theta = f(v)$ relationship clearly corresponds to a nonlinear system, it can be considered linear for small angle movement. In such circumstances, the joint can be modelled as a transfer function given by

$$G(s) = \frac{\Theta(s)}{V(s)} = \frac{1}{a_3 s^3 + a_2 s^2 + a_1 s + a_0},$$

where

$$a_3 = \frac{L_m}{EK_a}\left(I_m + E^2 I\right).$$

$$a_2 = \frac{1}{EK_a}\left(R_m I_m + R_m E^2 I + L_m B_m\right).$$

$$a_1 = \frac{1}{EK_a}\left(K_a K_b + R_m B_m - L_m E^2 MgLsen\theta_0\right).$$

$$a_0 = \frac{R_m E}{K_a} MgL\cos\theta_0.$$

For the shoulder flexion system, determine:

(a) The transfer function $G(s)$ for the values of the constants given below.
(b) The location of its zeros and poles.

The joint is controlled as a part of a feedback loop using a compensator defined by its transfer function $Q = 300$. For the continuous-time closed-loop system, obtain:

(c) The location of its zeros and poles.
(d) The step response.

Robotic arm mechanical parameters:

Mass: $M = 2.2$ kg.
Length: $L = 33$ cm.
Initial flexion shoulder angle: $\theta_0 = 45°$.
Gravitational constant: $g = 9.8\ \mathrm{m/s^2}$.
Moment of inertia: $I = ML^2$.

DC motor electrical parameters:

Resistance: $R_m = 3.05\ \Omega$.
Inductance: $L_m = 0.87$ mH.
Torque constant: $K_a = 62.2\ \mathrm{mN\ m/A}$.
Speed constant: $S_c = 154\ \mathrm{rpm/V}$.
Inverse speed constant: $K_b = 1/S_c$.

DC motor mechanical parameters:

Rotor inertia: $I_m = 67.9\ \mathrm{g\ cm^2}$.
No load current: $i_0 = 77.5$ mA.
No load speed: $\omega_0 = 7300$ rpm.
Damping coefficient: $B_m = K_a \cdot i_0/\omega_0$.
Gear reduction: $E = 1/300$.

Solution 2.2
(a) The shoulder flexion transfer function $G(s)$ can be obtained using the following MATLAB code:

```
%% Obtaining robotic arm transfer function

% Robotic arm mechanical parameters
M = 2.2; %Mass (kg)
L = 0.33; %Length (m)
th0 = degtorad(45); %Initial flexion shoulder angle
g = 9.8; %Gravitational constant (m/s^2)
I = M * L^2; %Moment of inertia
```

```
% DC motor electrical parameters
Rm = 3.05; %Electrical resistance (ohm)
Lm = 0.87E-3; %Electrical inductance (H)
Ka = 62.2E-3; %Torque constant (N·m/A)
Sc = 154;  %Speed constant (rpm/V)
Kb = 60/(2*pi)/Sc;  %Inverse speed constant (V / rad/s)

% DC motor mechanical parameters
Im = 67.9E-7; %Rotor inertia (kg·m^2)
i0 = 77.5E-3; %No load current (A)
w0 = 7300; %No load speed (rpm)
w0 = w0 * (2*pi/60); %No load speed (rad/s)
+ (kA + kL + kTA + kTL) s
E = 1/103; %Gear reduction

%Defining transfer function
s = tf('s'); %Defining ``s'' as the elementary transfer
function
a3 = Lm/(E*Ka)*(Im+E^2*I);
a2 = 1/(E*Ka)*(Rm*Im+Rm*E^2*I+Lm*Bm);
a1 = 1/(E*Ka)*(Ka*Kb+Rm*Bm-Lm*E^2*M*g*L*sin(th0));
a0 = (Rm*E/Ka)*M*g*L*cos(th0);
G = 1 / ( a3*s^3 + a2*s^2 + a1*s + a0 );
display(G);
```

The result obtained by executing this code is:

```
G =

                          1
   ---------------------------------------------------
   4.232e-05 s^3 + 0.1484 s^2 + 6.418 s + 2.395
```

that is

$$G(s) = \frac{1}{4.232 \times 10^{-5}s^3 + 0.1484s^2 + 6.418s + 2.395}.$$

(b) The zeros and poles can be obtained by adding the following MATLAB code:

```
%% Obtaining robotic arm zeros & poles
z = zero(G); %Zeros
p = pole(G); %Poles
```

```
display(z);
display(p);
```

The results obtained by executing this code are:

```
z =
   Empty matrix: 0-by-1

p =
  1.0e+03 *

   -3.4622
   -0.0434
   -0.0004
```

That is, the system has no zero and 3 poles, located at

$$p_1 = -3462.2; \quad p_2 = -43.4; \quad p_3 = -0.4$$

(c) The closed-loop system can be depicted as in Fig. 2.42.

So, the closed-loop transfer function is

$$M(s) = \frac{\Theta(s)}{R(s)} = \frac{Q(s) \cdot G(s)}{1 + Q(s) \cdot G(s)}.$$

This paragraph can be solved by adding the following MATLAB code:

```
%% Obtaining closed-loop zeros & poles

% Obtaining closed-loop transfer function
Q = 300; %Compensator
M = feedback(Q*G,1); %Closed-loop transfer function
```

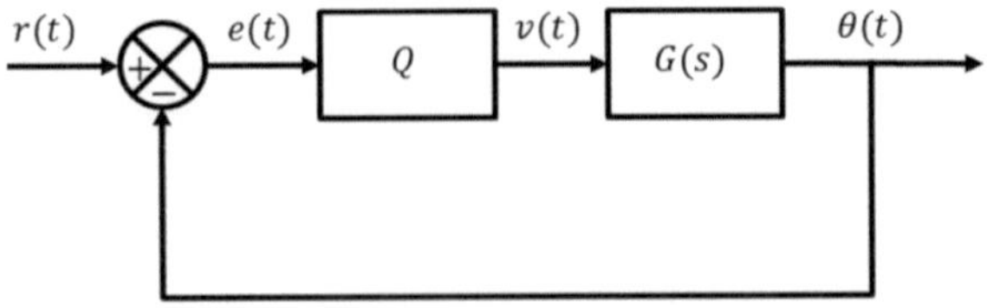

Fig. 2.42 Closed-loop system for the surgical robotic arm example

```
M = minreal(M); %Obtaining minimal realization of the
closed-
loop tf

% Obtaining closed-loop zeros & poles
zc = zero(M); %Zeros
pc = pole(M); %Poles
display(zc);
display(pc);
```

The results obtained by executing this code are:

```
zc =

    Empty matrix: 0-by-1

pc =
   1.0e+03 *

  -3.4628 + 0.0000i
  -0.0216 + 0.0400i
  -0.0216 - 0.0400i
```

That is, the closed-loop system has no zero and 3 poles, located at

$$p_1 = -3462.8; \quad p_2 = -21.6 + 40.0j; \quad p_3 = -21.6 - 40.0j$$

(d) To obtain the closed-loop step response, the following MATLAB code is added

```
%% Obtaining closed-loop step response
step(M);
```

which obtains the plot shown in Fig. 2.43.

Problem 2.3

Consider a catheter of those commonly used in the measurement of pressures in the cardiovascular system. The catheter receives the pressure $p(t)$ from the patient at the distal end and converts it into a measure of voltage $v(t)$ for its subsequent monitoring. The relationship between the pressure (input) and the voltage (output) constitutes a system whose impulse response is given by the equation:

$$g(t) = Ke^{\frac{-t}{\tau}}\sin(\omega t)$$

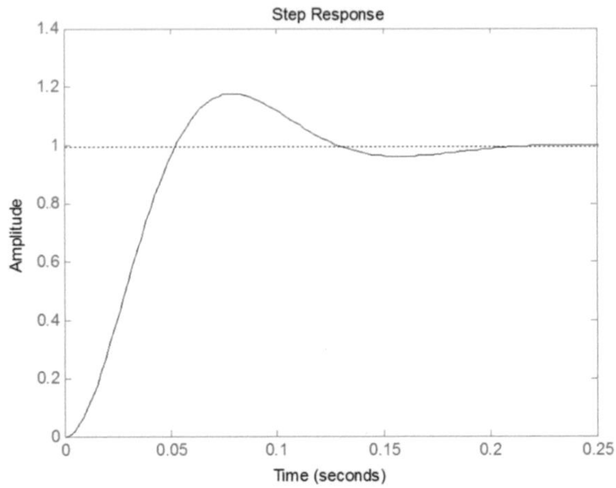

Fig. 2.43 Step response for the surgical robotic arm example

The constants of this equation are determined by the following expressions:

$$K = \frac{2\,A\,K_e}{\alpha b}; \quad \tau = \frac{2\,\rho\,A\,L}{b}; \quad \omega = \frac{\alpha b}{2\rho AL};$$

where A and L are the catheter's section and length, ρ the density of fluid within the catheter, c the lineal friction coefficient, b the viscous friction coefficient, K_e the proportionality constant between catheter displacement and electrical voltage and α a coefficient determined by the equation

$$\alpha = \sqrt{\frac{4\rho ALc}{b^2} - 1}$$

(a) Obtain the transfer function of the system.
(b) Determine the zeros and poles of the transfer function.
(c) Find the relationship among the constants to get a system with non-real poles.

Solution 2.3

(a) The system can be described according to Fig. 2.44.

In the s domain, the output $V(s) = G(s) \cdot P(s)$ or

$$G(s) = \frac{V(s)}{P(s)}$$

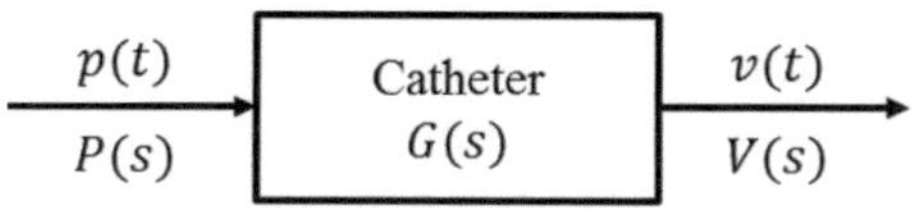

Fig. 2.44 Catheter diagram

By the definition, if the input is an unit impulse $p(t) = \delta(t)$, then the output will be the system's impulse response; in this case,

$$v(t) = g(t) = K\mathrm{e}^{\frac{-t}{\tau}} \sin(\omega t)$$

It can be written that

$$G(s) = \frac{V(s)}{P(s)} = \frac{\mathcal{L}[v(t)]}{\mathcal{L}[p(t)]} = \frac{\mathcal{L}[g(t)]}{\mathcal{L}[\delta(t)]} = \frac{\mathcal{L}[g(t)]}{1} = \mathcal{L}[g(t)]$$

Therefore,

$$G(s) = \mathcal{L}\left[K\mathrm{e}^{\frac{-t}{\tau}} \sin(\omega t)\right] = K\mathcal{L}\left[\mathrm{e}^{\frac{-t}{\tau}} \sin(\omega t)\right]$$

Looking up in the Laplace transform tables, it is found that

$$\mathcal{L}[\mathrm{e}^{-at} \sin(\omega t)] = \frac{\omega}{(s+a)^2 + \omega^2}$$

So,

$$G(s) = K\frac{\omega}{\left(s+\frac{1}{\tau}\right)^2 + \omega^2} = \frac{K\omega}{\left[\frac{1}{\tau}(\tau s+1)\right]^2 + \frac{1}{\tau^2}(\tau\omega)^2} = \frac{K\tau^2\omega}{(\tau s+1)^2 + (\tau\omega)^2}$$

Substituting the expressions of the constants provided in the problem statement,

$$G(s) = \frac{\left(\frac{2AK_e}{\alpha b}\right)\left(\frac{2\rho AL}{b}\right)^2\left(\frac{\alpha b}{2\rho AL}\right)}{\left(\frac{2\rho AL}{b}s+1\right)^2 + \left(\frac{2\rho AL}{b}\frac{\alpha b}{2\rho AL}\right)^2}$$

$$G(s) = \frac{2\rho AL\frac{2AK_e}{b^2}}{\left(\frac{2\rho AL}{b}s+1\right)^2 + \alpha^2} = \frac{4\rho A^2LK_e}{b^2\left(\frac{2\rho AL}{b}s+1\right)^2 + b^2\alpha^2}$$

$$G(s) = \frac{4\rho A^2LK_e}{(2\rho ALs+b)^2 + b^2\alpha^2}$$

$$G(s) = \frac{4\rho A^2LK_e}{(2\rho ALs+b)^2 + b^2\left(\frac{4\rho ALc}{b^2} - 1\right)}$$

$$G(s) = \frac{4\rho A^2 L K_e}{(2\rho A L s + b)^2 + 4\rho A L c - b^2}$$
$$G(s) = \frac{4\rho A^2 L K_e}{4\rho^2 A^2 L^2 s^2 + 4\rho A L b s + b^2 + 4\rho A L c - b^2}$$
$$G(s) = \frac{4\rho A^2 L K_e}{4\rho^2 A^2 L^2 s^2 + 4\rho A L b s + 4\rho A L c}$$
$$G(s) = \frac{4\rho A L (A K_e)}{4\rho A L (\rho A L s^2 + b s + c)}$$
$$G(s) = \frac{A K_e}{\rho A L s^2 + b s + c}$$

This paragraph can also be addressed using the following MATLAB code:

```
%% Obtaining the symbolic transfer function

% Defining symbolic constants and variables
syms A L; %Catheter section and length
syms rho; %Density of the fluid inside the catheter
syms c b; %Lineal and fluid friction constants
syms Ke; %Voltage-displacement proportionality constant
syms t;

% Defining the system's impulse response
alpha = sqrt(4*rho*A*L*c / b^2 - 1);
K = 2*A*Ke / (alpha*b);
tau = 2*rho*A*L / b;
omega = (alpha*b) / (2*rho*A*L);
g = K * exp(-t/tau) * sin(omega*t);

% Obtaining the system's transfer function
G = laplace(g);
G = collect(G);
pretty(G);
```

Executing this code, the same result showing the catheter transfer function $G(s)$ is obtained:

```
         A Ke
-------------------------
          2
A L rho s  + b s + c
```

(b) The numerator of the transfer function does not depend on s, so no s value makes it null. Therefore, the system has no zero.

To obtain the poles of the transfer function, let us determine the roots of the denominator in $G(s)$. For this purpose, let us solve the equation

$$\rho A L s^2 + bs + c = 0$$

So, the poles are at

$$s = \frac{-b \pm \sqrt{b^2 - 4\rho A L c}}{2\rho A L}$$

Alternatively, this result can be obtained by adding the following MATLAB code:

```
%% Obtaining the zeros and poles
syms s;

[numG,denG] = numden(G); %Obtaining num. & den. of G(s)
z = solve(numG,s); %Obtaining the zeros of G(s)
p = solve(denG,s); %Obtaining the poles of G(s)
pretty(z);
pretty(p);
```

Executing this code, the zeros and poles of $G(s)$ are obtained:

```
/                2                    \
|   b + sqrt(b  - 4 A L c rho) |
| -  ---------------------      |
|            2 A L rho         |
|                               |
|              2                |
|   b - sqrt(b  - 4 A L c rho) |
| -  ---------------------     |
\            2 A L rho        /
```

This result is the same as that previously derived.

(c) If the system has to have non-real poles, then the result of the square root must be an imaginary number or, which is equivalent,

$$b^2 - 4\rho ALc < 0$$
$$b^2 < 4\rho ALc$$
$$b < 2\sqrt{\rho ALc}$$

Problem 2.4

The establishment and spread of infectious diseases in a certain population is usually studied by so-called compartmental models in which the population is divided into compartments (subsets of the population) and individuals move from one compartment to another as the disease evolves along the time. The simplest of these models is the SIR model consisting on three compartments: the *S*, with people susceptible of contracting the disease; the *I*, with the infected people; and *R*, with those recovered from the disease or immunized. Upon infection, an individual goes from compartment *S* to compartment *I*, and recovering from the disease goes from *I* to *R*. In this model, it is assumed that there is no associated mortality and no new individuals are born during the development of the epidemic. The dynamics of the system is determined by the following system of differential equations:

$$\frac{\mathrm{d}S(t)}{\mathrm{d}t} = -\frac{\beta I(t)S(t)}{N}$$
$$\frac{\mathrm{d}I(t)}{\mathrm{d}t} = \frac{\beta I(t)S(t)}{N} - \gamma I(t)$$
$$\frac{\mathrm{d}R(t)}{\mathrm{d}t} = \gamma I(t)$$

In these equations, N is the number of individuals in the population; β is the contagious contact rate; and γ is the rate of recovery. The inverse values of these rates are, respectively, the mean time between contagious contacts T_c and the mean recovery time T_r. The following ratio is often called infectious contact number or basic reproduction ratio (ICN or σ)

$$\mathrm{ICN} \equiv \sigma = \frac{\beta}{\gamma} = \frac{T_r}{T_c}$$

In a population with one million inhabitants, an influenza epidemic develops with an average recovery time of 7 days and an infectious contact number of 4.

(a) Obtain a linearized model considering null initial conditions.
(b) Obtain the evolution of the epidemic when, initially, 0.1% of the population is infected and no individual is immune.
(c) Obtain the total number of individuals affected and the total duration of the infectious outbreak.
(d) Repeat the above computations for the case in which, by adopting prophylactic measures, infectious contact number is reduced to a value of 2.

(e) Obtain the dependence of the duration of the epidemic outbreak and the number of persons affected on the infectious contact number.

Solution 2.4

(a) Considering the second equation that describes the system in the following way

$$\frac{\beta I(t)S(t)}{N} = \frac{\mathrm{d}I(t)}{\mathrm{d}t} + \gamma I(t)$$

and substituting in the first equation

$$\frac{\mathrm{d}S(t)}{\mathrm{d}t} = -\frac{\mathrm{d}I(t)}{\mathrm{d}t} - \gamma I(t)$$

it is possible to describe the system with a set of two equations

$$\begin{aligned} F_1\left(\dot{S}, S, \dot{I}, I, \dot{R}, R\right) &= \frac{\mathrm{d}S(t)}{\mathrm{d}t} + \frac{\mathrm{d}I(t)}{\mathrm{d}t} + \gamma I(t) = 0 \\ F_2\left(\dot{S}, S, \dot{I}, I, \dot{R}, R\right) &= \frac{\mathrm{d}R(t)}{\mathrm{d}t} - \gamma I(t) = 0 \end{aligned}$$

This is a nonlinear model that can be linearized according to Eq. (2.10)

$$\begin{aligned} \frac{\partial F_1}{\partial \dot{S}}\delta\dot{S} + \frac{\partial F_1}{\partial S}\delta S + \frac{\partial F_1}{\partial \dot{I}}\delta\dot{I} + \frac{\partial F_1}{\partial I}\delta I + \frac{\partial F_1}{\partial \dot{R}}\delta\dot{R} + \frac{\partial F_1}{\partial R}\delta R = 0 \\ \frac{\partial F_2}{\partial \dot{S}}\delta\dot{S} + \frac{\partial F_2}{\partial S}\delta S + \frac{\partial F_2}{\partial \dot{I}}\delta\dot{I} + \frac{\partial F_2}{\partial I}\delta I + \frac{\partial F_2}{\partial \dot{R}}\delta\dot{R} + \frac{\partial F_2}{\partial R}\delta R = 0 \end{aligned}$$

Therefore, the linearized system can be expressed as

$$\begin{aligned} \delta\dot{S} + \delta\dot{I} + \gamma\delta I = 0 \\ -\gamma\delta I + \delta\dot{R} = 0 \end{aligned}$$

where the partial derivatives of F_1 and F_2 with regard to each independent variable are evaluated at the operating point, and whose validity is restricted to a small region around it, assuming the following definitions

$$\delta\dot{S}(t) = \dot{S}(t) - \dot{S}(t)\big|_{t=0} = \frac{\mathrm{d}\bar{S}(t)}{\mathrm{d}t}$$
$$\delta S(t) = S(t) - S(t)|_{t=0} = \bar{S}(t)$$
$$\delta\dot{I}(t) = \dot{I}(t) - \dot{I}(t)\big|_{t=0} = \frac{\mathrm{d}\bar{I}(t)}{\mathrm{d}t}$$
$$\delta I(t) = I(t) - I(t)|_{t=0} = \bar{I}(t)$$
$$\delta\dot{R}(t) = \dot{R}(t) - \dot{R}(t)\big|_{t=0} = \frac{\mathrm{d}\bar{R}(t)}{\mathrm{d}t}$$
$$\delta R(t) = R(t) - R(t)|_{t=0} = \bar{R}(t)$$

This way, the system description could be understood with the following equations

$$\frac{\mathrm{d}\bar{S}(t)}{\mathrm{d}t} + \frac{\mathrm{d}\bar{I}(t)}{\mathrm{d}t} + \gamma\bar{I}(t) = 0$$
$$-\gamma\bar{I}(t) + \frac{\mathrm{d}\bar{R}(t)}{\mathrm{d}t} = 0$$

Now, that we have a linearized system, it is possible to apply the Laplace transform, considering null initial conditions, yielding

$$s\bar{S}(s) + s\bar{I}(s) + \gamma\bar{I}(s) = 0$$
$$-\gamma\bar{I}(s) + s\bar{R}(s) = 0$$

which can be expressed as two transfer functions

$$s\bar{S}(s) = -\bar{I}(s)(s+\gamma) \rightarrow F_1(s) = \frac{\bar{S}(s)}{\bar{I}(s)} = -\frac{s+\gamma}{s}$$
$$s\bar{R}(s) = \gamma\bar{I}(s) \rightarrow F_2(s) = \frac{\bar{R}(s)}{\bar{I}(s)} = \frac{\gamma}{s}$$

(b) To determine the spread of the infectious disease, the number of persons in each compartment should be computed. For this purpose, the system of differential equations is solved using the following MATLAB code:

```
function main

%% Paragraph b)
Par = SetParameters;
Par.sigma = 4; %Infectious contact number or reproduction ratio
DrawEvolution(Par);
% Set the parameters' values
function Par = SetParameters
%Par: Structure containing model's parameters

% Parameters
Par.N = 1; %Population (in millions)
Par.Tr = 7; %Time until recovery (contagious period)
Par.sigma = nan; %Infectious contact number or reproduction ratio
Par.tf = 100; %Final time for computation

% Initial values
Par.rI0 = 0.001; % Initial Infected population ratio
Par.rR0 = 0.; % Initial Recovered population ratio
Par.I0 = Par.rI0 * Par.N;
Par.R0 = Par.rR0 * Par.N;
Par.S0 = Par.N - Par.I0 - Par.R0;


%Draw the evolution of the epidemic
function DrawEvolution(Par)
%Par: Structure containing model's parameters

% Solve the set of differential equations
[t,S,I,R] = ObtainInfectionEvolution(Par);

%Draw the results
plot(t,S,t,R,t,I);
xlabel('Time (days)');
ylabel('Population (millions)');
legend({'Susceptible','Recovered','Infected'},'Location','Best');
```

```
% Obtain the evolution of the infection solving the set of equations
function [t,S,I,R] = ObtainInfectionEvolution(Par)
%Par: Parameters
%t: time vector
%S: Susceptible population
%I: Infected population
%R: Recovered population

%Solve the compartmental model
tf = Par.tf; %Final time for computation
[t,X] = ode45... %Solve set of diferential equations
    (@(t,X)CompMod(t,X,Par),[0 tf],[Par.S0,Par.I0,Par.R0]);
S = X(:,1); %Susceptible population (element 1 in state vector X)
I = X(:,2); %Infected population (element 2 in state vector X)
R = X(:,3); %Recovered population (element 3 in state vector X)


% Compartmental model of the epidemic
function dX = CompMod(t,X,Par)
%t: time vector
%X: state vector (S,I,R)
%Par: Parameters

Tr = Par.Tr;
N = Par.N;
sigma = Par.sigma;

%Obtain auxiliary constants
gamma = 1/Tr; %Recovery rate
beta = sigma*gamma; %Contact rate

%Obtain state variables
S = X(1); %Susceptible population (element 1 in state vector X)
I = X(2); %Infected population (element 2 in state vector X)
R = X(3); %Recovered population (element 3 in state vector X)

%Compute the compartmental model
dS = -beta*I*S/N; %Differential of susceptible population
dI = beta*I*S/N - gamma*I; %Differential of infected population
dR = gamma*I; %Differential of recovered population

dX = [dS;dI;dR]; %Differential of state vector
```

The obtained result is shown in Fig. 2.45.

(c) To obtain the total number of people affected by the disease N_a, the final number of recovered people R_f will be computed and then compared to the initial number of people recovered or immune R_0, since

$$N_a = R_f - R_0$$

To obtain the value of R_f, the model's first and third differential equations are considered, so

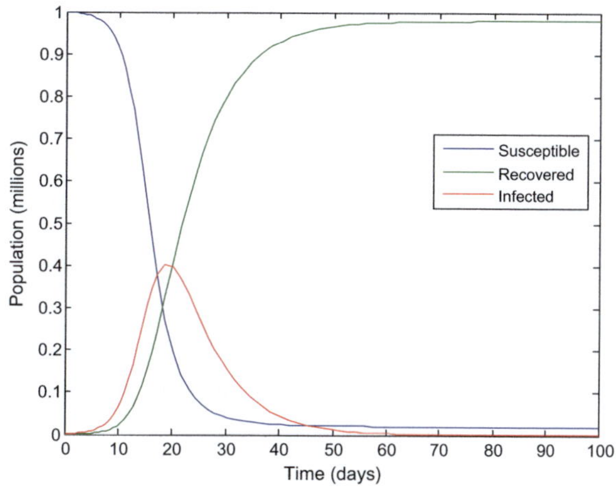

Fig. 2.45 Evolution of the susceptible, recovered and infected population over time

$$\frac{dS(t)}{dt} = -\frac{\beta I(t)S(t)}{N}; \quad \frac{dR(t)}{dt} = \gamma I(t)$$

Dividing the first over the third equation, it can be derived that

$$\frac{dS(t)}{dR(t)} = \frac{-\frac{\beta I(t)S(t)}{N}}{\gamma I(t)} = \frac{-\beta S(t)}{\gamma N} = -\sigma\frac{S(t)}{N}$$

$$\frac{dS(t)}{S(t)} = -\frac{\sigma}{N}dR(t)$$

$$\int\frac{dS(t)}{S(t)} = \int -\frac{\sigma}{N}dR(t)$$

$$LnS(t) = -\frac{\sigma}{N}R(t) + K_1$$

$$S(t) = e^{\left[-\frac{\sigma}{N}R(t)+K_1\right]} = e^{K_1}\cdot e^{-\frac{\sigma}{N}R(t)} = K_2\cdot e^{-\frac{\sigma}{N}R(t)}$$

At $t = 0$,

$$S(0) = K_2\cdot e^{-\frac{\sigma}{N}R(0)}$$

Therefore,

$$K_2 = \frac{S(0)}{e^{-\frac{\sigma}{N}R(0)}} = \frac{S_0}{e^{-\frac{\sigma}{N}R_0}} = S_0\cdot e^{\frac{\sigma}{N}R_0}$$

Thus,

$$S(t) = S_0 \cdot e^{\frac{\sigma}{N}R_0} \cdot e^{-\frac{\sigma}{N}R(t)} = S_0 \cdot e^{\frac{\sigma}{N}R_0 - \frac{\sigma}{N}R(t)}$$
$$S(t) = S_0 \cdot e^{\frac{\sigma}{N}[R_0 - R(t)]}$$

The final values of every variable are obtained at $t = \infty$,

$$S_f \equiv S(\infty) = S_0 \cdot e^{\frac{\sigma}{N}[R_0 - R(\infty)]}$$
$$S_f = S_0 \cdot e^{\frac{\sigma}{N}\left[R_0 - R_f\right]}$$

But at $t = \infty$, the number of infected people is $I_f = 0$, so

$$S_f + R_f + I_f = N$$
$$S_f + R_f = N$$

Then

$$S_f = N - R_f$$

Recalling the previous equation for S_f, it can be written that

$$S_f = N - R_f = S_0 \cdot e^{\frac{\sigma}{N}\left[R_0 - R_f\right]}$$

The final value R_f is obtained solving the following transcendental equation

$$R_f = N - S_0 \cdot e^{\frac{\sigma}{N}\left[R_0 - R_f\right]}$$

On the other hand, to obtain the duration of the epidemic, it will be considered the time taken for the number of recovered $R(t)$ going from the initial value R_0 to the final value R_f. In practice, the duration is usually measured as the time taken in going from 10 to 90% of the evolution of $R(t)$.

For both issues, the following code is added to the `function main`:

```
%% Paragraph c)
[De,Na] = EpidemicDuration(Par);
display(De);
display(Na);
```

and, additionally, the following MATLAB function is added:

```
% Compute the duration of the epidemic & the number of affected
persons
function [De,Na] = EpidemicDuration(Par)
%Par: Structure containing model's parameters
%De: Duration of epidemic
%Na: Number of affected persons

%Solve the set of differential equations
[t,S,I,R] = ObtainInfectionEvolution(Par);

%Compute the final number of recovered persons
R0 = Par.R0; %Initial numer of recovered
Rf = fzero(@(Rf) RfEquation(Rf,Par),[R0,1]);
Na = Rf-R0; % Number of affected persons

%Compute the duration of the epidemic
%   as the time needed for R rising from 10% to 90%
%   of its total increasing
Rmin = 0.1;
Rmax = 0.9;
%Discard values of t outside the [10%,90%] interval
t = t( R>=R0+Rmin*(Rf-R0) & R<=R0+Rmax*(Rf-R0) );
De = t(end) - t(1); %Duration of the epidemic


% Define de equation for the final value of recovered persons
function RfError = RfEquation(Rf,Par)
%Rf: Final number of recovered persons
%Par: Structure containing model's parameters
%RfError: Error in the computation of Rf (should be minimized)

% Initial values
N = Par.N; %Population (in millions)
sigma = Par.sigma; %Infectious contact number or reproduction ratio
R0 = Par.R0; %Initial numer of recovered
S0 = Par.S0; %Initial numer of susceptible

%Error in the equation for the final value of recovered persons
RfError = N - S0 * exp( (-sigma/N) * (Rf-R0)) - Rf;
```

The result obtained is the following:

```
De =

   20.2121


Na =

    0.9802
```

that is the duration of the epidemic is 20 days and the number of affected people is 98% of the total population.

(d) In this case, the following MATLAB code is added to the function main:

```
%% Paragraph d)
Par.sigma = 2; %Infectious contact number or reproduction ratio
DrawEvolution(Par);
[De,Na] = EpidemicDuration(Par);
display(De);
display(Na);
```

The obtained result is the following:

```
De =
    37.5000

Na =
      0.7972
```

that is the duration of the epidemic is 37.5 days and the number of affected people is 80% of the total population. The evolution of the epidemic is shown in Fig. 2.46.

(e) To obtain the dependence of the duration of the epidemic outbreak and the number of affected people on the infectious contact number, the following code is added to the function main:

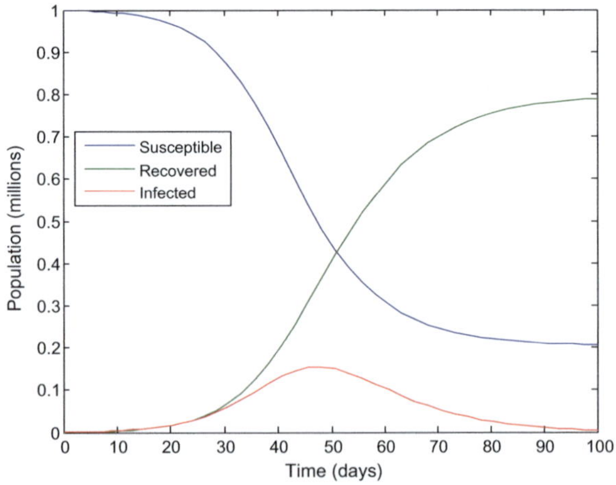

Fig. 2.46 Evolution of the susceptible, recovered and infected population over time, adopting prophylactic measures

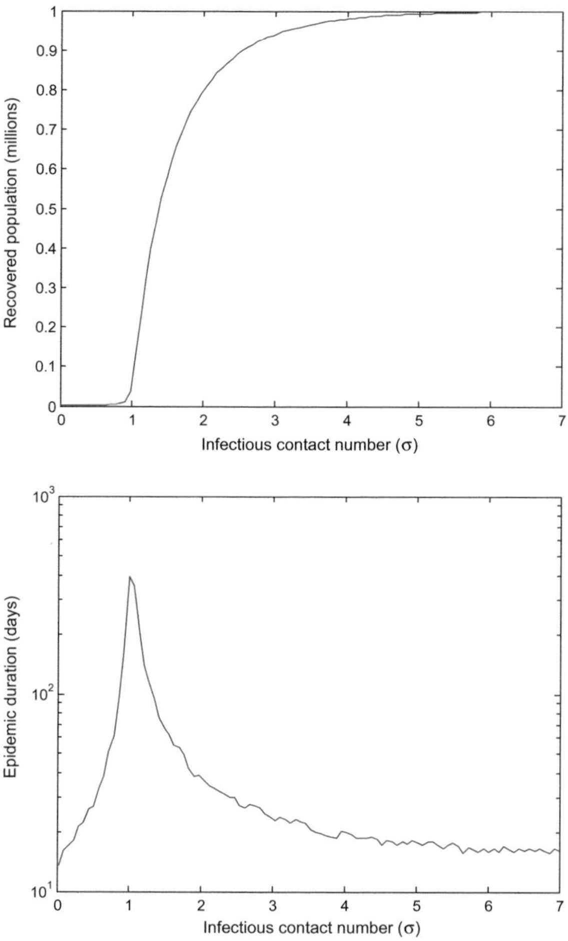

Fig. 2.47 (Top) Dependence of the number of affected people on the infectious contact number. (Bottom) Dependence of the duration of the epidemic outbreak on the infectious contact number

```
%% Paragraph e)
DrawEpidemicDuration(Par);
```

and, additionally, the following MATLAB function is added:

```
% Draw the epidemic duration vs. the infectious contact number
function DrawEpidemicDuration(Par)
%Par: Structure containing model's parameters

Par.tf = 1000; %Final time for computation
sigma = linspace(0,7); %Values for Infectious contact number
nsigma = length(sigma);
Na = zeros(1,nsigma);
De = zeros(1,nsigma);
for i = 1:nsigma
    Par.sigma = sigma(i);
    [De(i),Na(i)] = EpidemicDuration(Par);
end

figure;
plot(sigma,Na);
xlabel('Infectious contact number (\sigma)');
ylabel('Recovered population (millions)');

figure;
semilogy(sigma,De);
xlabel('Infectious contact number (\sigma)');
ylabel('Epidemic duration (days)');
```

The obtained result is shown in Fig. 2.47.

Time Domain Analysis of Continuous and Discrete Systems

3

This chapter analyses the output of systems, also known as response, when a particular input is applied. In this way, if system response is not adequate a redesign procedure can be made by adding a controller so as to make the controlled system to behave as desired.

The main characteristics of the response can be divided into two parts: the transient response and the steady-state response. The former entails the first response of the system, being the swiftness and smoothness of the response the key features, while the latter refers to the permanent or stationary value of the output.

In this chapter, we will study analytically the response of first- and second-order systems continuous systems and sketch the characteristic of the response of higher-order systems based on their components. Based on the values of response parameters, an identification procedure will be presented so as to derive the transfer function corresponding to the linear or linearized system.

In the same way, sampling and data reconstruction techniques will be presented so as to derive the pulse transfer concept or discretized model of a continuous system, whose transient response will be obtained starting from the continuous one by applying the *s*-*z*-plane mapping.

Both stability of continuous and discretized systems will be stated, and steady-state analysis of tracking error will be made when a closed-loop control configuration is under study both for continuous and for discrete-time control systems.

Continuous Transient Response Analysis

The way in which physical systems react against input stimuli depends, obviously, on their own nature and the given input. Considering biomedical systems, it becomes essential to understand how systems react against different inputs in order

J. Fernández de Cañete et al., *Automatic Control Systems in Biomedical Engineering*,
https://doi.org/10.1007/978-3-319-75717-9_3

to adapt them to fulfil the needed requirements in terms of speed, precision and stability.

Independently of the order of the system, system output is the combination of the *natural* response and the *forced* response. The natural response of a system is due to the initial conditions of the system, while the forced response is due to the input stimulus considering the initial conditions as null. For example, if we study the height of liquid in a reservoir when pumping liquid into it, the forced response will account only for the height change as a consequence of the input flow, while the natural response will be the height change due to the initial filling without considering the pumping effect.

Given a general differential equation that models a linear continuous system defined by

$$a_n y^n + a_{n-1} y^{n-1} + \cdots + a_1 \dot{y} + a_0 y(t) = b_m u^m + b_{m-1} u^{m-1} + \cdots + b_1 \dot{u} + b_0 u(t) \tag{3.1}$$

the Laplace transformation considering initial conditions will be given by

$$\begin{aligned} a_n\left(s^n Y(s) - s^{n-1} y(0) - \cdots - y^{n-1}(0)\right) + \cdots + a_1(sY(s) - y(0)) + a_0 Y(s) \\ = b_m\left(s^m U(s) - s^{m-1} u(0) - \cdots - u^{m-1)}(0)\right) + \cdots + b_1(sU(s) - u(0)) + b_0 U(s) \end{aligned} \tag{3.2}$$

Assembling terms, we will obtain the system response as the result of the sum of two components, let say $Y_1(s)$ the forced response and $Y_2(s)$ the natural response which only contains the values of the initial conditions grouped here in a polynomial $P(s)$, that is

$$Y(s) = \frac{b_m s^m + \cdots b_1 s + b_0}{a_n s^n + \cdots a_1 s + a_0} U(s) + \frac{P(s)}{a_n s^n + \cdots a_1 s + a_0} = Y_1(s) + Y_2(s) \tag{3.3}$$

Example 3.1
A thermometer used to measure the temperature of patients (Fig. 3.1) can be modelled through the differential equation:

$$C\frac{\mathrm{d}T}{\mathrm{d}t} = \frac{T_r(t) - T(t)}{R} \tag{3.4}$$

where C is the thermal capacitance of the thermometer, R is the thermal resistance, $T(t)$ is the measured temperature, and $T_r(t)$ is the reference temperature. Thus, this equation can be rewritten to match the general format as

Fig. 3.1 Schematic of a thermometer used in clinical practice

$$\frac{dT}{dt} + \frac{1}{RC}T(t) = \frac{T_r(t)}{RC} \tag{3.5}$$

being $T(t)$ the system output and $T_r(t)$ the reference input and with an initial value of $T(t) = T_0$.

Obtain the system response for $RC = 20\,\mathrm{s}$ starting from $T_0 = 25\ °\mathrm{C}$ for a step input signal $T_r(t)$ of amplitude 38 °C.

Solution

Applying the Laplace transformation, we obtain

$$T(s) = \frac{0.05}{s+0.05}T_r(s) + \frac{1.25}{s+0.05} = Y_1(s) + Y_2(s) \tag{3.6}$$

with $T_r(s) = \frac{38}{s}$ the input signal in Laplace domain.

Figure 3.2 shows some plots corresponding to the different parts of the response obtained by applying the MATLAB code:

```
≫s=zpk('s');
≫G=0.05/(s+0.05);
≫t=0:1:100;
≫yf=step(38*G,t)
≫plot(t,yf);grid

≫A=-0.05;B=0.05;C=1;D=0;sis=ss(A,B,C,D);
≫yn=initial(sis,25,t)
≫plot(t,yn);grid;
≫plot(t,yf+yn);grid;
```

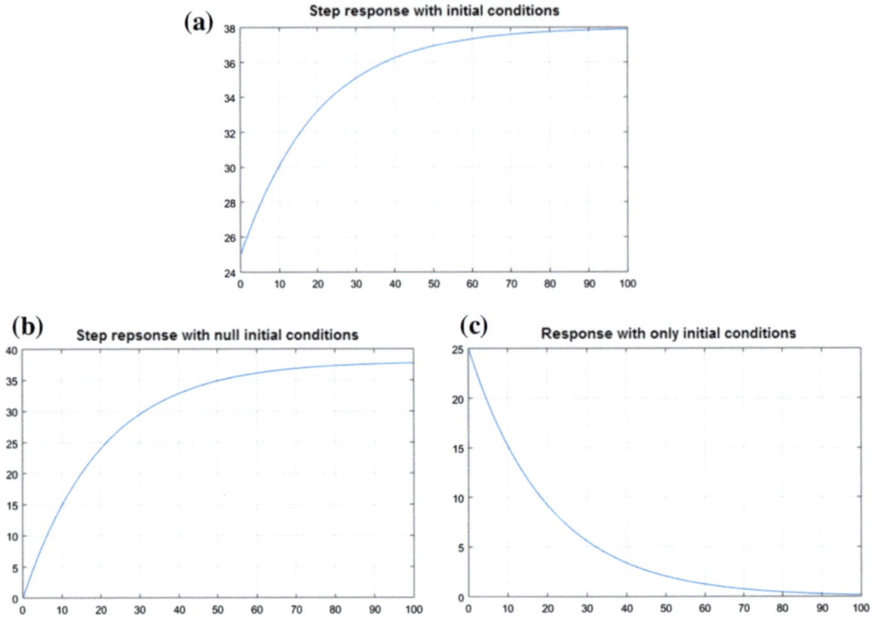

Fig. 3.2 Step response with initial conditions for a thermometer (**a**) Forced response (**b**) Natural response (**c**)

First-Order System Response

In general, first-order systems can be represented as differential equations in the form

$$\frac{\mathrm{d}y}{\mathrm{d}t} + a_0 y(t) = b_0 u(t) \tag{3.7}$$

In general, the transfer function of a first-order system obtained by means of the Laplace transformation of the former equation is given by

$$G(s) = \frac{b_0}{s + a_0} \tag{3.8}$$

The response of the system $Y(s)$ when a step input with amplitude A is applied (considering initial conditions as null) can be derived as follows:

$$G(s) = \frac{b_0}{s + a_0} \tag{3.9}$$

$$Y(s) = \frac{b_0}{s + a_0} U(s) = \frac{b_0}{s + a_0} \frac{A}{s} \tag{3.10}$$

Applying residues, this expression can be computed as

$$Y(s) = \frac{K_1}{s} + \frac{K_2}{s + a_0} \tag{3.11}$$

with

$$K_1 = \frac{b_0 A}{a_0}; \quad K_2 = -\frac{b_0 A}{a_0} \tag{3.12}$$

yielding

$$Y(s) = \frac{b_0 A}{a_0} \left(\frac{1}{s} - \frac{1}{s + a_0} \right) \xrightarrow{\mathcal{L}} y(t) = \frac{b_0 A}{a_0} (1 - \mathrm{e}^{-a_0 t}) u_s(t) \tag{3.13}$$

That is, the step response of a first-order system exhibits a negative exponential behaviour as shown in Fig. 3.3.

Notice that by assuming $a_0 > 0, \lim_{t \to \infty} \mathrm{e}^{-a_0 t} \to 0$, and thus, $\lim_{t \to \infty} y(t) = \frac{b_0 A}{a_0}$. That is, the steady-state or final value of a first-order response is given by $\frac{b_0 A}{a_0}$ and the term $\frac{b_0}{a_0}$ represents a factor that multiplies the amplitude of the input, so it is normally called *gain* and represented with K. Rewriting the system output using the gain, we have

$$y(t) = K(1 - \mathrm{e}^{-a_0 t}) u_s(t) \tag{3.14}$$

Also note that the higher the magnitude of a_0, the faster the system response achieves the steady state. We can define thus the *time constant* of a first-order

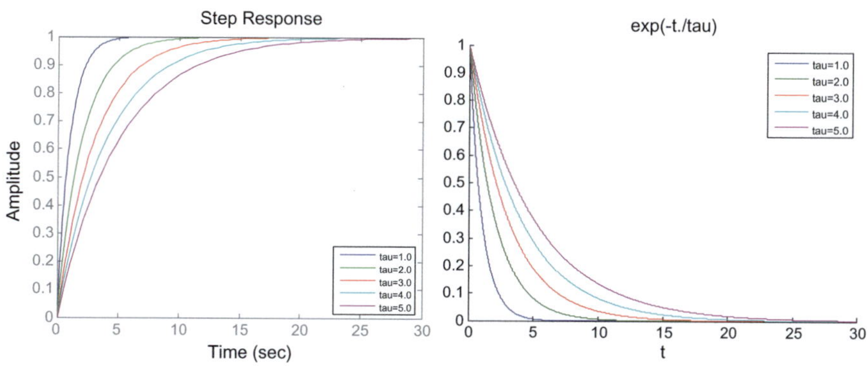

Fig. 3.3 System response of first-order systems for variables values of gain K and time constant τ

systems as $\tau = \frac{1}{a_0}$ which determines the moment in which the response reaches the 63.21% of the steady-state value, i.e. $y(\tau) = 0.6321\,KA$, since

$$y(\tau) = KA(1 - \mathrm{e}^{-a_0 \tau}) = KA\left(1 - \mathrm{e}^{-1}\right) = 0.6321\,KA \tag{3.15}$$

The effect of the time constant τ on a first-order system is shown in Fig. 3.3. Now, in terms of the parameters, K and τ, the transfer function of a first-order system can be rewritten in parametrical form as

$$G(s) = \frac{b_0}{s + a_0} = \frac{\frac{b_0}{a_0}}{\frac{s}{a_0} + 1} = \frac{K}{\tau s + 1} \tag{3.16}$$

that quickly identifies the system gain and the time constant.

In case of $\boldsymbol{a}_0 < 0$, the system is made unstable since the exponential term of the response would grow indefinitely, that is

$$\lim_{t \to \infty} y(t) = \lim_{t \to \infty} \frac{b_0 A}{a_0}(1 - \mathrm{e}^{a_0 t}) \to \infty \tag{3.17}$$

Example 3.2

An intravenous infusion system is used to administer a drug during medical treatments. Fig. 3.4 can be modelled through the differential equation

$$V_d \frac{\mathrm{d}C}{\mathrm{d}t} = Q(t) - K_s C(t) \tag{3.18}$$

where V_d is the diffusion volume of the drug, K_s is the drug transfer out of V_d, $C(t)$ is the concentration, and $Q(t)$ is the infusion flow of drug.

Derive the first-order transfer function by using the gain–time constant format as defined in Eq. (3.16).

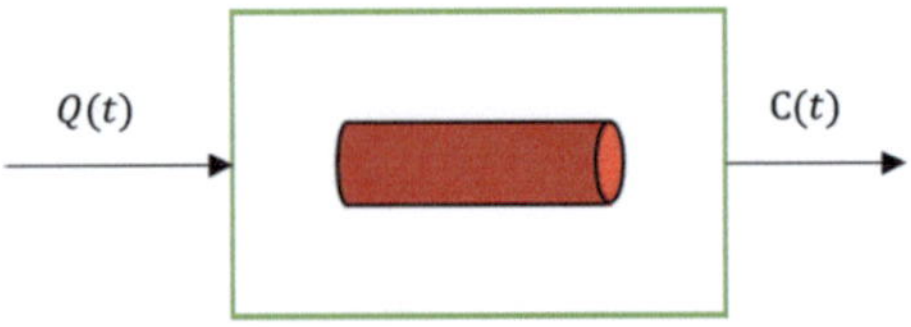

Fig. 3.4 Schematic of an intravenous infusion system

Solution

Rearranging (3.18) and after applying Laplace transform, we have

$$\begin{aligned} V_d s C(s) &= Q(s) - K_s C(s) \\ G(s) &= \frac{C(s)}{Q(s)} = \frac{1}{V_d s + K_s} \end{aligned} \tag{3.19}$$

Then, according to (3.16) we obtain

$$G(s) = \frac{\frac{1}{K_s}}{\frac{V_d}{K_s} s + 1} = \frac{K}{\tau s + 1} \tag{3.20}$$

Therefore, the system gain is inversely proportional to K_s and the time constant depends on the ratio $\frac{V_d}{K_s}$, so the greater the V_d (lesser V_d), the greater the τ.

Second-Order System Response

In general, second-order systems can be represented as differential equations in the form

$$\frac{\mathrm{d}^2 y}{\mathrm{d}t^2} + a_1 \frac{\mathrm{d}y}{\mathrm{d}t} + a_0 y(t) = b_0 u(t) \tag{3.21}$$

An example of a second-order system can be shown in Fig. 3.5. The passive force that develops in an elongating tissue is the sum of inertial, elastic, and viscous effects so that a human tissue can be represented by a simple mechanical dynamic equation as

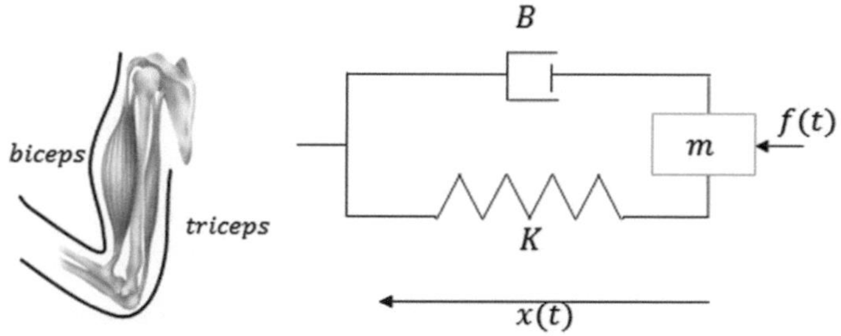

Fig. 3.5 Schematic of a simplified muscle model

$$m\frac{d^2x}{dt^2} + D\frac{dx}{dt} + Kx(t) = F_t(t) \quad (3.22)$$

where m represents the mass of the tissue, D the viscosity constant, K the elastic constant with displacement $x(t)$, and force $F(t)$ as output and input of the system, respectively.

In general, the transfer function for a second-order system can be considered as

$$G(s) = \frac{b_0}{s^2 + a_1 s + a_0} \quad (3.23)$$

where the characteristic equation (the denominator of the function) is of second-order type and whose response to an input step of amplitude A is applied (considering initial conditions as null) can be derived as follows:

$$Y(s) = \frac{b_0}{s^2 + a_1 s + a_0} U(s) = \frac{b_0}{s^2 + a_1 s + a_0} \frac{A}{s} \quad (3.24)$$

Applying residues, we can separate this expression in factors as

$$Y(s) = \frac{K_0}{s} + \frac{K_1}{s + s_1} + \frac{K_2}{s + s_2} \quad (3.25)$$

According to the nature of the three roots of the characteristic equation, we can distinguish three cases:

- s_1 and s_2 are both real and $s_1 \neq s_2$
- s_1 and s_2 are both real and $s_1 = s_2$
- s_1 and s_2 are conjugate complex

In the following, we can discuss the particular responses for each case.

Second-Order System with Distinct Real Roots

In this case, applying the inverse transformation of Laplace to Eq. (3.25) we obtain the general response expression for second-order systems when their poles are different and real, so that

$$y(t) = (K_0 + K_1 e^{-s_1 t} + K_2 e^{-s_2 t}) u_s(t) \quad (3.26)$$

Notice that this expression is similar to the response of first-order system but adding a new exponential term. The steady-state value for the response will be K_0 since the two exponential terms will tend to zero as long as the time tends to infinite as presented in the previous section (assuming both s_1 and s_1 are positive). The systems showing this type of behaviour are called overdamped.

Example 3.3
Consider the simplified muscle system model of second-order type system given by (3.22) with $F(t) = 2.5u_s(t)$ as input step signal and parameters $m = 1$, $D = 3$ and $K = 2$. Obtain the transfer function, and calculate the system response.

Solution
Applying Laplace transform in Eq. (3.22), we have

$$G(s) = \frac{X(s)}{F_t(s)} = \frac{1}{ms^2 + Ds + K} = \frac{1}{s^2 + 3s + 2} \tag{3.27}$$

The time response is calculated as

$$X(s) = G(s)F_t(s) = \frac{1}{s^2 + 3s + 2}\frac{2.5}{s} \tag{3.28}$$

$$X(s) = \frac{K_0}{s} + \frac{K_1}{s+1} + \frac{K_2}{s+2} \tag{3.29}$$

The values for the residues K_0, K_1 and K_2 can be easily calculated by using the MATLAB command `residue`. Given the numerator and denominator of an expression, this command returns the ordered list of roots and their residues

```
>> [r,p]=residue(2.5,conv([1 0],(conv([1 1],[1 2]))))
r =
    1.2500
   -2.5000
    1.2500
p =
    -2
    -1
     0
```

That is, in this case $K_0 = 1.25$, $K_1 = -2.5$ and $K_2 = 1.25$ so that the output for this system is given by

$$x(t) = \left(1.25 - 2.5e^{-t} + 1.25e^{-2t}\right)u_s(t) \tag{3.30}$$

The output response for this system is shown in Fig. 3.6, whose shape is similar to the response of first-order systems, but a little bit slower at the beginning, appearing a characteristic S-shape, which is normal due the addition of an extra

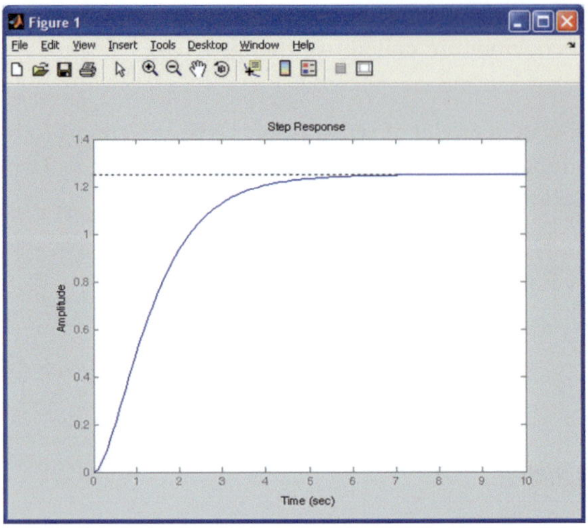


```
>> s=tf('s');
>> G=2.5/( (s+1)*(s+2))

Transfer function:
      2.5
-------------
s^2 + 3 s + 2

>> step(G)
```

Fig. 3.6 Second-order system response for the simplified muscle model for the overdamped case

exponential term. Also, the magnitude of the poles is responsible for the time needed to achieve the steady state; that is, the smaller they are, the quicker the output reaches the final value.

Second-Order System with Repeated Real Roots

This case is very similar to the previous one with a step response represented by

$$Y(s) = \frac{K_0}{s} + \frac{K_1}{s+s_1} + \frac{K_2}{(s+s_1)^2} \tag{3.31}$$

$$y(t) = (K_0 + K_1 \mathrm{e}^{-s_1 t} + K_2 t \mathrm{e}^{-s_1 t}) u_s(t) \tag{3.32}$$

The expression changes the exponential term by an exponential term multiplied by t, which will also tend to zero as long as time tends to infinite (assuming s_1 is positive). The systems showing this type of behaviour are called critically damped.

Example 3.4
Consider a modification of the previous example whereby $m = 1$, $D = 2$ and $K = 1$. Obtain the transfer function and calculate the system response when $F(t) = 2.5u_s(t)$.

Solution
Similar to Example 3.3, the transfer function would be this time:

$$X(s) = G(s)F_t(s) = \frac{1}{s^2 + 2s + 1}\frac{2.5}{s} \tag{3.33}$$

$$X(s) = \frac{K_0}{s} + \frac{K_1}{s+1} + \frac{K_2}{(s+1)^2} \tag{3.34}$$

The calculation of residues via MATLAB renders the values $K_0 = 2.5$, $K_1 = -2.5$ and $K_2 = -2.5$ as

```
≫ [r,p]=residue(2.5,conv([1 0],(conv([1 1],[1 1]))))
r =
  -2.5000
  -2.5000
   2.5000
p =
    -1
    -1
     0
```

so that

$$x(t) = (2.5 - 2.5e^{-t} - 2.5te^{-t})u_s(t) \tag{3.35}$$

Regarding the response shape, it exhibits again similarities with respect to the case of two different poles, i.e. a *S*-shape at the beginning Fig. 3.7. In this example, the final value is 2.5, i.e. the gain of the system is 2.5, as can be checked by applying the final value theorem.

Second-Order System with Conjugate Complex Roots

In case of complex conjugates poles located in $s_1, s_2 = \sigma \pm j\omega$, the decomposition of the step input response of a second-order system is given by

$$Y(s) = \frac{K_0}{s} + \frac{K_1}{s + \sigma - j\omega} + \frac{K_2}{s + \sigma + j\omega} \tag{3.36}$$

which cannot be directly transformed into a temporal expression through the Laplace transformation table. By means of a simple algebraic manipulation, we can sum up the terms that involve the complex conjugates poles as

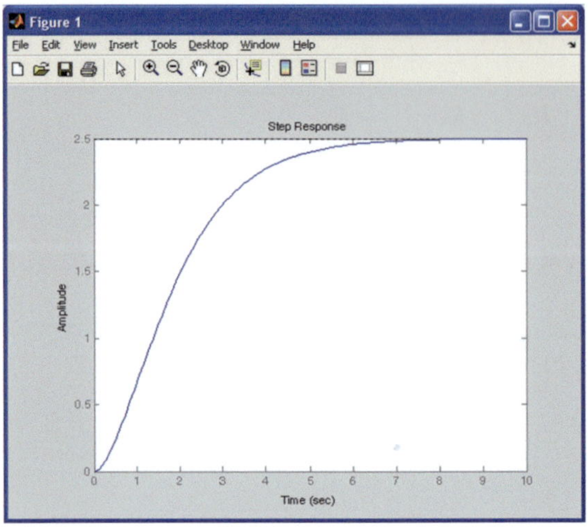


```
>> G=2.5/( (s+1)*(s+1))

Transfer function:
     2.5
-------------
s^2 + 2 s + 1

>> step(G)
```

Fig. 3.7 Second-order system response for the simplified muscle model for the critically damped case

$$Y(s) = \frac{K_0}{s} + \frac{K_1^*(s+\sigma) + K_2^*\omega}{(s+\sigma)^2 + \omega^2}$$
$$K_1^* = 2\,\mathrm{Re}\,(K_1), \quad K_2^* = -2\,\mathrm{Im}\,(K_2) \tag{3.37}$$

In this manner, we can obtain an equivalent expression in which each term can be directly transformed into

$$y(t) = \left(K_0 + K_1^*\mathrm{e}^{-\sigma t}\cos(\omega t) + K_2^*\mathrm{e}^{-\sigma t}\sin(\omega t)\right)u_s(t) \tag{3.38}$$

Analysing this expression we can note the presence of oscillations in the output response due to the sinusoidal terms. Also note that such oscillations will disappear as long as the time tends to infinite given they are multiplied by negative exponential terms. In general, it can be said that attenuation of the oscillations will depend on the magnitude of the real part of the poles (σ) and their frequency will be caused by the conjugate part (ω). The systems showing this type of behaviour are called underdamped.

Example 3.5
Consider again a modification of the previous example whereby $m = 1$, $D = 4$ and $K = 13$. Obtain the transfer function, and calculate the system response when $F(t) = 2.5u_s(t)$.

Solution

$$X(s) = G(s)F_t(s) = \frac{1}{s^2+4s+13}\frac{2.5}{s} \tag{3.39}$$

$$X(s) = \frac{K_0}{s} + \frac{K_1}{s+2+3j} + \frac{K_2}{s+2-3j} \tag{3.40}$$

The residues can be computed as

```
≫ [r,p]=residue(2.5,conv([1 0],[1 4 13]))
r =
  -0.0962 + 0.0641i
  -0.0962 - 0.0641i
   0.1923
p =
  -2.0000 + 3.0000i
  -2.0000 - 3.0000i
        0
```

That is, $K_0 = 0.1923$, $K_1 = -0.0962 + 0.0641j$ and $K_2 = -0.0962 - 0.0641j$. Therefore, $K_1^* = -0.1924$ and $K_2^* = -0.1282$. With these values, the response function is given by (Fig. 3.8)

$$x(t) = \left(0.1923 - 0.1924\mathrm{e}^{-2t}\cos(3t) - 0.1282\mathrm{e}^{-2t}\sin(3t)\right)u_s(t) \tag{3.41}$$

Second-Order System Response in Parametric Form

Although the analysis and the representation used in the previous development is sufficient for understanding the behaviour of second-order systems, a different representation in parametric form is widely employed. The initial expression (3.23) is replaced by

$$G(s) = \frac{b_0}{s^2+a_1s+a_0} \rightarrow G(s) = \frac{K\omega_n^2}{s^2+2\xi\omega_n s+\omega_n^2} \tag{3.42}$$

Since they are both equivalent, it is clear that

$$K\omega_n^2 = b_0 \quad 2\xi\omega_n = a_1 \quad \omega_n^2 = a_0 \tag{3.43}$$

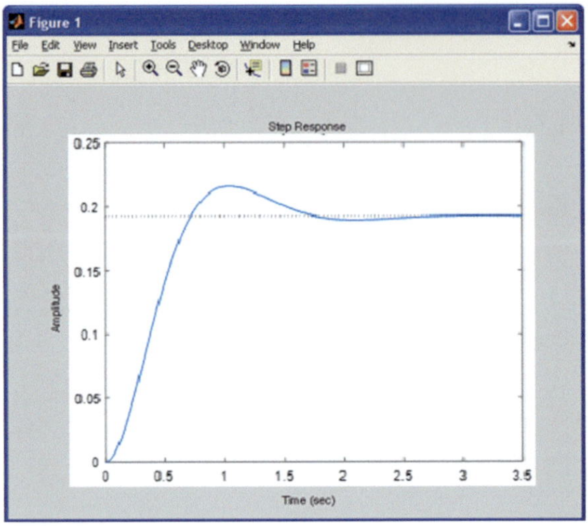


```
>> G=2.5/(s^2+4*s+13)

Transfer function:
      2.5
--------------
s^2 + 4 s + 13

>> step(G)
```

Fig. 3.8 Second-order system response for the simplified muscle model for the underdamped case

In this representation, K is the gain of the system, ω_n is called the undamped natural frequency of the system, and ξ is known as damping ratio. The roots of the characteristic equation (the poles of the system) are obtained by solving

$$\begin{aligned} &s^2 + 2\xi\omega_n s + \omega_n^2 = 0 \\ &s_1, s_2 = -\xi\omega_n \pm j\omega_n\sqrt{1-\xi^2} = -\sigma \pm j\omega_d \end{aligned} \tag{3.44}$$

where ω_d is called the damped natural frequency since it relates the damping ratio and the natural frequency.

Figure 3.9 shows the geometric representation of these parameters. Thus, given a second-order system expressed as in (3.42) one can guess the position of its poles quickly according to the values of ξ and ω_n distinguishing the three cases previously commented

1. $\xi > 1$

   When $\xi > 1$, the expression (3.44) turns into two real numbers; that is, the poles of the system are real and different, and thus, we have an *overdamped* second-order system.
2. $\xi = 1$
   When $\xi = 1$, the radical of expression (3.44) is null and thus the poles are real and equal; thus, we have a *critically damped* second-order system.
3. $0 < \xi < 1$

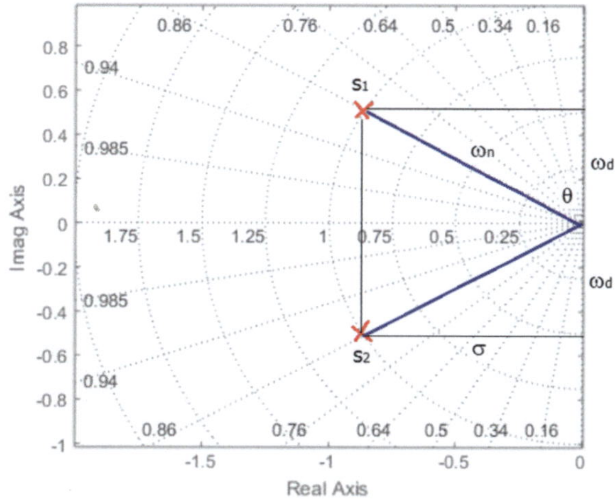

Fig. 3.9 Graphical meaning of the parameters $\xi, \omega_n, \sigma(\xi\omega_n)$ and $\omega_d\left(\omega_n\sqrt{1-\xi^2}\right)$ in the *s*-plane

When $0<\xi<1$, the solution of (3.44) is complex conjugates, and the system is called *underdamped* second-order system.

4. $\xi = 0$

   This is an extreme case where there is no damping effect and the poles are located at the imaginary axis. In this situation, the response of the system is oscillatory, since (3.42) is converted to

$$G(s) = \frac{K\omega_n^2}{s^2 + \omega_n^2} \tag{3.45}$$

and the step response after substituting $\xi = 0$ in (3.38) will be given by

$$y(t) = \left(K_0 + K_1^* \cos(\omega_n t) + K_2^* \sin(\omega_n t)\right) u_s(t) \tag{3.46}$$

To sum up, the transient response of continuous second-order systems will depend on the location of the pair of poles of Eq. (3.44). In fact, ξ-constant loci and ω_n-constant loci can be drawn in *s*-plane, the intersection of which gives the precise location of such poles (Fig. 3.10).

In the same way, the output response will be depend also on the relative value of the pair (ξ, ω_n) (see Fig. 3.10). Figure 3.11 shows a family of curves corresponding to the step response.

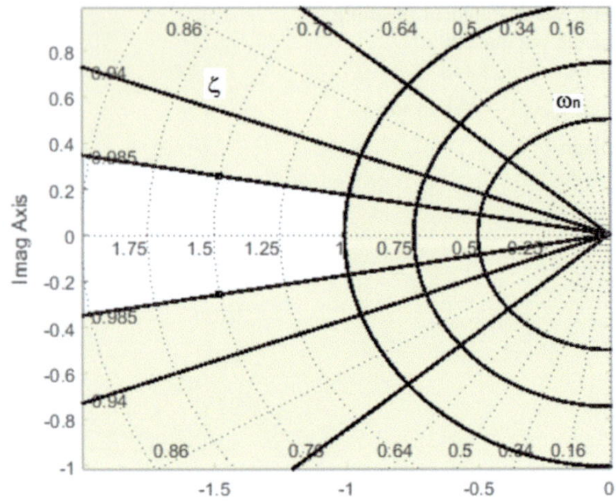

Fig. 3.10 ξ-constant loci and ω_n-constant loci in s-plane

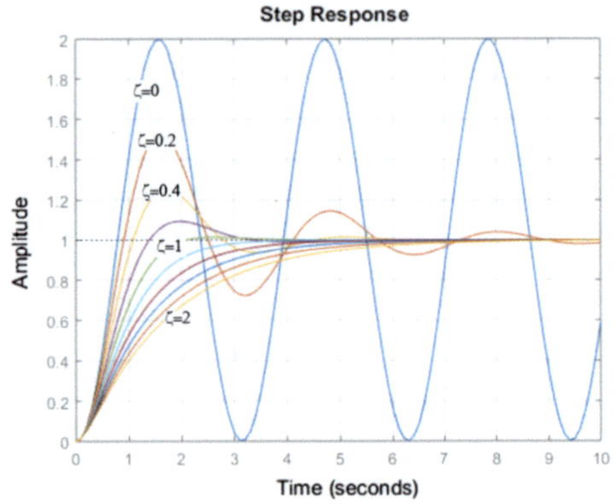

Fig. 3.11 Step responses for second-order systems with distinct (ξ, ω_n)

Higher-Order System Response

Systems dynamics are in general more complex than the first- or second-order systems, and the study of their responses cannot be analysed in the same way it was done in the previous section. Therefore, the order of the system (i.e. the number of poles) and the number of zeros will determine the transient response. However, a generalized study can be derived starting from a general n-order-system represented by

$$G(s) = \frac{K(s+z_1)(s+z_2)\ldots(s+z_m)}{(s+p_1)(s+p_2)\ldots(s+p_n)} \tag{3.47}$$

Without loss of generality, it can be assumed that the n poles are real and different and that the input signal $u(t)$ is a unitary step, and no initial conditions are present. With these assumptions, the system response in s-plane $Y(s)$ can be separated as

$$Y(s) = G(s)U(s) = \frac{G(s)}{s} = \frac{K_0}{s} + \sum_{i=1}^{n} \frac{K_i}{(s+p_i)} \tag{3.48}$$

which yields the time response

$$y(t) = \left(K_0 + \sum_{i=1}^{n} K_i e^{-p_i t} \right) u_s(t) \tag{3.49}$$

Note that the response of a second-order system with real and different poles is a particular case of expression (3.26). That is, by assuming $p_i > 0$, the system output will tend to the final stationary value K_0. The transient response will depend on the location of the poles in the s-plane and the relative magnitude of the coefficients K_i.

In case of k repeated real poles, an expression similar to (3.39) would be obtained this time with power time expressions t^k accompanying the exponential term $e^{-p_i t}$, whose dynamics tend to be dominated by the exponential negative terms. In the same way when conjugate complex poles $s_1, s_2 = -\sigma \pm j\omega_d$ are present, then the same expression (3.39) would suffice, and these poles would also contribute to damped oscillatory behaviour in system response.

When analysing high-order systems, the concept of *dominance of poles* should be taken into account. It can be said that a pole p_j dominates pole p_i if $\frac{\mathrm{Re}(p_j)}{\mathrm{Re}(p_i)} > 5$.

Example 3.6

A blood pressure transducer (Fig. 3.12) is modelled as a second-order system whose dynamics is given by real poles located at $-p_1$ and $-p_2$ and constant K_s as

$$H(s) = \frac{K_s}{(s+p_1)(s+p_2)} \tag{3.50}$$

Determine which pole is the dominant one for $K_s = 10$, $p_1 = 1$ and $p_2 = 10$.

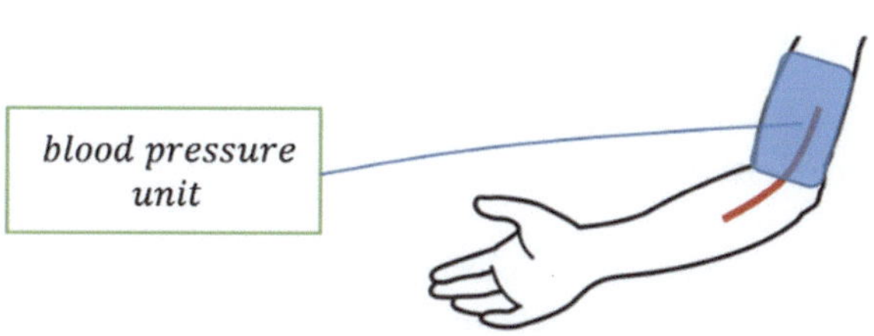

Fig. 3.12 Schematic diagram of a blood pressure sensor

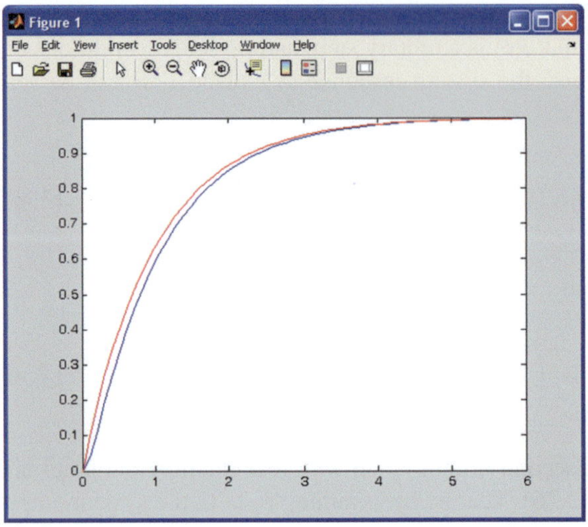


```
>> s=tf('s');
>> G=10/( (s+1)*(s+10))
>> G2=1/(s+1)
>> [x,y]=step(G);
>> [x2,y2]=step(G2);
>> plot(y,x)
>> hold on
>> plot(y2,x2,'r')
```

Fig. 3.13 Dominance of poles. The response of the second-order system (blue) is quite similar to the one of a first-order system with a single dominant pole at p_1 (red)

Solution

As $\frac{\mathrm{Re}(p_2)}{\mathrm{Re}(p_1)} = 10 > 5$, then p_2 dominates p_1 and therefore the effect of the latter on the response can be insignificant. Figure 3.13 shows the step response of the original system and the approximated dominant (without p_1 and same gain). In this way, the second-order system can be considered as a first-order one by neglecting p_2, thus reducing the order of systems and facilitating its analysis.

In the general case of dealing with conjugate complex poles, similar considerations are applicable. Figure 3.14 shows the contribution of individual poles in the response of a high-order system according to their relative position. The final output will be the sum of their contribution along the time.

The transient response is also influenced by the existence of zeros in the general expression (3.47). While zeros do not affect exponential terms $e^{-p_i t}$, they determine the relative value of residues K_i. In general, it can be stated that addition of zeros makes the system output faster and more stable while the addition of poles makes the opposite effect.

Systems Identification

Systems identification refers to the procedure to obtain the transfer function of a system by only considering its output response to a given input signal (Fig. 3.15). This is a useful technique when as it happens in biomedical engineering it is not

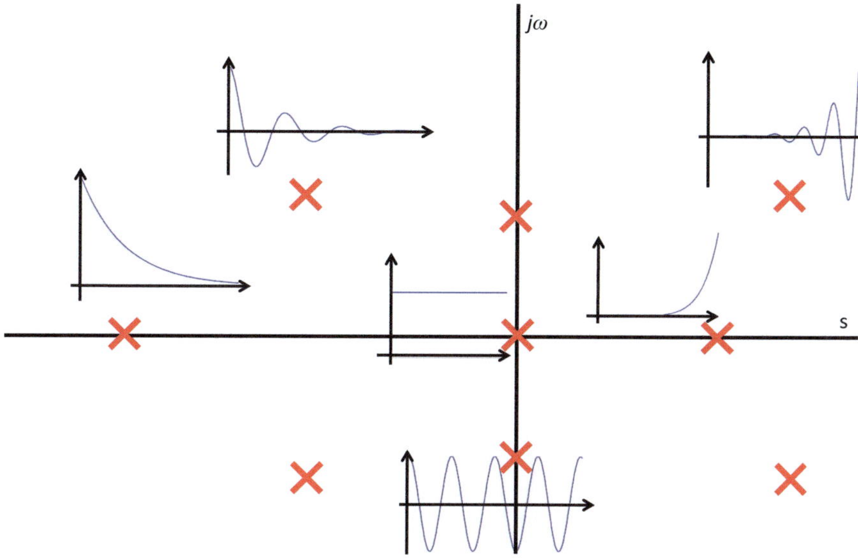

Fig. 3.14 Contribution of poles according to their relative position at the s-plane. Notice that poles located at the right part of the s-plane produce unstable contributions

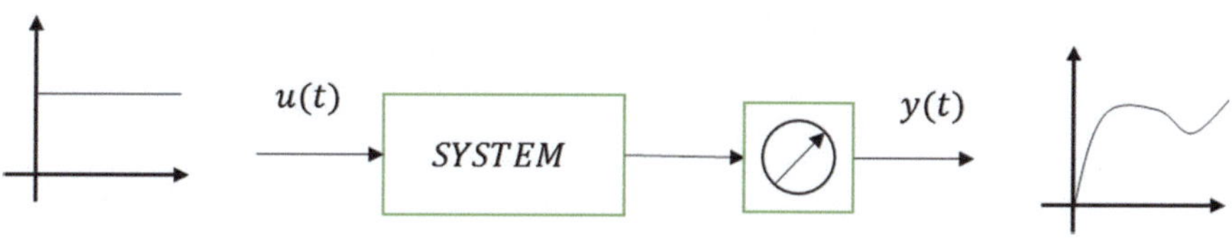

Fig. 3.15 Schematic of parametric system identification

possible to make test on a patient but on a patient model instead. In this way, a black box model is derived stating from input–output operation data.

In this book, we will only focus on parametric system identification of first- and second-order systems considering step signals as inputs.

First-Order System Identification

As was described formerly, the parametric representation of a first-order system was constituted by a gain K and a time constant τ. Thus, when a system is identified assuming it is of first-order type the goal will be to find the values of both K and τ. Therefore, in order to obtain K a simple amplitude relation between the stationary values of output and input signals will be necessary to get it as $K = \frac{y(\infty)}{u(\infty)}$, while time constant τ will be obtained by searching the value of t which makes the output $y(t) = 0.6321\, y(\infty)$.

Example 3.7
Considering the intravenous infusion system already presented in Example 3.2, it is depicted in Fig. 3.16 the output response (concentration) when an input step (flow) with amplitude of 2 is applied. Obtain its transfer function and the parameter values K_s and V_d.

Solution
Recalling that the transfer function of the intravenous infusion system was given by

$$G(s) = \frac{\frac{1}{K_s}}{\frac{V_d}{K_s}s + 1} = \frac{K}{\tau s + 1} \tag{3.51}$$

the system gain $K = \frac{1}{K_s}$ can be computed as the relation between the final value of the output and the final value of the input as $K = \frac{y(\infty)}{u(\infty)} = \frac{0.5}{2} = 0.25$ and then $K_s = \frac{1}{0.25} = 4$. Remembering that time constant is defined by $y(\tau) = 0.6321$

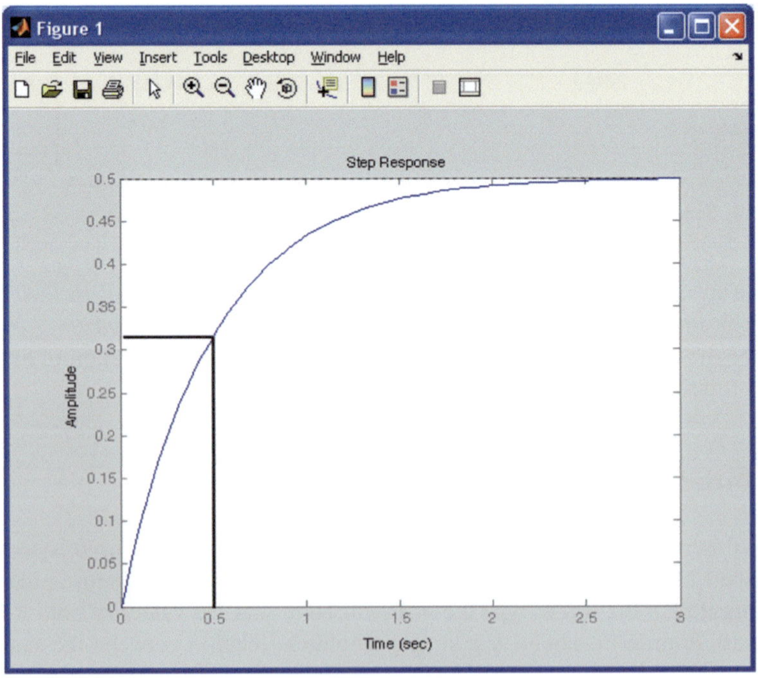

Fig. 3.16 System response of the intravenous system after application of step infusion flow with amplitude of 2

$y(\infty) = 0.315$, by inspection (see Fig. 3.16) one can conclude that $\tau = 0.5$ s and consequently $\frac{V_d}{K_s} = 0.5 \rightarrow V_d = 0.5\, K_s = 2$, so that the identified transfer function is

$$G(s) = \frac{K}{\tau s + 1} = \frac{0.25}{0.5s + 1} = \frac{0.5}{s + 2} \tag{3.52}$$

Second-Order System Identification

The process of system identification becomes similar when the system exhibits a second-order underdamped system dynamics. If it is the case, the goal will be to find the values of gain K, damping coefficient ξ and undamped natural frequency ω_n. Nevertheless, these two former parameters cannot be obtained by inspection as it happened with first-order systems. Therefore, a subset of step input response features (Fig. 3.17) are defined in order to determine the value of ξ and ω_n, as K will be computed in a similar fashion as the one described for first-order type. These output response features are defined next:

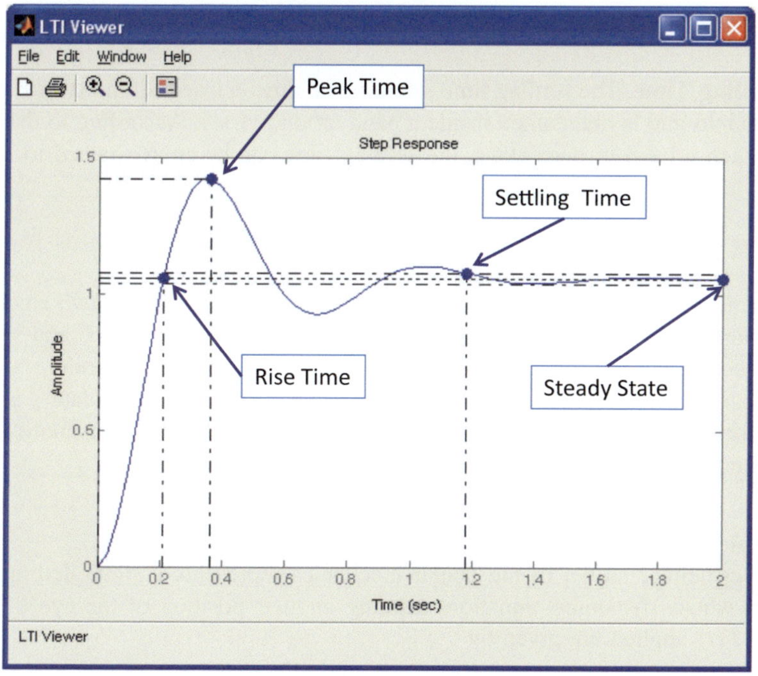

Fig. 3.17 Step response features of an underdamped second-order system visualized through the LTI viewer application

- **Peak Time**. It corresponds to the instant when first peak of the output is reached. This is an easily measurable parameter which is computed as

$$t_p = \frac{\pi}{\omega_d} \tag{3.53}$$

- **Rise Time**. It refers to the time to reach the stationary value $y(\infty)$ for the first time starting form initial conditions (0–100%) and is calculated as

$$t_p = \frac{\pi - \beta}{\omega_d} \tag{3.54}$$

where $\beta = \mathrm{a}\cos(\xi)$.

- **Overshoot**. The overshoot represents the percentage in which the system output exceeds the stationary value $y(\infty)$. Mathematically, it can be expressed as

$$O_s = \frac{y(t_p) - y(\infty)}{y(\infty)} (\text{in } \%) \tag{3.55}$$

which is also an easily measurable parameter computed as

$$O_s = \mathrm{e}^{-\frac{\sigma\pi}{\omega_d}} = \mathrm{e}^{-\frac{\xi\pi}{\sqrt{1-\xi^2}}} \tag{3.56}$$

- **Settling Time**. The settling time refers to the time elapsed till the system output gets into and is maintained inside a band around $y(\infty)$. According to the width of such a band in percentage, the settling time can be approximated to

$$t_s = \frac{4}{\sigma}(2\%) \quad t_s = \frac{3}{\sigma}(5\%) \tag{3.57}$$

In general, the identification process can be completed by selecting two among the four output response features above defined, so as to determine ξ and ω_n and consequently the transfer function of the second-order system. Therefore, there is not a unique way to proceed since we have several choices to calculate ξ and ω_n. Nevertheless, as long as it is possible it will be easier using the overshoot together with the peak time.

Example 3.8

The Westheimer model of the ocular mechanics represents a simplified approximation whose dynamical equations relating angular position of the eye $\theta(t)$ and torque $T(t)$ applied are given by

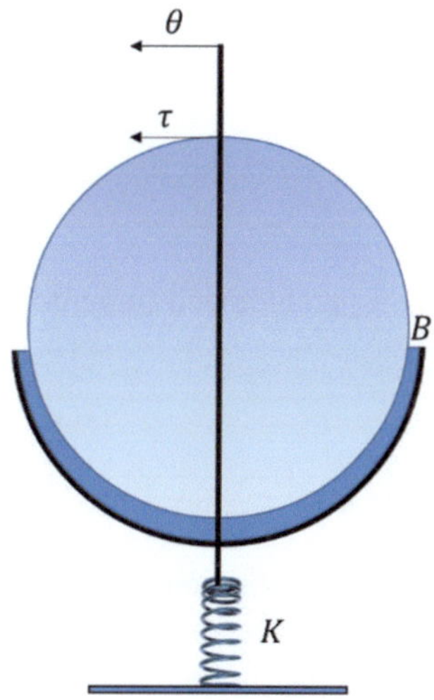

Fig. 3.18 Schematic diagram of the Westheimer ocular model

$$J\frac{d^2\theta}{dt^2} + B\frac{d\theta}{dt} + K\theta(t) = T(t) \tag{3.58}$$

where J represents the moment of inertia, B the friction constant and K the friction constant (Fig. 3.18).

The angular position evolution is plotted in Fig. 3.19 when a unitary step input torque is applied. By inspection, we get and overshoot $O_s = 0.254(25.4\%)$, peak time $t_p = 0.34\,\mathrm{s}$ and stationary value $\theta(\infty) = 1.5$. Determine by parametric identification the transfer function.

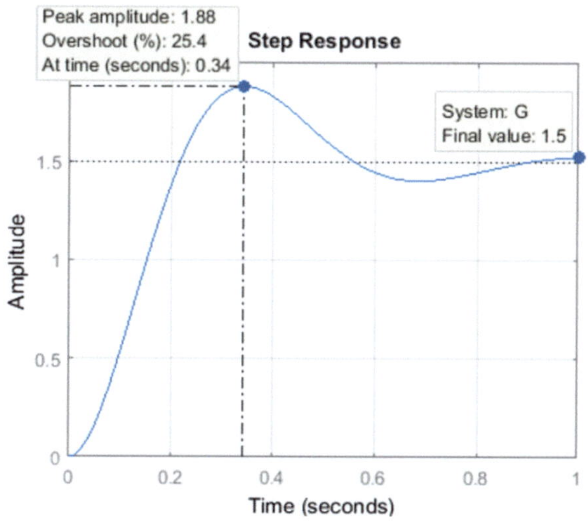

Fig. 3.19 System response of Westheimer model of the ocular mechanics after application of unitary step torque

Solution

From Eq. (3.58) after applying Laplace transformation we have

$$G(s) = \frac{\theta(s)}{T(s)} = \frac{1}{Js^2 + Bs + K} = \frac{\frac{1}{J}}{s^2 + \frac{B}{J}s + \frac{K}{J}} \tag{3.59}$$

so a parametrized underdamped second-order transfer function to be identified as

$$G(s) = \frac{K_s\omega_n^2}{s^2 + 2\xi\omega_n s + \omega_n^2} = \frac{\frac{1}{J}}{s^2 + \frac{B}{J}s + \frac{K}{J}} \tag{3.60}$$

In this way, once are determined the gain K_s, ξ and ω_n from the output response features (overshoot, peak time and stationary value), then it is straightforward to obtain the mechanical set of parameters J, B and K.

As $O_s = 0.254$, then by applying Eq. (3.56) or its inverse, we get

$$\xi = \sqrt{\frac{(\ln O_s)^2}{(\ln O_s)^2 + \pi^2}} = 0.4 \tag{3.61}$$

As $t_p = 0.34$ s then by applying Eq. (3.53)

$$\begin{aligned} t_p &= 0.34 = \frac{\pi}{\omega_d} = \frac{\pi}{\omega_n\sqrt{1-\xi^2}} \\ \omega_n &= \frac{\pi}{t_p\sqrt{1-\xi^2}} = \frac{\pi}{0.34\sqrt{1-0.16}} = 10 \end{aligned} \tag{3.62}$$

Finally, the gain of this system is calculated from $K_s = \frac{\theta(\infty)}{T(\infty)} = \frac{1.5}{1} = 1.5$, so the transfer function will be

$$G(s) = \frac{K_s\omega_n^2}{s^2 + 2\xi\omega_n s + \omega_n^2} = \frac{1.5 \times 10^2}{s^2 + 2 \cdot 0.4 \cdot 10s + 10^2} = \frac{150}{s^2 + 8s + 100} \tag{3.63}$$

and the mechanical parameters are $J = \frac{1}{150}$, $K = 100J$ and $B = 8J$, respectively.

Discrete Transient Response Analysis

A discrete-time system is one whose operation is described by a difference equation relating sampled input and output signals. To understand the operation of digital control systems is necessary to understand the effects of sampling continuous signals and data reconstruction techniques, so as to derive a discretized model of a continuous system, whose transient response will be further analysed. This will enable us to analyse the combined continuous-discrete structure of a digital control system.

A mathematical model of the combined sampling-reconstruction operation is developed, which will allow us the use of the Z transform for the calculation of the response of discrete control systems.

Data Sampling

The process of sampling consists of taking a sequence of n samples $x(kT) = \{x(0), x(T), \ldots x(nT)\}$ of a continuous signal $x(t)$, where T is the sampling time (Fig. 3.20). Obviously, the sampling operation gives no information about $x(t)$ between samples. Hence, two signals $x_1(t)$ and $x_2(t)$ sharing the same values in the sampling instants will give rise to the same sampled signal $x_1(kT) = x_2(kT)$; therefore, this sampling operation will not be ever represented as a transfer function.

Data Reconstruction

A data reconstruction (data hold) device generates a continuous signal $h(t)$ from a discrete sequence $x(kT)$ such that $h(t)$ is approximated in the interval $kT \leq t \leq (k+1)T$ by a polynomial in terms of τ such as

$$h(t) = h(kT+\tau) = a_n\tau^n + a_{n-1}\tau^{n-1} + \cdots + a_1\tau + a_0 \quad 0 \leq \tau \leq T \tag{3.64}$$

In case of $n = 0$, we have the zero-order hold ($H0$) where the actual value of $x(kT)$ is used and maintained so that

$$h(t) = h(kT+\tau) = x(kT) \quad 0 \leq \tau \leq T \tag{3.65}$$

and for n-order hold (Hn) $n+1$ past values of $x(kT)$ are employed to generate $h(t)$, concretely $x(kT), x((k-1)T), \ldots, x((k-1)T)$. As n increases, accuracy in approximation improves, but large delays are also produced rendering instability

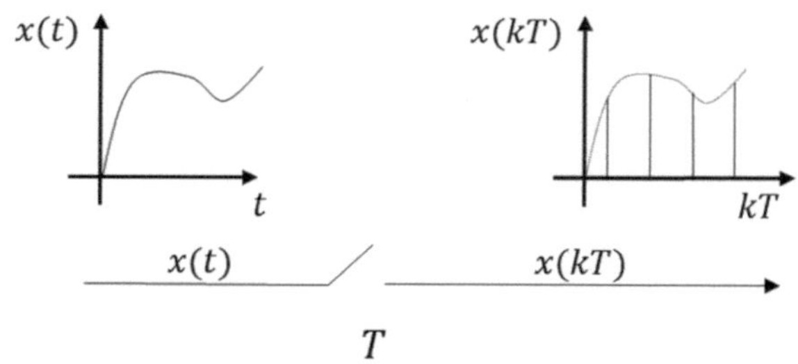

Fig. 3.20 Sampling operation of a continuous signal

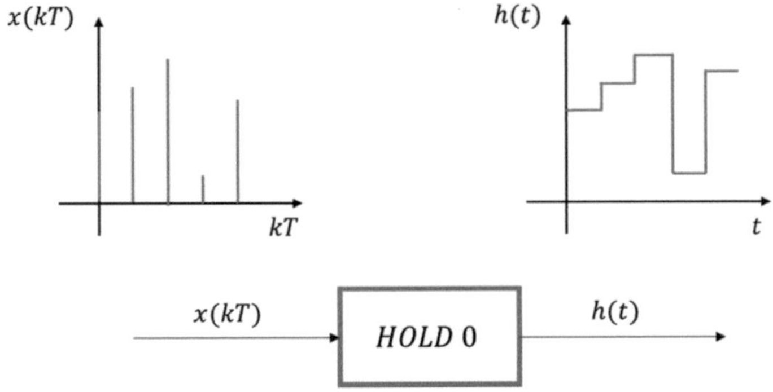

Fig. 3.21 Data reconstruction operation of a discrete signal with a zero-order hold

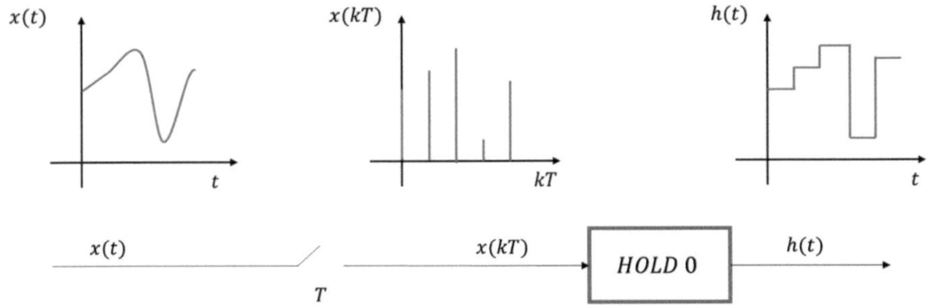

Fig. 3.22 Combined data sampling and zero-order hold operation

when control operation is demanded. The zero order holds the simpler data reconstruction device and is the most common used by far, by clamping its output signal to a value equal to that of the sample signal at the sampling instant (Fig. 3.21).

A mathematical model of the combined sampling-$H0$ operation is developed (Fig. 3.22), which will allow the use of the Z transform for the calculation of the response of discrete control systems by substituting the real sampling-$H0$ operation by its ideal model without altering its input–output characteristics.

The reconstructed signal $h(t)$ can be expressed as a sum of pulses of duration T with amplitude equal to that of $x(kT)$, that is.

$$\begin{aligned} h(t) &= x(0)(u_s(t) - u_s(t-T)) + x(T)(u_s(t-T) - u_s(t-2T)) + \cdots \\ h(t) &= \sum_{k=0}^{\infty} x(kT)(u_s(t-kT) - u_s(t-(k+1)T)) \end{aligned} \tag{3.66}$$

The Laplace transform of $h(t)$ is given by

$$\begin{aligned} H(s) &= \sum_{k=0}^{\infty} x(kT)\frac{1}{s}\left(\mathrm{e}^{-kTs} - \mathrm{e}^{-(k+1)Ts}\right) \\ &= \left(\sum_{k=0}^{\infty} x(kT)\mathrm{e}^{-kTs}\right)\left(\frac{1-\mathrm{e}^{-Ts}}{s}\right) \end{aligned} \tag{3.67}$$

The first factor of (3.67) depends on the samples of $x(t)$ and is denoted as $X^*(s)$ called the starred Laplace transform and denotes an ideal sampler. In fact, $x^*(t)$ will be

$$x^*(t) = \mathcal{L}^{-1}(X^*(s)) = x(0)\delta(t) + x(T)\delta(t-T) + \cdots \tag{3.68}$$

which represents a train of impulse functions with amplitudes equal to that of the samples of $x(t)$ (Fig. 3.23).

The second factor of (3.67) does not depend on the samples of $x(t)$ and is properly a transfer function denoted as $G_{H0}(s)$ called the ideal zero-order hold. Neither the ideal sampler nor the ideal zero-order hold models physical devices, but the combination accurately models the real sample and zero-order hold operation (Fig. 3.24). Besides, the only function allowed for $G_{H0}(s)$ is the train of impulse functions $x^*(t)$.

It is highly noticed the similarity between the starred transform and the z transform; in fact, it can be stated that $X^*(s) = X(z)|_{z=\mathrm{e}^{sT}}$ and it represents the link between the s-plane and the z-plane. In fact, the ideal sampling allows to analyse the sampled continuous systems through the Z transform.

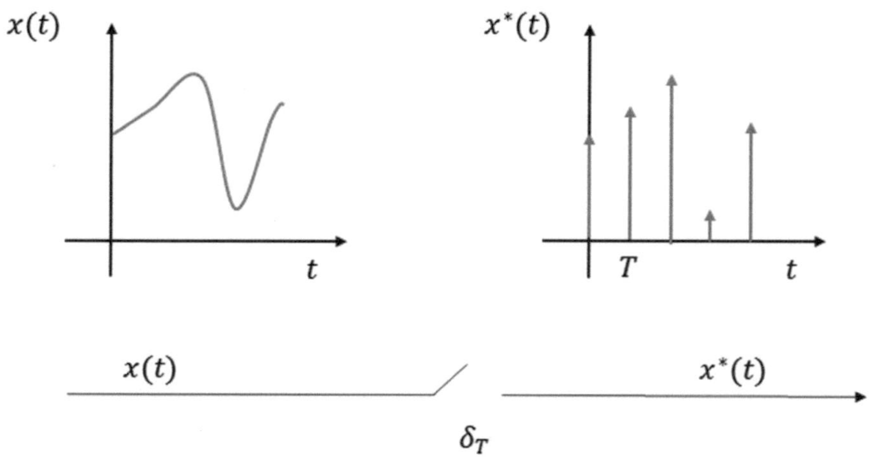

Fig. 3.23 Representation of the ideal sampling operation

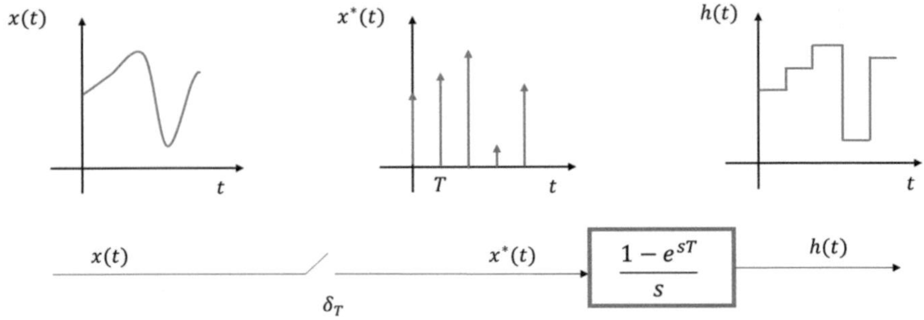

Fig. 3.24 Combined ideal sampler and ideal zero-order hold

The starred transform has an alternate form equivalent to the one defined in (3.67), related to the Laplace transform that can be expressed as

$$X^*(s) = \frac{1}{T}\sum_{n=-\infty}^{n=\infty} X(s + jn\omega_s) + \frac{x(0)}{2} \tag{3.69}$$

being $\omega_s = \frac{2\pi}{T}$ the sampling frequency, which will prove to be useful when selecting a proper value for sampling period T. In fact, if the frequency spectrum of $x^*(t)$ is evaluated by making $s = j\omega$, then we have

$$\begin{aligned} X^*(j\omega) &= |X^*(j\omega)|e^{j\angle X^*(j\omega)} \\ &= \frac{1}{T}(\cdots + X(j\omega + j\omega_s) + X(j\omega) + X(j\omega - j\omega_s) + \cdots \end{aligned} \tag{3.70}$$

Hence, the ideal sampling produces a replication of the original spectrum of $x(t)$, so the signal content is distorted when ideal sampling is applied (Fig. 3.25). Therefore, an ideal filter with T gain should be employed to restore the original content of $x(t)$ provided that $\frac{\omega_s}{2} \geq \omega_1$ which is basically the Shannon's sampling theorem. Otherwise, an aliasing effect will appear so that there will be not possible to reconstruct the original signal starting from its samples (Fig. 3.26).

It must be underlined that the ideal filter cannot be implemented physically. In its place, a zero-order hold is usually used instead, whose frequency response can be evaluated by making $s = j\omega$ in $G_{H0}(s)$ so that

$$G_{H0}(j\omega) = \frac{1 - e^{-j\omega T}}{j\omega} \tag{3.71}$$

The amplitude and phase of $G_{H0}(j\omega)$ are represented in Fig. 3.27 in comparison with that of the ideal hold. It can be seen that even though there appear high-frequency ripples and phase delay, both effects can be attenuated as long as ω_s is increased beyond that of the Shannon's sampling theorem establishes.

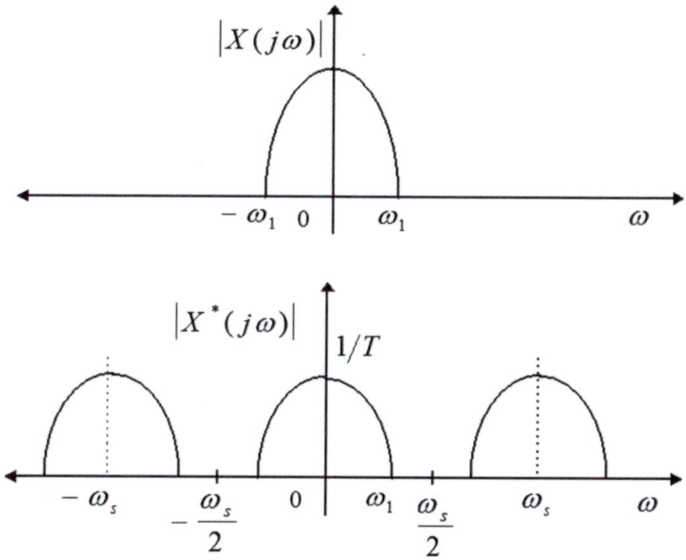

Fig. 3.25 Frequency spectrum in amplitude for $\boldsymbol{x}(t)$ and $\boldsymbol{x}^*(t)$. It is supposed that $\boldsymbol{x}(t)$ is band-limited with highest frequency of ω_1

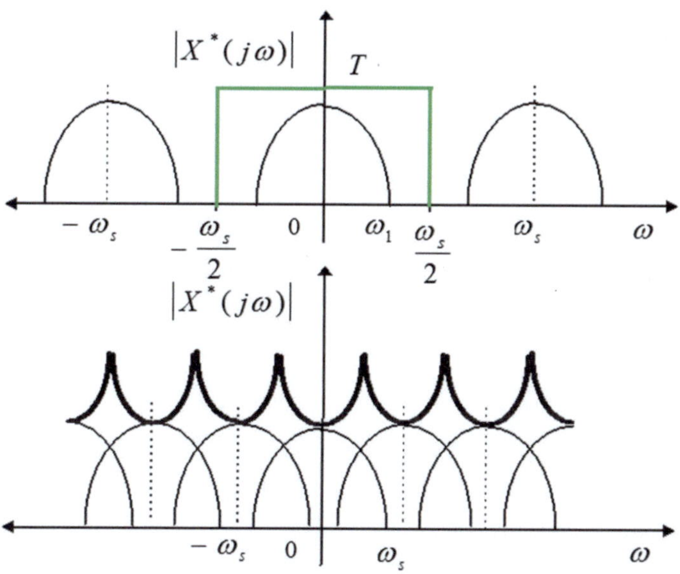

Fig. 3.26 Frequency spectrum in amplitude $x^*(t)$ with $\frac{\omega_s}{2} \geq \omega_1$ and reconstruction with filter (up) and with $\frac{\omega_s}{2} < \omega_1$ (down) with aliasing effect

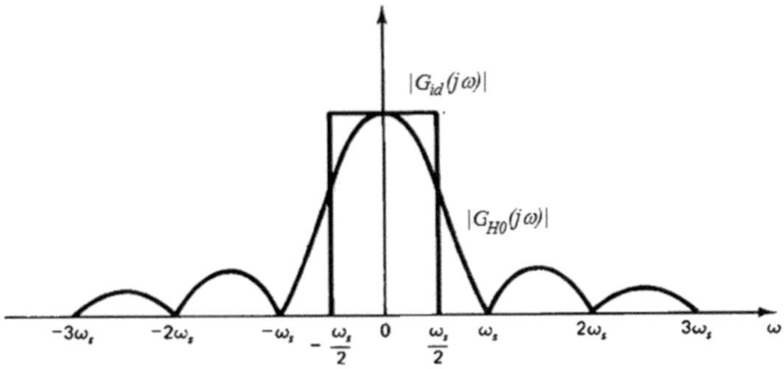

Fig. 3.27 Comparative frequency response of ideal hold and zero-order hold

Pulse Transfer Function

The pulse transfer function allows the analysis of the sampled continuous systems with the Z transform and is defined as the relationship between the Z transform of the output at the sampling times and the Z transform of the sampled input.

To develop the pulse transfer function for an open-loop sampled data system we are going to consider the system shown in Fig. 3.28, which consists of an ideal sampler, a zero-order hold in series with the continuous plant $G_p(s)$, being $e(t)$ and $y(t)$ the input and output, respectively.

The open-loop transfer function is given by the serial system

$$\begin{aligned} G(s) &= \frac{Y(s)}{E^*(s)} = \frac{1 - \mathrm{e}^{-Ts}}{s} G_p(s) \\ Y(s) &= G(s)E^*(s) \end{aligned} \tag{3.72}$$

In order to determine the pulse transfer function, the output $y(t)$ at sampling times should be taken by means of an ideal sampler (Fig. 3.29), thus

$$\begin{aligned} Y^*(s) &= (G(s)E^*(s))^* = \frac{1}{T} \sum_{n=-\infty}^{n=\infty} G(s + jn\omega_s)E^*(s + jn\omega_s) \\ Y^*(s) &= G^*(s)E^*(s) \end{aligned} \tag{3.73}$$

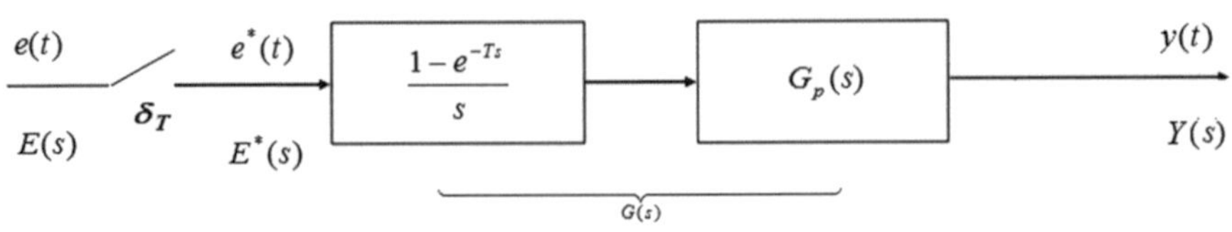

Fig. 3.28 Open-loop sampled data system

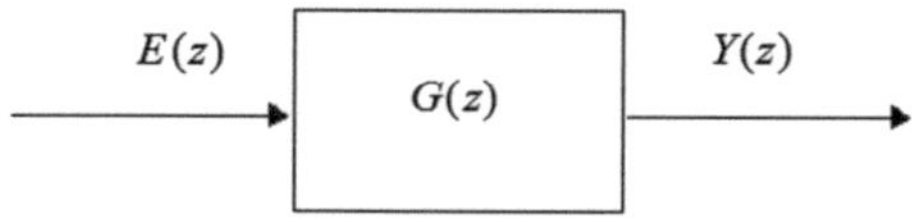

Fig. 3.29 Pulse transfer function of $G(s)$

since ideal sampling of $e^*(t)$ is the same signal. Then, by recalling the relation between starred transform and Z transform given by $z = e^{sT}$, then

$$Y(z) = G(z)E(z) \tag{3.74}$$

where $G(z)$ is the so-called pulse transfer function which given no information about $y(t)$ between sampling instants. In this way, it represents a discretized model of the continuous system (Fig. 3.29).

In order to derive $G^*(s)$, the open-loop transfer function $G(s)$ is expressed as

$$G(s) = \left(1 - e^{-Ts}\right)\frac{G_p(s)}{s} = F^*(s)H(s) \tag{3.75}$$

being $F^*(s)$ and $H(s)$ the first and second factor of (3.72). Then $G^*(s)$ can be easily obtained since

$$\begin{aligned} G^*(s) &= F^*(s)H^*(s) \\ G(z) &= F(z)H(z) \end{aligned} \tag{3.76}$$

with $F(z) = 1 - z^{-1}$ while $H(z) = Z(H(s))$, which is obtained by making the partial fraction expansion of $H(s)$ and after applying the equivalence between Laplace transform table and Z transform table for each partial fraction.

Example 3.9

Obtain the pulse transfer function $G(z) = \frac{Y(z)}{E(z)}$ of the thermometer of Example 1.1 (Fig. 3.30) with transfer function $G_p(s) = \frac{0.05}{s+0.05}$ for $T = 0.1$

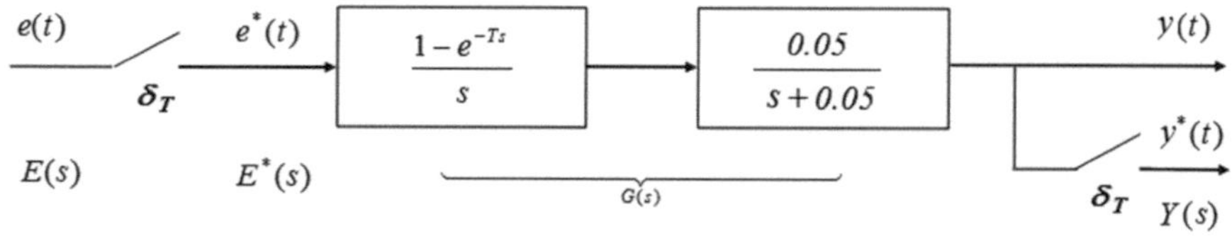

Fig. 3.30 Open-loop sampled data system corresponding to thermometer

Solution

The open-loop transfer function is defined by

$$G(s) = \left(1 - e^{-Ts}\right)\frac{G_p(s)}{s} = \underbrace{\left(1 - e^{-Ts}\right)}_{F^*(s)}\underbrace{\frac{0.05}{s(s+0.05)}}_{H(s)} \tag{3.77}$$

Therefore, the pulse transfer function (discretized model) is given by

$$G(z) = Z(G(s)) = F(z)Z(H(s)) = \left(1 - z^{-1}\right)Z\left(\frac{0.05}{s(s+0.05)}\right) \tag{3.78}$$

Partial fraction expansion of $H(s)$ is computed as

$$\begin{aligned} H(s) &= \frac{0.05}{s(s+0.05)} = \frac{1}{s} - \frac{1}{s+0.05} \\ Z(H(s)) &= \frac{z}{z-1} - \frac{z}{z-e^{-0.05T}} = \frac{0.005z}{(z-1)(z-0.995)} \end{aligned} \tag{3.79}$$

and from (3.78) we have

$$G(z) = \frac{z-1}{z}\frac{0.005z}{(z-1)(z-0.995)} = \frac{0.05}{z-0.995} \tag{3.80}$$

By applying the MATLAB code

```
≫ s= zpk('s');
≫ G=0.05/(s+0.05);
≫ T=0.1;Gz=c2d(G,T,'zoh')

Gz =

  0.005
  ---------
  (z-0.995)

Sample time: 0.1 seconds
≫ step(G,Gz)
```

By comparing the unit step response of the continuous thermometer system $G(s)$ and of the discretized model $G(z)$, it can be seen that both responses are equal at sampling instants (Fig. 3.31).

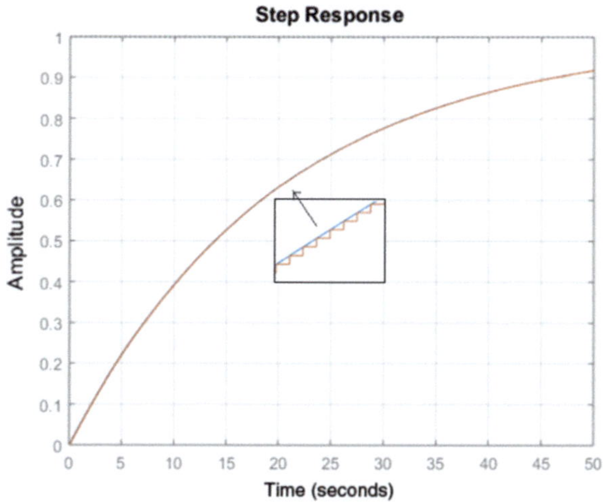

Fig. 3.31 Step responses of continuous and discretized system corresponding to thermometer

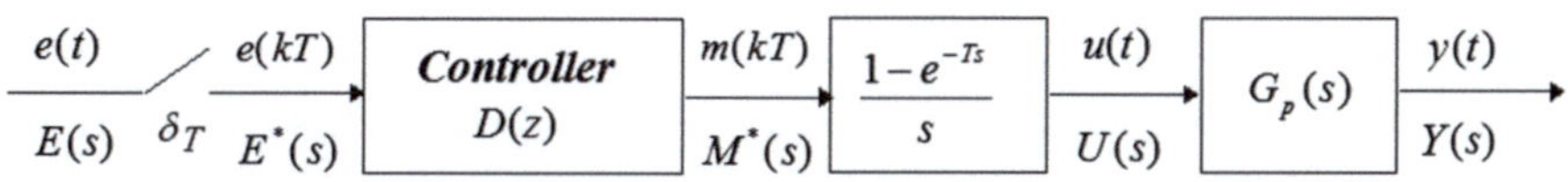

Fig. 3.32 Open-loop sampled data system with discrete controller

We extend the pulse transfer function determination to cover the case that the open-loop sampled continuous system contains a discrete controller (Fig. 3.32).

Applying the relation between starred transform and Z transform, we have for the discrete controller

$$\begin{aligned} M(z) &= D(z)E(z) \\ M^*(s) &= D^*(s)E^*(s) \end{aligned} \tag{3.81}$$

Taking the samples of $y(t)$ by means of an ideal sampler and recalling (3.73), we have

$$\begin{aligned} Y^*(s) &= G^*(s)M^*(s) = G^*(s)D^*(s)E^*(s) \\ Y(z) &= G(z)D(z)E(z) \end{aligned} \tag{3.82}$$

Therefore, the pulse transfer function of this configuration is computed by the serial product of the discrete controller transfer function $D(z)$ and of the discretized system transfer function $G(z)$ (Fig. 3.33) calculated as indicated in (3.76). It is important to highlight that the sampling theorem condition will be applied to the input signal $e(t)$ even if the controller interposes between the sampler and the

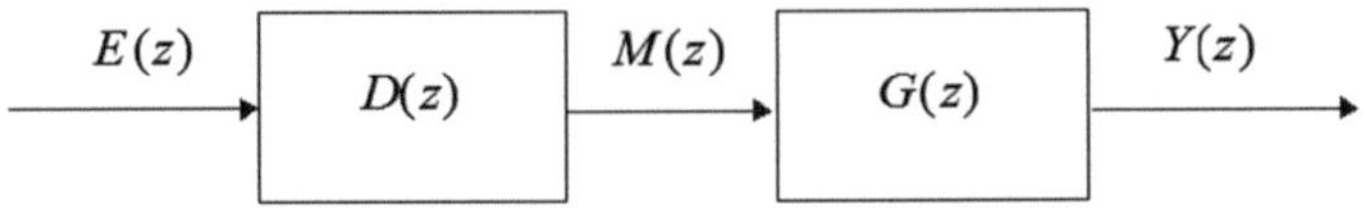

Fig. 3.33 Pulse transfer function of discrete controller $D(z)$ in series with discretized $G(s)$

zero-order hold, since the band limitation in $e(t)$ guarantees band limitation in $m(t)$ as frequency spectrum of $D(z)$ is periodic with period $\frac{\omega_s}{2}$.

Considering only the pulse transfer function or discretized plant $G(z)$, the discrete-time response in z-plane will consist of the result of the sum of forced response $Y_1(z)$ dependent of the input $U(z)$ and natural response $Y_2(z)$ influenced by the initial conditions polynomial $P(z)$ in the same way as described for continuous systems, that is

$$\begin{aligned} Y(z) = Y_1(z) + Y_2(z) = \frac{K(z-z_1)(z-z_2)\ldots(z-z_m)}{(z-p_1)(z-p_2)\ldots(z-p_n)}U(z) \\ + \frac{P(z)}{(z-p_1)(z-p_2)\ldots(z-p_n)} \end{aligned} \tag{3.83}$$

where it has been assumed that the discretized plant contains n poles and m zeros. Considering the input signal $u(kT)$ as the unitary step and no initial conditions, the system response in z-plane $Y(z)$ can be separated in case of distinct poles as

$$Y(z) = G(z)U(z) = \frac{K_0 z}{z-1} + \sum_{i=1}^{n} \frac{K_i}{(z-p_i)} \tag{3.84}$$

which yields the time response

$$y(k) = K_0 + \sum_{l=1}^{n} K_i (p_i)^l u_s(k) \tag{3.85}$$

where each pole p_i contributes with a power $(p_i)^l$; thus by assuming $|p_i| < 1$, the discrete system output will tend to the final stationary value K_0. The transient response will depend on the location of the poles in z-plane and the relative magnitude of the coefficients K_i in the same way as described for continuous systems.

Mapping from *s*-Plane to *z*-Plane

It has been determined that variables s and z are related through the mapping $z = e^{sT}$ through the relation established between starred transform and z transform.

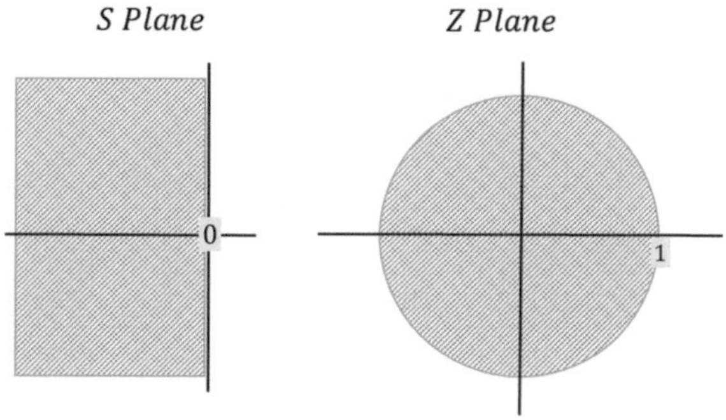

Fig. 3.34 Mapping from s-plane to z-plane

Therefore, if the effects of pole positioning on the s-plane are known, the corresponding effects of pole positioning on the z-plane can be also determined.

Considering first the mapping of a general point in s-plane $s = \sigma + j\omega$ into z-plane then

$$z = e^{(\sigma + j\omega)T} = e^{\sigma T} e^{j\omega T} = e^{\sigma T} e^{j(\omega + k\omega_s)T}, \quad k = 0, 1, 2\ldots \tag{3.86}$$

as $e^{j\omega_s kT} = e^{j2\pi k} = 1$. Thus poles and zeros in s whose frequencies ω differ by multiples of sampling frequency ω_s are mapped in the same locations in z; that is, the s-z mapping is not unique. Along the $s = j\omega$ axis, $z = e^{j\omega T} = 1\angle\omega T$; hence, pole locations on the unit circumference in z-plane are equivalent to pole locations on the imaginary axis in s-plane, which in both cases represents the marginally stable systems (Fig. 3.34). Therefore, for a discrete system to be stable, the poles of the system must be located inside the unit circle.

Besides, each strip of width ω_s in s-plane is mapped into the circle unit. The first strip is called the primary strip, while the rest are referred to as complementary strips. This proves the uniqueness of the s-z mapping (Fig. 3.35).

The effect of a given poles location in z-plane on the transient response can be seen by examining the correspondence between the poles location in s-plane and the poles in z-plane through the mapping s-z (Fig. 3.36). In this way, it will not be necessary to tackle the discrete-time system response by analysing first-order, second-order, or higher-order discrete system dynamics in the way it was done in s-plane.

In Fig. 3.37, it shows the contribution of individual poles in the response of a high-order system according to their relative position both in s-plane and in z-plane comparatively. The discrete output response will be the sum of their contribution along time kT.

The discrete transient response also depends on T, since the location of poles in z-plane will be given in general by $z = e^{(\sigma + j\omega)T} = e^{\sigma T} e^{j\omega T}$. The value of T must satisfy the sampling theorem, since it would not be possible having poles $s_1 = \sigma_1 + j\omega_1$ of the characteristic equation with $\omega_1 \geq \frac{\omega_s}{2} = \frac{\pi}{T}$, since there would be overlapping and the original poles and zeros location in s-plane would be changed

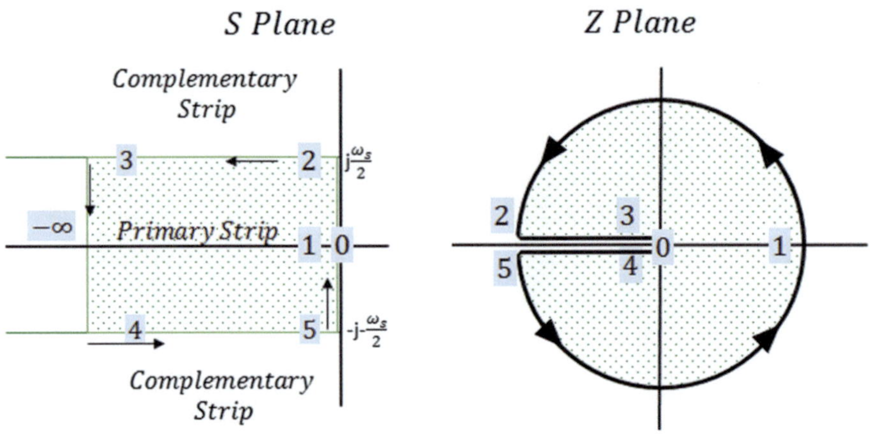

Fig. 3.35 Mapping primary and complementary strips from s-plane to z-plane

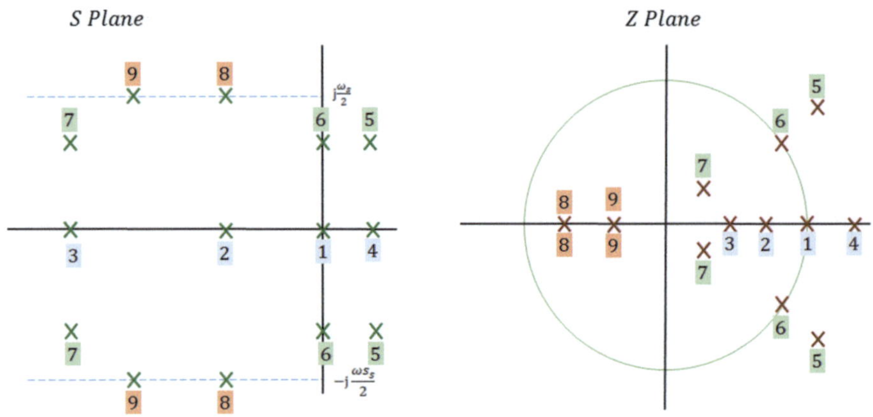

Fig. 3.36 Mapping primary and complementary strips from s-plane to z-plane

in z-plane. In that case, T must be decreased so as to fulfil $\omega_1 < \frac{\pi}{T}$. In Fig. 3.37, it also shows the contribution of each individual pole and the changing effect with T.

Example 3.10
Derive for the intravenous infusion system of Example 3.2 the discretized model $G(z)$ starting form the first-order continuous transfer function as defined by

$$G(s) = \frac{C(s)}{Q(s)} = \frac{1}{V_d s + K_s} \tag{3.87}$$

with $V_d = 10$ and $K_s = 2$ for $T = T_1 = 0.1$ and $T = T_2 = 0.5$. Also obtain the step response for each case.

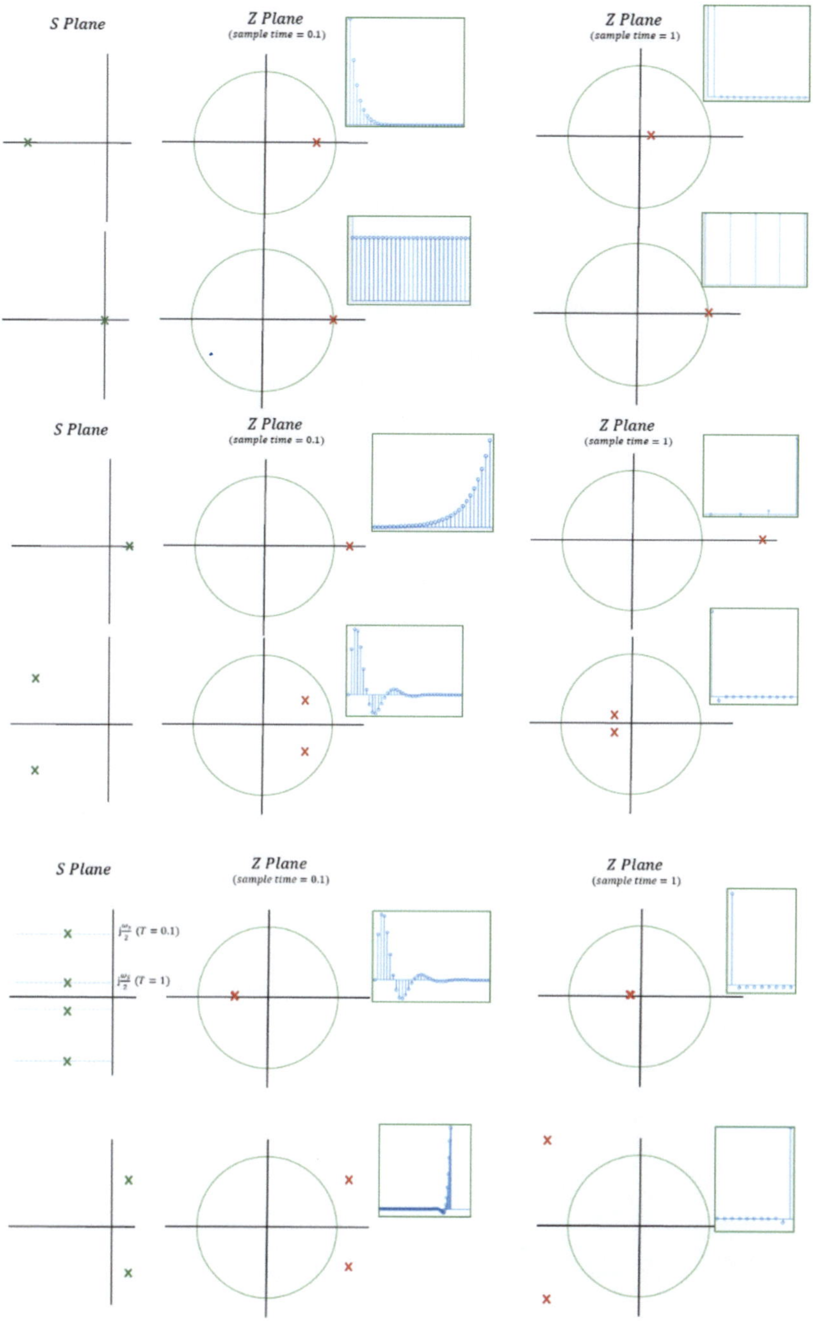

Fig. 3.37 Contribution of poles according to their relative position at the *s*-plane and its correspondence in *z*-plane for $T = 0.1$ and $T = 1$. Notice that poles located outside the unit circle in *z*-plane produce unstable contributions

Solution

Substituting V_d and K_s in (3.87), it is possible to calculate the pulse transfer function as it was done in (3.77) so as to get

$$\begin{aligned} G(z) &= \left(1-\mathrm{e}^{-Ts}\right)\frac{G(s)}{s} = \left(1-\mathrm{e}^{-Ts}\right)\frac{1}{s(10s+2)} \\ G(z) &= \left(1-z^{-1}\right)Z\left(\frac{0.1}{s(s+0.2)}\right) \end{aligned} \tag{3.88}$$

and following the same approach as was described in Example 3.9 it is obtained

$$\begin{aligned} G(z) &= \left(\frac{z-1}{z}\right)0.5\left(\frac{z}{z-1}-\frac{z}{z-\mathrm{e}^{-0.2T)}}\right) \\ G(z) &= 0.5\frac{1-\mathrm{e}^{-0.2T}}{z-\mathrm{e}^{-0.2T}} \end{aligned} \tag{3.89}$$

By applying MATLAB, the two pulse transfer functions and their step responses (Fig. 3.38) can be easily obtained both for $T = T_1 = 0.1$ and for $T = T_2 = 0.5$

```
≫ s=zpk('s');
≫ G=1/(10*s+2);
≫ T=0.1;
≫ Gz1=c2d(G,T,'zoh')

Gz1 =

    0.0099007
    ----------
```

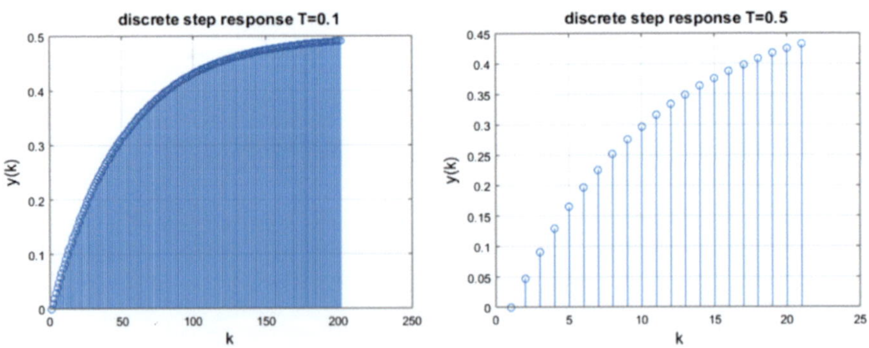

Fig. 3.38 Step response of discretized infusion system for $T_1 = 0.1$ (left) and for $T_2 = 0.5$ (right)

```
   (z-0.9802)

Sample time: 0.1 seconds
Discrete-time zero/pole/gain model.

≫ T=0.5;
≫ Gz2=c2d(G,T,'zoh')

Gz2 =

   0.047581
  ----------
  (z-0.9048)

Sample time: 0.5 seconds
Discrete-time zero/pole/gain model.

≫ y1=step(Gz1,10);stem(y1)
≫ y2=step(Gz2,10);stem(y2)
```

To sum up, the transient response of discrete-time systems will depend on the location of poles of characteristic equation of discretized transfer function $G(z)$. In the same way as it happened with second-order continuous-time systems, ξ-constant loci and ω_n-constant loci can be drawn in z-plane just by applying the mapping $z = e^{sT}$ with poles inside the primary strip where sampling theorem is fulfilled (Fig. 3.39).

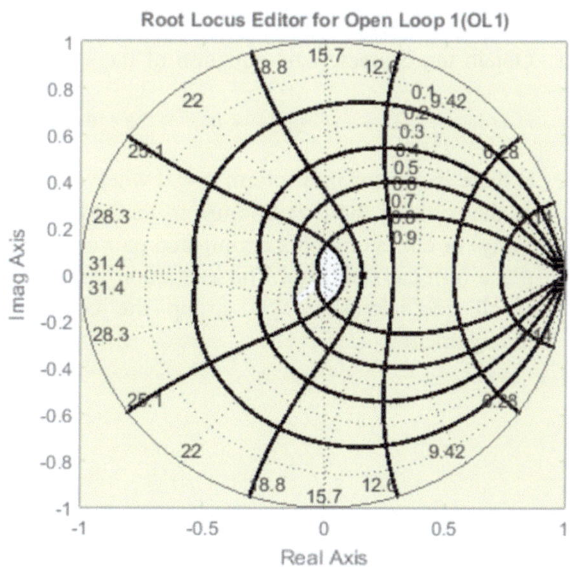

Fig. 3.39 ξ-constant loci and ω_n-constant loci in z-plane

Continuous and Discrete Systems Stability

The stability concept refers to a desirable dynamical feature whereby a system is stable when in face of any bounded input signal it produces always an output bounded signal. It is important to highlight that the stability is a system property no matter if the system under analysis is continuous or discrete and it does not depend on the input signal either. Far from mathematical expressions, the stability concept is widely employed in an uninformed way.

Firstly, considering continuous-time systems from a mathematical point of view, we have already referred to system stability when we analysed the system response. Remember that in the general case the system step response considering both multiple real and conjugate complex poles in $p_i = \sigma_i \pm j\omega_i$ in general was given by

$$y(t) = \left(K_0 + \sum_{i=1}^{n} \frac{K_i}{(n-1)!} t^{n-1} e^{-p_i t} \right) u_s(t) \tag{3.90}$$

so the system output $y(t)$ will be bounded as long as every exponential term fulfils $\lim_{t\to\infty} e^{-p_i t} = 0$. Therefore, stability is assured when $p_i > 0 \,\forall\, i$, in other words, when all the roots of the characteristic equation are located on the left side of the s-plane. Thus, the most direct way to check the stability of a system is to obtain the characteristic equation and calculate the location of the system poles.

As it can be observed, when studying system stability, sometimes it is not needed to know where the poles are exactly placed, but to know whether all the poles are on the left part of the s-plane, and this task can be easily done in MATLAB. Nevertheless, there are some simple procedures as the *Routh–Hurwitz* criterion, which yields the number of poles that are placed in the right part of the s-plane. The steps to easily check the stability of a system are given by:

1. Obtain the characteristic equation of the system (assuming $a_0 \neq 0$)

$$Q(s) = a_n s^n + \cdots + a_1 + a_0 = 0 \tag{3.91}$$

2. If there is one (or more) term $a_i < 0$ when another term $a_j > 0$, then one at least of the roots of the equation are in the right part of the s-plane positive (instability) or else a complex conjugated pair of roots are located in the $j\omega$ axis of s-plane (critical stability). This condition is necessary but not sufficient.
3. If all coefficients $a_i > 0$, group the coefficients in the following matrix arrangement

$$\begin{array}{c|cccc} s^n & a_n & a_{n-2} & \cdots & a_0 \\ s^{n-1} & a_{n-1} & a_{n-3} & \cdots & a_1 \\ s^{n-2} & b_{n-2} & \cdots & b_0 & \\ s^{n-3} & c_{n-3} & \cdots & & \\ \cdots & \cdots & \cdots & & \\ s^2 & e_2 & e_0 & & \\ s^1 & f_1 & & & \\ s^0 & g_0 & & & \end{array} \tag{3.92}$$

where the coefficients b_i are computed as

$$b_{n-2} = \frac{a_{n-1}a_{n-2} - a_n a_{n-3}}{a_{n-1}} \tag{3.93}$$

$$b_{n-4} = \frac{a_{n-1}a_{n-4} - a_n a_{n-5}}{a_{n-1}} \tag{3.94}$$

and in the same way, the coefficients $c_i, d_i, e_i \ldots$ are evaluated by continuing the process until the last row is completed.

The Routh–Hurwitz theorem assures that the number of poles located at the right part of the s-plane coincides with the changes of signs of the elements of the first column of the table. Thus, a system is stable if and only such elements are all positive or all negative.

Example 3.11
In Fig. 3.40, it shows a feedback control system designed to regulate the concentration of a drug in the circulatory system $q(t)$ by varying the input dose $i(t)$, by using a sensor to measure concentration with gain K_m and a pump with gain K_p as actuator.

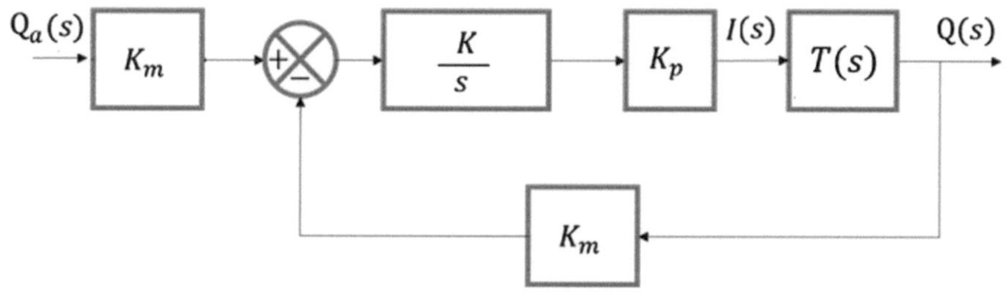

Fig. 3.40 Closed-loop control system for regulation of concentration of drug

Assuming a transfer function model (see Example 2.3)

$$T(s) = \frac{0.1}{s^2 + 0.17s + 0.006} \tag{3.95}$$

with gains $K_p = 0.1$, $K_m = 1$, determine the range of K of the integral controller $\frac{K}{s}$ which makes the closed-loop system stable.

Solution

In first place, we have to obtain the closed-loop transfer function and its characteristic equation in particular. Therefore, applying block reduction techniques we have

$$\begin{aligned} G_{cl}(s) &= \frac{Q(s)}{Q_a(s)} = \frac{K_m \frac{K}{s} K_p T(s)}{1 + K_m \frac{K}{s} K_p T(s)} \\ G_{cl}(s) &= \frac{\frac{0.1 \cdot 0.1K}{s(s^2 + 0.17s + 0.006)}}{1 + \frac{0.1 \cdot 0.1K}{s(s^2 + 0.17s + 0.006)}} \end{aligned} \tag{3.96}$$

where $G_{cl}(s)$ is the resulting closed-loop transfer function, given by

$$G_{cl}(s) = \frac{0.01K}{s^3 + 0.17s^2 + 0.006s + 0.01K} \tag{3.97}$$

The characteristic equation is given by the denominator of $G_{cl}(s)$ with $n = 3$ as

$$Q(s) = a_n s^n + \cdots + a_1 + a_0 = s^3 + 0.17s^2 + 0.006s + 0.01K \tag{3.98}$$

where it should be assumed that $K \neq 0$. From the second rule of the method, as all coefficients of $Q(s)$ are positive, then it follows that $K > 0$.

In order to check the stability of the closed-loop system then the Routh–Hurwitz matrix should be completed

$$\begin{array}{c|cc} s^3 & 1 & 0.006 \\ s^2 & 0.17 & 0.01K \\ s^1 & \frac{0.17 \cdot 0.06 - 0.01K}{0.17} & 0 \\ s^0 & 0.01K & \end{array} \tag{3.99}$$

where

$$b_1 = \frac{a_2 a_1 - a_3 a_0}{a_2} = \frac{0.17 \cdot 0.06 - 0.01K}{0.17} \tag{3.100}$$

$$c_1 = \frac{b_1 a_0 - b_2 a_2}{b_1} = \frac{\frac{0.17 \cdot 0.06 - 0.01K}{0.17} 0.01K}{\frac{0.17 \cdot 0.06 - 0.01K}{0.17}} = 0.01K \tag{3.101}$$

Therefore, from rule 3 it should be fulfilled that no sign changes should occur, that is $0.01K > 0$ (same as rule 2) and also $\frac{0.17 \cdot 0.06 - 0.01K}{0.17} > 0 \rightarrow K < 0.102$; thus, stability is guaranteed when $0 < K < 0.102$.

Certainly, when $K = 0.102$ the roots of the characteristic equation are at the limit of stability as there are two roots at the imaginary axis in s-plane, as it proven in MATLAB

```
≫ K=0.102;
≫ poly=[1 0.17 0.006 0.01*K]

poly =

   1.0000   0.1700   0.0060   0.0010

≫ roots(poly)

ans =

  -0.1700 + 0.0000i
  -0.0000 + 0.0775i
  -0.0000 - 0.0775i
```

Concerning discrete-time systems and realizing an analysis similar to that of continuous systems, it can be stated that the order of the system and the number of zeros in z-plane will determine the transient response, so for a general n-order-discretized system represented by

$$G(s) = \frac{K(z - z_1)(z - z_2)\ldots(z - z_m)}{(z - p_1)(z - p_2)\ldots(z - p_n)} \tag{3.102}$$

Without loss of generality, it can be assumed that the n poles are real and different and that the input signal $u(kT)$ is a unitary step, and no initial conditions are present. With these assumptions, the system response in z-plane $Y(z)$ can be separated as

$$Y(z) = G(z)U(z) = \frac{G(z)z}{z - 1} = \frac{K_0 z}{z - 1} + \sum_{i=1}^{n} \frac{K_i z}{(z - p_i)} \tag{3.103}$$

which yields the discrete-time response

$$y(kT) = \left(K_0 + \sum_{j=1}^{n} K_i(p_i)^j\right) u_s(kT) \tag{3.104}$$

Considering poles located at $p_i = |p_i|\angle p_i$ in general, then the system output $y(kT)$ will be bounded as long as every power term fulfils $\lim_{k\to\infty}(|p_i|)^k = 0$. Therefore, stability is assured whenever $|p_i| < 1 \forall i$, in other words, when all the roots of the discrete characteristic equation are located inside the unit circle in z-plane. Thus, the most direct way to check the stability of a system is to obtain the characteristic equation of $G(z)$ and calculate the location of the system poles. Nevertheless, location of poles of the discretized system will be directly related to that of poles of the continuous system as $z_i = e^{s_i T}$.

It is also possible to deduce the discrete system stability by applying simple procedures to determine whether all the poles are inside the circle unit of z-plane, as the Jury test and the *Routh–Hurwitz* criterion also, which is not directly applicable in z-plane to the characteristic equation of $G(z)$. Indeed, it must be applied before the so-called bilinear transformation which maps z-plane to w-plane, as follows

$$z = \frac{1 + \frac{T}{2}w}{1 - \frac{T}{2}w} \rightarrow w = \frac{2}{T}\frac{z-1}{z+1} \tag{3.105}$$

transforming the unit circle in z-plane into the imaginary axis in the w-plane so as the Routh criterion could be applied to the characteristic equation of $G(w)$ (Fig. 3.41).

The transformation of unit circle in z-plane $z = e^{j\omega T}$ to w-plane is given by

$$w = \frac{2}{T}\frac{e^{j\omega T} - 1}{e^{j\omega T} + 1} = j\frac{2}{T}\tan\left(\frac{\omega T}{2}\right) = jv \tag{3.106}$$

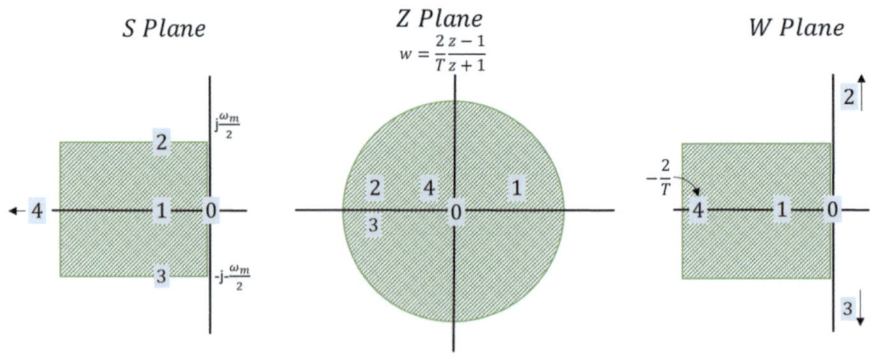

Fig. 3.41 Transformation from s-plane to z-plane and from z-plant to w-plane (bilinear)

Thus, the relation between the transformed frequency v and the original frequency ω in w-plane and s-plane, respectively, is given by

$$v = \frac{2}{T}\tan\left(\frac{\omega T}{2}\right) \cong \omega \tag{3.107}$$

which holds for low values of ω as $\frac{\omega T}{2} < \frac{\pi}{10}$, that is, $\omega < \frac{\omega_s}{10}$ so that stability of discrete-time system can be analysed by using the Routh–Hurwitz criterion as long as $\omega < \frac{\omega_s}{10}$ which represents a more restrictive statement of the sampling theorem.

Example 3.12
In Fig. 3.42 it shows a feedback discrete control system designed to regulate the concentration of a drug in the circulatory system $q(t)$ by varying the input dose $i(t)$, by using a sensor to measure concentration with gain K_m and a pump with gain K_p as actuator. Considering a digital integral controller $D(z) = K$ and using the same transfer function $T(s)$ as described in Example 3.11, determine the range K of the digital controller which makes the closed-loop system stable, assuming $K_p = 0.1$, $K_m = 1$ and sampling time $T = 0.5$.

Solution
In first place, we have to obtain the closed-loop transfer function and its characteristic equation in z-plane. In order to determine the pulse transfer function $G(z) = \frac{Q(z)}{U(z)}$, the output $q(t)$ at sampling times should be taken by means of an ideal sampler as it was done in (3.75) resulting into

$$G(z) = \frac{Q(z)}{U(z)} = Z\{G_{H0}(s)K_pT(s)\} = \left(1 - z^{-1}\right)Z\left\{K_p\frac{T(s)}{s}\right\} \tag{3.108}$$

As the transfer function model $T(s)$ is given by

$$T(s) = \frac{0.1}{s^2 + 0.17s + 0.006} \tag{3.109}$$

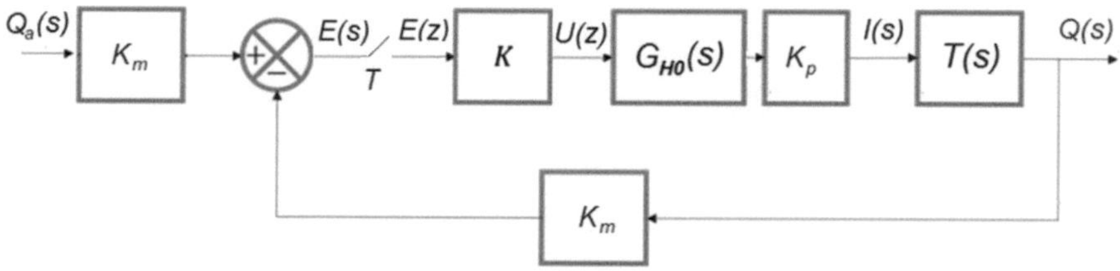

Fig. 3.42 Sampled continuous feedback control system for regulation of concentration of drug

and $K_p = 0.1$ then it is straightforward computing $G(z)$ by following the same approach as described in Example 3.10, then

$$\begin{aligned} G(z) &= \left(\frac{z-1}{z}\right) Z\left\{\frac{0.1 \cdot 0.1}{s(s^2+0.17s+0.006)}\right\} \\ G(z) &= \left(\frac{z-1}{z}\right) Z\left\{\frac{0.1 \cdot 0.1}{s(s^2+0.17s+0.006)}\right\} \\ G(z) &= \frac{Q(z)}{U(z)} = \frac{0.001215z+0.001181}{z^2-1.917z+0.9185} \end{aligned} \tag{3.110}$$

The discrete-time controller transfer function is given by a gain

$$D(z) = \frac{U(z)}{E(z)} = K \tag{3.111}$$

Then computing the Laplace transform of the error signal as

$$\begin{aligned} E(s) &= K_m(Q_a(s) - Q(s)) \\ E^*(s) &= K_m\left(Q_a^*(s) - Q^*(s)\right) \end{aligned} \tag{3.112}$$

and recalling the relation between starred transform and Z transform, we have

$$E(z) = K_m(Q_a(z) - Q(z)) \tag{3.113}$$

Therefore, the sampled continuous feedback control system of Fig. 3.42 is transformed into a discrete-time closed-control loop (Fig. 3.43) in the z domain.

After applying block reduction techniques, the closed-loop transfer function results into

$$G_{cl}(z) = \frac{Q(z)}{Q_a(z)} = \frac{K_m D(z) T(z)}{1 + K_m D(z) T(z)} \tag{3.114}$$

whose characteristic equation is given by the denominator of $G_{cl}(z)$

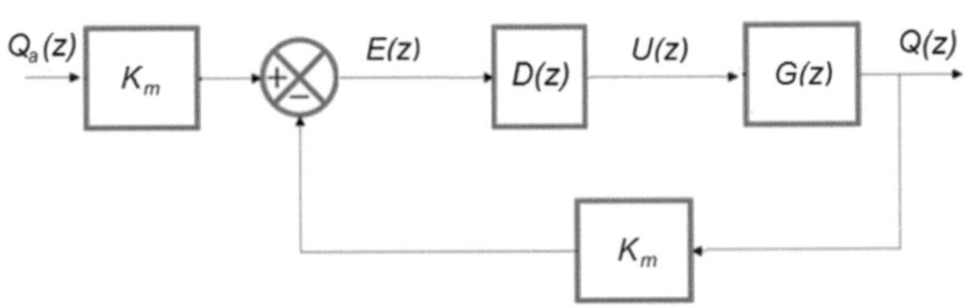

Fig. 3.43 Transformed discrete feedback control system for regulation of concentration of drug

$$\begin{aligned}Q(z) &= 1 + K_m D(z)T(z) = 1 + K\frac{0.001215z + 0.001181}{(z^2 - 1.917z + 0.9185)} \\ Q(z) &= \left(z^2 - 1.917z + 0.9185\right) + K(0.001215z + 0.001181) \\ Q(z) &= z^2 + (0.001215K - 1.917)z + (0.9185 + 0.001181K)\end{aligned} \tag{3.115}$$

In order to check the stability of the discrete closed-loop system through the application of the Routh–Hurwitz criterion, the bilinear transformation has to be applied, thus

$$\begin{aligned}Q(w) = Q(z)|_{z=\frac{1+0.25w}{1-0.25w}} &= \left(\frac{1+0.25w}{1-0.25w}\right)^2 \\ &+ (0.001215K - 1.917)\left(\frac{1+0.25w}{1-0.25w}\right) + (0.9185 + 0.001181K) = 0\end{aligned} \tag{3.116}$$

$$\begin{aligned}Q(w) &= (1+0.25w)^2 + (0.001215K - 1.917)(1+0.25w)(1-0.25w) \\ &+ (0.9185 + 0.001181K)(1 - 0.25w^2) = 0\end{aligned} \tag{3.117}$$

Then, arranging the polynomial $Q(w)$ and applying the Routh–Hurwitz criterion, the roots of the characteristic equation are at the limit of stability in z-plane when $K = 69$, as there are two roots at the unit circumference as it proven in MATLAB:

```
≫ K=69;
≫ poly=[1 0.001215*K-1.917 0.9185+0.001181*K]
≫ poly

poly =

    1.0000  -1.8332  1.0000

≫ roots(poly)

ans =

   0.9166 + 0.3998i
   0.9166 - 0.3998i
```

```
>> abs(ans)

ans =

    1.0000
    1.0000
```

Continuous and Discrete System Stationary Response

In previous sections, we were dealing with characteristics of the transient response of systems, paying attention to issues such as maximum overshoot, peak times. Although such features are highly relevant for real systems, it is desirable that a control system be accurate when tracking a reference command signal. In this way, the *steady-state error* of a closed-loop control system is defined as the difference between the reference command signal and the stationary system output (Fig. 3.44).

The study and analysis of this parameter is of paramount importance for designing reliable control systems. Let us think, for example, in administration of a medicament to a patient. Obviously, any inherent error in the administered dose can lead to the patient to death in case of inaccurate dose.

Considering continuous-time systems, we are going to focus on closed-loop control system comprising system, actuator, sensor and controller with corresponding $G(s)$, $A(s)$, $H(s)$ and $C(s)$ transfer functions. For the sake of simplicity, it will be assumed $C(s) = 1$, as we are not going to dealt with control design at present.

Let $r(t)$ be the reference command signal and $b(t)$ be measured system output through the sensor, then the steady-state error can be defined as

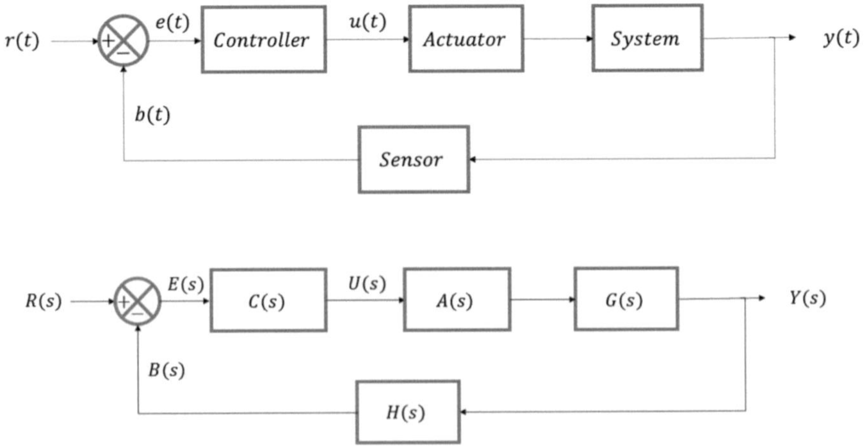

Fig. 3.44 Closed-loop configuration for continuous steady-state error analysis

$$e_{ss} = \lim_{t\to\infty} e(t) = \lim_{t\to\infty} (r(t) - b(t)) \tag{3.118}$$

In the complex domain, the error signal $e(t)$ can be expressed and expanded as

$$\begin{aligned} E(s) &= R(s) - B(s) = R(s) - H(s)Y(s) \\ E(s) &= R(s) - H(s)\frac{A(s)G(s)}{1 + A(s)G(s)H(s)}R(s) \\ E(s) &= R(s)\left(1 - \frac{A(s)G(s)H(s)}{1 + A(s)G(s)H(s)}\right) \\ E(s) &= \frac{1}{1 + A(s)G(s)H(s)}R(s) \end{aligned} \tag{3.119}$$

Using this last equation together with the final value theorem, the steady-state error can be defined as

$$e_{ss} = \lim_{t\to\infty} e(t) = \lim_{s\to 0} sE(s) = \lim_{s\to 0} \frac{s}{1 + A(s)G(s)H(s)}R(s) \tag{3.120}$$

Note that the error obviously depends on the open-loop dynamics at $s = 0$ and on the reference command $r(t)$ to be tracked. That is, regarding the stationary response, a system will behave differently according to the given reference command signal.

In the following, we will particularize the study of the steady-state error for different input reference commands and types of system dynamics at $s = 0$. For that reason, the concept of *type of a system* is to be considered and defined as the number of integrators N of the combined open-loop dynamics $A(s)G(s)H(s)$, then rewriting it in general as

$$A(s)G(s)H(s) = \frac{K(\tau_{c1}s + 1)(\tau_{c2}s + 1)\ldots(\tau_{cm}s + 1)}{s^N(\tau_{p1}s + 1)(\tau_{p2}s + 1)\ldots(\tau_{pn}s + 1)} \tag{3.121}$$

so it is said that the open-loop $A(s)G(s)H(s)$ is a *N-type*. Next, the dependence of the steady-state error on different particular input reference commands is analysed.

Steady-State Error for Step Reference

In case of input reference command $r(t)$ with amplitude A, then e_{ss} is given by

$$e_{ss} = \lim_{s\to 0} sE(s) = \lim_{s\to 0} \frac{s}{1 + A(s)G(s)H(s)}\frac{A}{s} = \frac{A}{1 + K_p} \tag{3.122}$$

where $K_p = \lim_{s\to 0} A(s)G(s)H(s)$ is the static position error. Notice that as long as K_p tends to infinite, the steady-state error tends to zero, and this is only achieved when open-loop system is type-1 or greater $(N \geq 1)$.

Steady-State Error for Ramp Reference

Similarly, when the reference command signal is a ramp with amplitude A, then e_{ss} is given by

$$\begin{aligned} e_{ss} &= \lim_{s\to 0} sE(s) = \lim_{s\to 0} \frac{s}{1 + A(s)G(s)H(s)} \frac{A}{s^2} \\ e_{ss} &= \lim_{s\to 0} \frac{A}{s + sA(s)G(s)H(s)} = \frac{A}{K_v} \end{aligned} \tag{3.123}$$

where $K_v = \lim_{s\to 0} sA(s)G(s)H(s)$ is the static velocity error. Notice that as long as K_v tends to infinite, the steady-state error tends to zero, and this is only achieved when open-loop system is type-2 or greater $(N \geq 2)$.

Example 3.13
For the feedback control system designed to regulate the concentration of a drug in the circulatory system already defined in Example 3.11, determine the actuator gain K_p so as the control system can track a unit-ramp concentration $q_a(t)$ with a steady-state error less than 0.1, assuming $K = 0.1$ for the integral controller.

Solution
Computing the error signal $E(s)$ by following a similar analysis as described in Example 3.11 (observe the common sensor gain in Fig. 3.42) and including the integral controller into the open-loop dynamics, we have

$$E(s) = K_m(Q_a(s) - Q(s)) = \frac{K_m}{1 + C(s)A(s)T(s)H(s)} Q_a(s) \tag{3.124}$$

As $Q_a(s) = \frac{1}{s^2}$, it follows that

$$\begin{aligned} e_{ss} &= \lim_{s\to 0} sE(s) = \lim_{s\to 0} \frac{K_m}{sC(s)A(s)T(s)H(s)} \\ e_{ss} &= \frac{K_m}{K_v} \end{aligned} \tag{3.125}$$

being $K_v = \lim_{s\to 0} sC(s)A(s)G(s)H(s) = \lim_{s\to 0} \frac{K}{s} K_p \frac{0.1s}{s^2 + 0.17s + 0.006} K_m = \frac{0.1K_pK_m}{0.006}$.

Then, it follows that

$$e_{ss} = \frac{K_m}{\frac{0.1K_pK_m}{0.006}} < 0.1 \rightarrow K_p > \frac{0.006}{0.01} = 0.6 \tag{3.126}$$

It is clear that the results obtained are consistent since the open-loop transfer function of type 1 (i.e. it has an integrator), and thus, it will follow that the feedback control systems will not accurately respond to a unit-ramp reference while no errors in concentration output will be shown when a unit step reference is applied.

The same reasoning can be done for other input reference commands, for instance, for parabolic reference command $(s) = \frac{A}{s^3}$; then, the error will be calculated as

$$\begin{aligned} e_{ss} &= \lim_{s\to 0} sE(s) = \lim_{s\to 0} \frac{s}{1 + A(s)G(s)H(s)} \frac{A}{s^3} \\ e_{ss} &= \lim_{s\to 0} \frac{A}{s^2 + s^2A(s)G(s)H(s)} = \frac{A}{K_a} \end{aligned} \tag{3.127}$$

where $K_a = \lim_{s\to 0} s^2A(s)G(s)H(s)$ is the static acceleration error. Notice that as long as K_a tends to infinite, the steady-state error tends to zero, and this is only achieved when open-loop system is type-3 or greater $(N \geq 3)$.

It is important to highlight that in case of non-unitary sensor gain K_h in $H(s)$ it is compulsory to include an equal gain K_h after the reference command $R(s)$ so as to correctly compute the error signal (Fig. 3.45), that is

$$e_{ss} = \lim_{t\to\infty} e(t) = \lim_{s\to 0} sE(s) = \lim_{s\to 0} \frac{s}{1 + A(s)G(s)H(s)} K_hR(s) \tag{3.128}$$

Concerning discrete-time systems, we are going to follow a similar approach to that of continuous closed-loop control systems. Therefore, a sampled continuous feedback control systems is going to be analysed comprising system, actuator, sensor and discrete controller with $G(s)$, $A(s)$, $H(s)$ and $D(z)$ corresponding transfer functions, along with interface devices as sampler and zero-order data hold $G_{H0}(s)$ (Fig. 3.46). For the sake of simplicity it will be assumed again, $D(z) = 1$, as we are not going to deal with control design at present.

We have to extend the pulse transfer function determination to cover this particular case when the closed-loop sampled continuous system contains both a

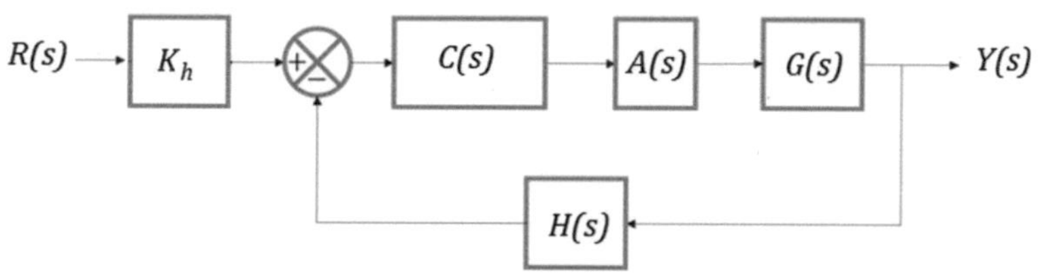

Fig. 3.45 Consideration of sensor gain K_h in steady-state error analysis

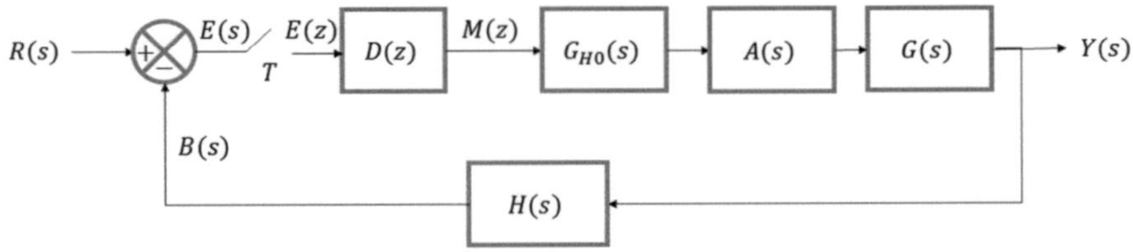

Fig. 3.46 Closed-loop configuration for sampled continuous steady-state error analysis

discrete controller and a sensor. Applying the relation between starred transform and Z transform, we have for the discrete controller

$$\begin{aligned} M(z) &= D(z)E(z) \\ M^*(s) &= D^*(s)E^*(s) \end{aligned} \tag{3.129}$$

Taking the samples of $y(t)$ by means of an ideal sampler and considering the serial product $G_s(s) = G_{HO}(s)A(s)G(s)$, we have

$$\begin{aligned} Y^*(s) &= G_s^*(s)M^*(s) = G_s^*(s)D^*(s)E^*(s) \\ Y(z) &= G_s(z)D(z)E(z) \end{aligned} \tag{3.130}$$

where $G_s(z)$ is defined as

$$G_s(z) = Z(G_s(s)) = Z(G_{HO}(s)A(s)G(s)) = (1 - z^{-1})Z\left(\frac{A(s)G(s)}{s}\right) \tag{3.131}$$

As stated in (3.118) then by applying the sampling operation, we have

$$E^*(s) = R^*(s) - B^*(s) = R^*(s) - G_{sH}^*(s)D^*(s)E^*(s) \tag{3.132}$$

with $G_{sH}^*(s) = (G_s(s)H(s))^*$, then rearranging with $D^*(s) = D(z) = 1$ and applying the relation between starred transform and Z transform we have

$$\begin{aligned} E^*(s) &= \frac{1}{1 + G_{sH}^*(s)}R^*(s) = \frac{1}{1 + (G_s(s)H(s))^*}R^*(s) \\ E(z) &= \frac{1}{1 + \overline{G_sH}(z)}R(z) \end{aligned} \tag{3.133}$$

with $\overline{G_sH}(z) = (1 - z^{-1})Z\left(\frac{A(s)G(s)H(s)}{s}\right)$.

Using this last equation together with the final value theorem, the discrete steady-state error can be defined as

$$e_{ssd} = \lim_{k \to \infty} e(kT) = \lim_{z \to 1} \frac{(z-1)}{z} E(z) = \lim_{z \to 1} \frac{(z-1)}{1 + \overline{G_s H}(z)} R(z) \tag{3.134}$$

Note that the discrete error depends on the open-loop dynamics at $z = 1$ and on the reference command $r(kT)$ to be tracked. That is, regarding the stationary response, a system will behave differently according to the given reference command signal.

In the following, we will particularize the study of the discrete steady-state error for different input reference commands and types of system dynamics at $z = 1$. For that reason, the concept of *type of a discrete system* is to be considered and defined as the number of "discrete" integrators N (poles at $z = 1$) of the combined open-loop dynamics $\overline{G_s H}(z)$, then rewriting it in general as

$$\begin{aligned} \overline{G_s H}(z) &= Z(G_{H0}(s)A(s)G(s)H(s)) \\ &= \frac{K_{GH}(z - z_1)(z - z_2)\ldots(z - z_m)}{(z-1)^N (z - p_1)(z - p_2)\ldots(z - p_n)} \end{aligned} \tag{3.135}$$

so it is said that the open-loop $\overline{G_s H}(z)$ is a *N-type*. Next, the dependence of the discrete steady-state error on different particular input reference commands is analysed.

Steady-State Discrete Error for Step Reference

In case of input reference command $r(kT)$ with amplitude A, then e_{ssd} is given by

$$e_{ssd} = \lim_{z \to 1} \frac{(z-1)}{z} \frac{1}{1 + \overline{G_s H}(z)} \frac{Az}{(z-1)} = \frac{A}{1 + K_{pd}} \tag{3.136}$$

where $K_{pd} = \lim_{z \to 1} \overline{G_s H}(z)$ is the static position error. Notice that as long as K_{pd} tends to infinite, the discrete steady-state error tends to zero, and this is only achieved when open-loop system is type-1 or greater ($N \geq 1$).

Steady-State Discrete Error for Ramp Reference

Similarly, when the reference command signal is a ramp with amplitude A, then e_{ssd} is given by

$$e_{ssd} = \lim_{z \to 1} \frac{(z-1)}{z} \frac{1}{1 + \overline{G_s H}(z)} \frac{ATz}{(z-1)^2} = \frac{A}{K_{vd}} \tag{3.137}$$

where $K_v = \lim_{z \to 1} \frac{(z-1)}{zT} \overline{G_s H}(z)$ is the static velocity error. Notice that as long as K_{vd} tends to infinite, the discrete steady-state error tends to zero, and this is only achieved when open-loop system is type-2 or greater $(N \geq 2)$.

Example 3.14
For the feedback control system designed to regulate the concentration of a drug in the circulatory system already defined in Example 3.12, determine the sensor gain K_m so as the control system can track a unit step concentration $q_a(t)$ with a steady-state error less than 0.1, assuming $K = 2$ for the proportional discrete controller with $T = 0.5$.

Solution
Starting from Fig. 3.42, we have

$$E(z) = K_m(Q_a(z) - Q(z)) = \frac{K_m}{1 + D(z)G_s(z)K_m} Q_a(z) \tag{3.138}$$

As $Q_a(z) = \frac{z}{z-1}$ and $D(z) = K$ it follows that

$$\begin{aligned} e_{ssd} &= \lim_{z \to 1} \frac{z-1}{z} E(z) = \lim_{z \to 1} \frac{K_m}{1 + D(z)G_s(z)K_m} \\ e_{ssd} &= \frac{K_m}{1 + KG_s(1)K_m} \end{aligned} \tag{3.139}$$

As the discretized system $G(z)$ has been obtained in (3.109) then by substituting, we get

$$\begin{aligned} G(z) &= \frac{Q(z)}{U(z)} = \frac{0.001215z + 0.001181}{z^2 - 1.917z + 0.9185} \\ G(1) &= \frac{0.001215 + 0.001181}{1 - 1.917 + 0.9185} = 1.6 \end{aligned} \tag{3.140}$$

Then, it follows that

$$e_{ssd} = \frac{K_m}{1 + 2 \cdot 1.6K_m}$$

$$e_{ssd} = \frac{K_m}{1 + 3.2K_m} < 0.1 \rightarrow K_m < \frac{0.1}{1 - 0.32} = 0.1471 \tag{3.141}$$

The same reasoning can be done for other input reference commands, for instance, for parabolic reference, whose discrete steady-state error tends to zero as long as the open-loop system is type-3 or greater $(N \geq 3)$.

It is important to highlight that in case of non-unitary sensor gain K_H in $H(z)$, it is compulsory to include an equal gain K_H after the reference command $R(z)$ so as to correctly compute the discrete error signal as it was commented in Fig. 3.45 for continuous systems.

Solved Problems

Problem 3.1

Consider an autoclave, as it is shown in Fig. 3.47, which is used in the sterilization process of clinical material. In order to measure the temperature inside the autoclave, a thermoelectric sensor (a thermometer) is used, which behaves as a first-order system, with static gain K and time constant τ.

(a) Determine the differential equation governing the behaviour of the system.
(b) Define in the s domain the relationship between the actual temperature of the autoclave $T_a(s)$ and the temperature measured by the thermometer $T_m(s)$. Note that the initial conditions (initial temperatures) are not necessarily zero.
(c) In the process of calibrating the thermometer, this is initially at the ambient temperature T_1 and is introduced into the autoclave where the temperature is T_2. Obtain the expression for the temperature measured by the thermometer when a sudden change occurs, both in the spectral domain $T_m(s)$ and in the time domain $T_m(t)$.
(d) In the previous case, the thermometer takes 15 s to cover 95% of the change required for the measurement to go from its initial value T_1 to its final value T_2. Determine the values of the static gain K and the time constant τ.
(e) Draw the evolution of the temperature measurement in the calibration process when $T_1 = 25$ °C and $T_2 = 135$ °C.

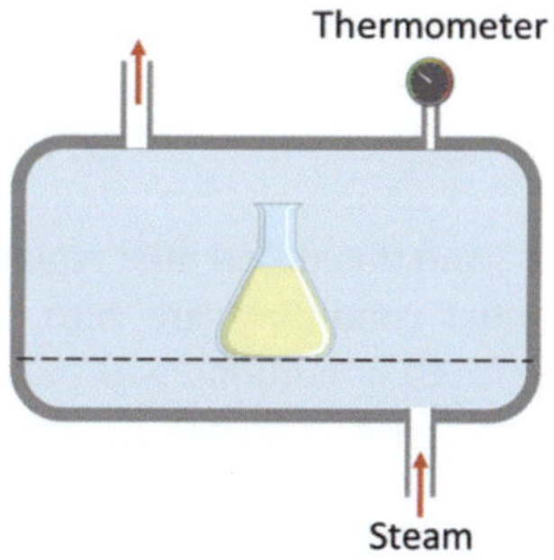

Fig. 3.47 Autoclave used in the sterilization process of clinical material

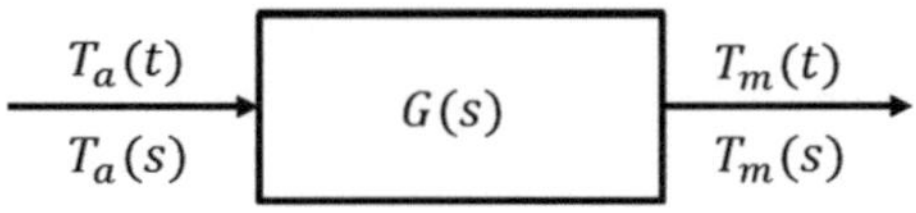

Fig. 3.48 Thermometer diagram for the autoclave problem

Solution 3.1

(a) According to the problem statement, the thermometer can be modelled as a first-order system $G(s)$ whose input is the temperature $T_a(s)$ inside the autoclave, and its output is the temperature $T_m(s)$ as measured by the sensor (see Fig. 3.48).

The thermometer transfer function $G(s)$ will then correspond to a first-order system

$$G(s) = \frac{K}{1 + \tau s}.$$

To obtain the differential equation governing the behaviour of the system, we must take into account that this equation is the same regardless of the initial conditions. If the system had zero initial conditions, the output could be calculated by the expression $T_m(s) = G(s)T_a(s)$. In that case:

$$\begin{aligned}
&T_m(s) = \frac{K}{1 + \tau s} T_a(s). \\
&(1 + \tau s) T_m(s) = K\, T_a(s). \\
&\tau s\, T_m(s) + T_m(s) = K\, T_{\mathrm{a}}(s). \\
&\mathcal{L}^{-1}[\tau s\, T_m(s) + T_m(s)] = \mathcal{L}^{-1}[K\, T_a(s)]. \\
&\tau \mathcal{L}^{-1}[s\, T_m(s)] + \mathcal{L}^{-1}[T_m(s)] = K \mathcal{L}^{-1}[T_a(s)]. \\
&\tau \frac{\mathrm{d}T_m(t)}{\mathrm{d}t} + T_m(t) = K\, T_a(t).
\end{aligned}$$

And this differential equation, although it has been obtained considering that the initial conditions are zero, is also valid in any other circumstances. The initial nonzero conditions will be taken into account in solving the differential equation, but do not influence its structure.

(b) Again, if the system had zero initial conditions, the output could be calculated by the expression $T_m(s) = G(s)T_a(s)$. Otherwise, we proceed as follows:

$$\mathcal{L}\left[\tau\frac{dT_m(t)}{dt}+T_m(t)\right]=\mathcal{L}[K\,T_a(t)].$$

$$\tau\mathcal{L}\left[\frac{dT_m(t)}{dt}\right]+\mathcal{L}[T_m(t)]=K\mathcal{L}[T_a(t)].$$

Recalling that

$$\mathcal{L}\left[\frac{df(t)}{dt}\right]=s\,F(s)-f(0).$$

$$\mathcal{L}\left[\frac{dT_m(t)}{dt}\right]=s\,T_m(s)-T_m(0).$$

Therefore,

$$\tau[s\,T_m(s)-T_m(0)]+T_m(s)=K\,T_a(s).$$

$$\tau s\,T_m(s)-\tau T_m(0)+T_m(s)=K\,T_a(s).$$

$$(1+\tau s)T_m(s)-\tau T_m(0)=K\,T_a(s).$$

$$T_m(s)=\frac{K}{1+\tau s}T_a(s)+\frac{\tau}{1+\tau s}T_m(0).$$

$$T_m(s)=G(s)T_a(s)+\frac{\tau}{1+\tau s}T_m(0).$$

Calling

$$G_0(s)\equiv\frac{\tau}{1+\tau s},$$

it can be written that

$$T_m(s)=G(s)T_a(s)+G_0(s)T_m(0).$$

(c) In the calibration process, the sensor is exposed to a sudden temperature change which instantaneously brings it from an initial temperature T_1 to a final temperature T_2. This rapid change can be modelled as a step between both temperatures determined by

$$T_a(t)=T_1+(T_2-T_1)\cdot u(t),$$

where $u(t)$ is the unit step function. In the s domain, the input can be stated as

$$T_a(s) = \mathcal{L}[T_a(t)] = \mathcal{L}[T_1 + (T_2 - T_1) \cdot u(t)].$$
$$T_a(s) = \mathcal{L}[T_1] + (T_2 - T_1)\mathcal{L}[u(t)].$$
$$T_a(s) = T_1 \frac{1}{s} + (T_2 - T_1)\frac{1}{s}.$$
$$T_a(s) = T_2 \frac{1}{s}.$$

In this case, the initial condition (temperature) of the system is not zero, but of value $T_m(0)$. Assuming that the thermometer is initially in a stable state (not changing its measurement), the derivatives of the temperatures will be null, so it can be written that

$$\tau \left.\frac{\mathrm{d}T_m(t)}{\mathrm{d}t}\right|_{t=0} + T_m(t)|_{t=0} = K\,T_a(t)|_{t=0}.$$
$$T_m(0) = K\,T_a(0) = K\,T_1.$$

To express the temperature measured by the sensor in the spectral domain, the result obtained in the previous section will be used, so it can be written as

$$T_m(s) = G(s)T_a(s) + G_0(s)T_m(0).$$
$$T_m(s) = \frac{K}{1+\tau s} T_2 \frac{1}{s} + \frac{\tau}{1+\tau s} K\,T_1.$$
$$T_m(s) = \frac{K\,T_2}{s(1+\tau s)} + \frac{\tau K\,T_1}{1+\tau s}.$$

To express the temperature measured by the sensor in the temporal domain, the inverse Laplace transform of the above expression will be derived as

$$T_m(t) = \mathcal{L}^{-1}[T_m(s)] = \mathcal{L}^{-1}\left[\frac{K\,T_2}{s(1+\tau s)} + \frac{\tau K\,T_1}{1+\tau s}\right].$$
$$T_m(t) = \mathcal{L}^{-1}\left[\frac{K\,T_2}{\tau s\left(\frac{1}{\tau}+s\right)} + \frac{K\,T_1}{\frac{1}{\tau}+s}\right].$$
$$T_m(t) = \mathcal{L}^{-1}\left[\frac{K\,T_2}{\tau s\left(\frac{1}{\tau}+s\right)}\right] + \mathcal{L}^{-1}\left[\frac{K\,T_1}{\frac{1}{\tau}+s}\right].$$
$$T_m(t) = \frac{K\,T_2}{\tau}\mathcal{L}^{-1}\left[\frac{1}{s\left(\frac{1}{\tau}+s\right)}\right] + K\,T_1\mathcal{L}^{-1}\left[\frac{1}{\frac{1}{\tau}+s}\right].$$

Recalling that

$$e^{-at} \overset{\mathcal{L}}{\leftrightarrow} \frac{1}{s+a}; \quad \frac{1}{a}(1 - e^{-at}) \overset{\mathcal{L}}{\leftrightarrow} \frac{1}{s(s+a)},$$

$$T_m(t) = \frac{K\,T_2}{\tau}\left[\frac{1}{\frac{1}{\tau}}\left(1 - e^{-\frac{1}{\tau}t}\right)\right] + K\,T_1\left[e^{-\frac{1}{\tau}t}\right].$$

$$T_m(t) = K\,T_2\left(1 - e^{\frac{-t}{\tau}}\right) + K\,T_1 e^{\frac{-t}{\tau}}.$$

$$T_m(t) = K\,T_2 + K(T_1 - T_2)e^{\frac{-t}{\tau}}.$$

An alternative way to obtain the transfer function is by using the following MATLAB code

```
%% Obtaining sensor response to a sudden temperature change

% Defining symbolic constants and variables
syms K; %DC gain
syms tau; %Time constant
syms T1 T2; %Initial & final temperature
syms t s;

% Obtaining system's transfer function
G = K / (1 + tau*s); %System's transfer function
G0 = tau / (1 + tau*s); %Considering initial conditions

% Modelling input
u(t) = heaviside(t); %Unit step
Ta_t = T1 + (T2-T1) * u(t); %Step input (t domain)
Ta_s = laplace(Ta_t); %Step input (s domain)

% Obtaining step response (s domain)
Tm_s = G * Ta_s + G0 * K * T1; %Step response (s domain)
Tm_s = expand(Tm_s);
pretty(Tm_s);

% Obtaining step response (t domain)
Tm = ilaplace(Tm_s); %Step response (t domain)
pretty(Tm);
```

Executing this code, both the spectral and the temporal domain expressions of the temperature measured by the sensor are obtained

```
K T1 tau          K T2
--------- + -------------
s tau + 1   s (s tau + 1)


   /    t  \
exp| - --- | (K T1 - K T2) + K T2
   \   tau /
```

(d) During the calibration process, as it was previously stated, the sensor is exposed to a sudden change which instantaneously brings it from an initial temperature $T_a(0) = T_1$ to a final temperature $T_a(\infty) = T_2$. The measurements obtained by the sensor are son $T_m(0) = T_1$ and $T_m(\infty) = T_2$. These values are enough to determine the static gain because

$$\begin{aligned} T_m(t)\big|_{t=\infty} &= K\,T_2 + K(T_1 - T_2)e^{\frac{-t}{\tau}}\Big|_{t=\infty}. \\ T_m(\infty) &= K\,T_2 + K(T_1 - T_2)e^{\frac{-\infty}{\tau}}. \\ T_2 &= K\,T_2. \end{aligned}$$

It follows that the value of the static gain is $K = 1$. This result could also have been reached in the spectral domain since

$$T_2 = T_m(\infty) = \lim_{t\to\infty} T_m(t) = \lim_{s\to 0} s\,T_m(s) = \lim_{s\to 0} s\left[\frac{K\,T_2}{s(1+\tau s)} + \frac{\tau K\,T_1}{1+\tau s}\right].$$

$$T_2 = \lim_{s\to 0}\left[\frac{K\,T_2}{(1+\tau s)} + s\frac{\tau K\,T_1}{1+\tau s}\right] = \lim_{s\to 0}\left[\frac{K\,T_2}{(1+\tau s)}\right] + \lim_{s\to 0}\left[s\frac{\tau K\,T_1}{1+\tau s}\right].$$

$$T_2 = K\,T_2 + 0.$$

Hence, again, it follows that the value of the static gain (also called DC gain) is $K = 1$. This value can also be obtained by adding the following MATLAB code:

```
%% Obtaining DC gain from the steady state (ss) equation

eq_ss = T2 == limit(s*Tm_s,s,0); %Defining ss equation
K = solve(eq_ss,K); %Solving steady state equation (symbolic value)
K = double(K); %Obtaining DC gain (numeric value)
display(K);
```

The result obtained confirms the previous calculation:

```
K =

     1
```

In order to determine the time constant, it will be considered the problem statement that at time $t_s = 15$ s, the value of $T_m(t_s)$ has covered an $R = 95\%$ of the total change between T_1 and T_2, that is,

$$T_m(t_s) = T_1 + R(T_2 - T_1) = K\,T_2 + K(T_1 - T_2)\mathrm{e}^{\frac{-t}{\tau}}\Big|_{t=t_s}.$$
$$T_1 + R(T_2 - T_1) = K\,T_2 + K(T_1 - T_2)\mathrm{e}^{\frac{-t_s}{\tau}}.$$

Recalling that $K = 1$, then

$$T_1 + R(T_2 - T_1) = T_2 + (T_1 - T_2)\mathrm{e}^{\frac{-t_s}{\tau}}.$$
$$R(T_2 - T_1) = (T_2 - T_1) - (T_2 - T_1)\mathrm{e}^{\frac{-t_s}{\tau}}.$$
$$R(T_2 - T_1) = (T_2 - T_1)\left(1 - \mathrm{e}^{\frac{-t_s}{\tau}}\right).$$
$$R = \left(1 - \mathrm{e}^{\frac{-t_s}{\tau}}\right).$$
$$\mathrm{e}^{\frac{-t_s}{\tau}} = 1 - R.$$
$$\frac{-t_s}{\tau} = \mathrm{Ln}\,(1 - R).$$
$$\tau = \frac{t_s}{-\mathrm{Ln}\,(1 - R)} = \frac{15\,\mathrm{s}}{-\mathrm{Ln}(1 - 0.95)} = 5\,\mathrm{s}.$$

This value can also be obtained by adding the following MATLAB code:

```
%% Obtaining time constant from the transient response (tr) equation

% Defining constants
ts = 15; %Settling time
R = 0.95; %Settling time value

eq_tr = T1 + R*(T2-T1) == K*T2 - K*(T2-T1)*exp(-ts/tau); %tr equation
tau = solve(eq_tr,tau); %Solving tr equation (symbolic value)
tau = double(tau); %Obtaining time constant (numeric value)
display(tau);
```

Executing this code, the same result for the time constant is obtained:

```
tau =
    5.0071
```

(e) To draw the evolution of the temperature measurement during the calibration process, the following MATLAB code will be added

```
%% Plotting step response indicating settling time & time constant

% Restoring symbolic K and tau
Kn = K; %Saving numeric value of K
taun = tau; %Saving numeric value of tau
syms K; %Restoring symbolic K
syms tau; %Restoring symbolic tau

% Defining step response
Tm_p = Tm * u (t); %Step response for t>0
Tm_n = T1 * u(-t); %Step response for t<0
Tm = Tm_p + Tm_n; %Step response for any value of t

% Substituting values of constants
T1n = 25; %Initial temperature in step input (numeric)
T2n = 135; %Final temperature in step input (numeric)
Tm = subs(Tm,{K,tau,T1,T2},{Kn,taun,T1n,T2n});

% Plotting step response
tRange = [-10,25]; %Defining plot temporal range
TempRange = [0,175]; %Defining plot temperature range
ezplot(Tm,tRange); %Plotting step response
hold on; %Retain current graph when adding new graphic elements
plot([0,tRange(2)],[T1n,T1n],'k:'); % Plotting Tm=T1 reference line
plot([0,0],[T1n,TempRange(2)],'k:'); % Plotting t=0 reference line
ylim(TempRange); %Defining plot vertical range
title('Step response'); %Defining plot title

% Plotting settling time
Tm_ts = T1n + R *(T2n-T1n); %Temperature at settling time
plot(ts,Tm_ts,'bo','MarkerFaceColor','b'); %Plotting point
    % Plotting temperature range at settling time
plot([ts,ts],[T1n,Tm_ts],'k:'); %Plotting range (line)
yText = (Tm_ts-T1n) / 2 + T1n; %Vertical position for text
RText = ['R=' , num2str(100*R) , '%']; %Defining text for R
text(ts,yText,RText); %Plotting range (text)
    % Plotting time range at settling time
plot([0,ts],[Tm_ts,Tm_ts],'k:'); %Plotting range (line)
xText = ts/2; %Horizontal position for text
```

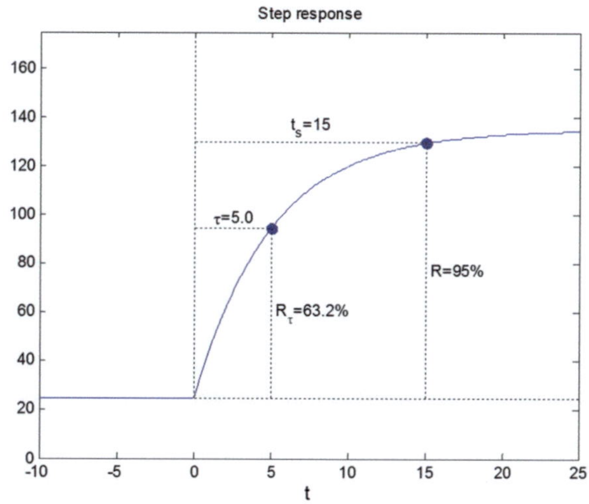

Fig. 3.49 Step response for the autoclave system

```
tsText = ['t_s=' ,num2str(ts)]; %Defining text
text(xText,Tm_ts,tsText,'HorizontalAlignment','Center',...
        'VerticalAlignment','Bottom'); %Plotting range (text)

% Plotting time constant
Tm_tau = double ( subs(Tm,t,taun) ); %Temperature at time constant
plot(taun,Tm_tau,'bo','MarkerFaceColor','b'); %Plotting point
    % Plotting temperature range at settling time
plot([taun,taun],[T1n,Tm_tau],'k:'); %Plotting range (line)
Rtau = (Tm_tau-T1n) / (T2n-T1n); %Vertical position for text
yText = (Tm_tau-T1n) / 2 + T1n; %Vertical position for text
RText = ['R_\tau=' , ...
          num2str(100*Rtau , '%3.1f') , '%']; %Defining text for R
text(taun,yText,RText); %Plotting range (text)
    % Plotting time range at settling time
plot([0,taun],[Tm_tau,Tm_tau],'k:'); %Plotting range (line)
xText = taun/2; %Horizontal position for text
tauText = ['\tau=' ,num2str(taun,'%3.1f')]; %Defining text
text(xText,Tm_tau,tauText,'HorizontalAlignment','Center',...
        'VerticalAlignment','Bottom'); %Plotting range (text)
```

The result of executing this code is shown in Fig. 3.49.

Problem 3.2

Consider a catheter of those commonly used in the measurement of pressures in the cardiovascular system. The catheter receives the pressure $p(t)$ from the patient at the distal end and converts it into a measure of voltage $v(t)$ for its subsequent

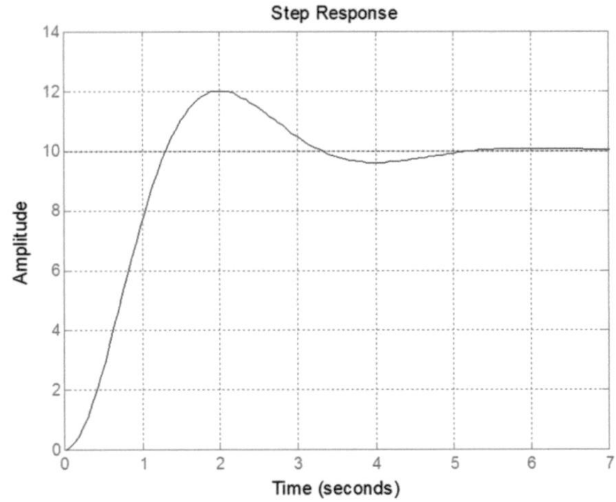

Fig. 3.50 Step system response

monitoring. The relationship between the pressure (input) and the voltage (output) constitutes a system which can be described by its transfer function

$$G(s) = \frac{A\,K_e}{\rho A\,L\,s^2 + b\,s + c},$$

where A and L are the catheter's section and length, ρ the density of fluid within the catheter, c the lineal friction coefficient, b the viscous friction coefficient, and K_e the proportionality constant between catheter displacement and electrical voltage. It is known the values of the constants $A = 0.25$, $L = 5$ and $\rho = 0.6$.

(a) Obtain the transfer function of the system using the values of the constants defined above.
(b) Determine the system's static gain, natural frequency and damping ratio.
(c) Obtain the values of K_e, b and c if the unit step system response is depicted in Fig. 3.50.

Solution 3.2

(a) The system transfer function, once the known values of the constants have been substituted, is

$$G(s) = \frac{A\,K_e}{\rho A\,L\,s^2 + b\,s + c} = \frac{0.25K_e}{0.6 \cdot 0.25 \cdot 5s^2 + b\,s + c}$$

$$G(s) = \frac{0.25K_e}{0.75s^2 + b\,s + c}$$

The same result can be obtained using the following MATLAB code

```
%% Obtaining the symbolic transfer function

% Defining symbolic constants and variables
syms c b; %Lineal and fluid friction constants
syms Ke; %Voltage-displacement proportionality constant
syms s;
A = sym(0.25); %Catheter section
L = sym(5); %Catheter length
rho = sym(0.6); %Density of the fluid inside the catheter

% Obtaining the system's transfer function
G = A*Ke / ( rho*A*L*s^2 + b*s +c);
pretty(G);
```

The execution of this code obtains the same transfer function, as it is shown below:

```
             Ke
--------------------
/     2              \
| 3 s                |
| ---- + b s + c | 4
\   4                /
```

This transfer function can also be expressed in its normalized form, dividing it by the coefficient in the denominator's highest order term. In this case

$$G(s) = \frac{\frac{0.25}{0.75} K_e}{s^2 + \frac{b}{0.75} s + \frac{c}{0.75}}$$

(b) A second-order system can be expressed, in its normalized form, as

$$G(s) = \frac{K\omega_n^2}{s^2 + 2\xi\omega_n s + \omega_n^2}$$

Comparing this expression with the result obtained in the previous paragraph, the natural frequency ω_n is derived as

$$\omega_n^2 = \frac{c}{0.75}$$

$$\omega_n = \sqrt{\frac{c}{0.75}}$$

To obtain the damping ratio ξ, let us compare the denominator's first-order term

$$2\xi\omega_n = \frac{b}{0.75}$$
$$\xi = \frac{b}{2\omega_n 0.75} = \frac{b}{1.5\sqrt{\frac{c}{0.75}}}$$

Finally, to derive the static gain K let us compare both numerators:

$$K\omega_n^2 = \frac{0.25}{0.75}K_e$$
$$K = \frac{0.25\,K_e}{0.75\,\omega_n^2} = \frac{0.25\,K_e}{0.75\frac{c}{0.75}} = 0.25\frac{K_e}{c}$$

These results can also be reached by adding the following MATLAB code:

```
%% Obtaining the gain, natural frequency and damping ratio

% Obtaining the denominator's coefficients in standard form
[numG,denG] = numden(G); %Obtaining num. & den. of G(s)
CdenG = coeffs(denG,s); %Obtaining the denominator's coefficients
d2 = CdenG(3); %Obtaining the coefficient for the s^2 term
CdenG = CdenG/d2; %Modifying the den. to have a one in the s^2 term
d1 = CdenG(2); %Coefficient for the denominator's s term
d0 = CdenG(1); %Denominator's independent term

% Obtaining the numerator's coefficients in standard form
n0 = numG/d2; %Numerator independent term

% Obtaining the standard form's coefficients
wn = sqrt(d0); %Obtaining natural frequency
xi = d1 / (2*wn); %Obtaining damping ratio
K = n0 / wn^2; %Obtaining the gain
pretty(K);
pretty(wn);
pretty(xi);
```

The execution of this code obtains the required standard parameters, which are the same that those derived above

```
Ke
---
4 c

   / 4 c \
sqrt| --- |
   \  3  /

     2 b
-------------
     / 4 c \
3 sqrt| --- |
     \  3  /
```

(c) In a second-order system given in its standard form, when it is excited by a unit step input, the response reaches a steady-state value given by

$$y_{ss} = \lim_{t \to \infty} y(t) = \lim_{s \to 0} s\,Y(s) = \lim_{s \to 0} s[G(s)X(s)] = K$$

$$y_{ss} = \lim_{s \to 0} s\,G(s)\frac{1}{s} = \lim_{s \to 0} G(s) = \lim_{s \to 0} \frac{K\omega_n^2}{s^2 + 2\xi\omega_n s + \omega_n^2} = \frac{K\omega_n^2}{\omega_n^2} = K$$

In the problem statement, it is clear that the system response reaches its steady state at the value of $y_{ss} = 10$, so the system static gain is $K = 10$.

On the other hand, if the system shows some overshoot, as it is the case, the value of this overshoot is given by

$$O_s = e^{\frac{-\xi\pi}{\sqrt{1-\xi^2}}}$$

From the problem statement, the value of the overshoot can be derived

$$O_s \equiv \frac{y_p - y_{ss}}{y_{ss}} = \frac{12 - 10}{10} = 0.2$$

Therefore, the damping ratio can be derived from

$$O_s = e^{\frac{-\xi\pi}{\sqrt{1-\xi^2}}}$$
$$\frac{-\xi\pi}{\sqrt{1-\xi^2}} = \mathrm{Ln}\, O_s$$
$$-\xi\pi = (\mathrm{Ln}\, O_s)\sqrt{1-\xi^2}$$
$$\xi^2\pi^2 = (\mathrm{Ln}\, O_s)^2(1-\xi^2) = (\mathrm{Ln}\, O_s)^2 - (\mathrm{Ln}\, O_s)^2\xi^2$$
$$\left[\pi^2 + (\mathrm{Ln}\, O_s)^2\right]\xi^2 = (\mathrm{Ln}\, O_s)^2$$
$$\xi^2 = \frac{(\mathrm{Ln}\, O_s)^2}{\pi^2 + (\mathrm{Ln}\, O_s)^2}$$
$$\xi = \frac{|\mathrm{Ln}\, O_s|}{\sqrt{\pi^2 + (\mathrm{Ln}\, O_s)^2}}$$

In this case,

$$\xi = \frac{|\mathrm{Ln}\, 0.2|}{\sqrt{\pi^2 + (\mathrm{Ln}\, 0.2)^2}} = 0.4559$$

Finally, in a system response with overshoot, the peak response is reached at the peak time

$$t_p = \frac{\pi}{\omega_n\sqrt{1-\xi^2}}$$

For the problem statement, it can easily be observed that the peak time is $t_p = 2$. Therefore, the natural frequency can be derived from the above equation:

$$\omega_n = \frac{\pi}{t_p\sqrt{1-\xi^2}}$$
$$\omega_n = \frac{\pi}{2\sqrt{1-0.4559^2}} = 1.7649$$

Alternatively, those values can also be obtained by adding the following MATLAB code:

```
%% Identifying the system

% System response to a unit step
yss = 10; %Steady state value
yp = 12; %Peak value
tp = 2; %Peak time
```

```
Os = (yp-yss) / yss; %Overshoot

% Obtaining experimental gain, natural frequency and damping ratio
K = yss; %Obtaining gain
xi = abs(log(Os)) / sqrt( pi^2 + log(Os)^2 ); %Obtaining damping ratio
wn = pi / ( tp * sqrt(1-xi^2) ); %Obtaining natural frequency
display (K);
display(xi);
display(wn);
```

By executing this code, the system parameters are obtained

```
K =
    10

xi =
    0.4559

wn =
    1.7649
```

From these results, the remaining system parameters can be obtained. Recalling from the above paragraph that

$$\omega_n = \sqrt{\frac{c}{0.75}},$$

it can be derived the value of the constant

$$c = 0.75\ \omega_n^2 = 2.3362$$

Also in the above paragraph it was shown that

$$K = 0.25\frac{K_e}{c}$$

So

$$K_e = \frac{K\,c}{0.25} = \frac{10 \cdot 2.3362}{0.25} = 93.4492$$

Finally, also recalling paragraph b,

$$\xi = \frac{b}{1.5\sqrt{\frac{c}{0.75}}}$$

Therefore,

$$b = \xi \cdot 1.5\sqrt{\frac{c}{0.75}} = 1.2071$$

Once again, it is possible to reach these results adding a new fragment of MATLAB code

```
% Obtaining the remaining system's constants
c = 0.75 * wn^2;
b = xi * 1.5 * sqrt(c/0.75);
Ke = 10*c/0.25;
display(Ke);
display(b);
display(c);
```

The execution of this code obtains the values of the constants

```
Ke =
   93.4492

b =
   1.2071

c =
   2.3362
```

To validate the above results, the system's step response can be obtained by adding the following MATLAB code:

```
%% Validating the result

% Converting symbolic to numeric constants
A = double(A);
L = double(L);
rho = double(rho);

% Generating the system's transfer function
s = tf('s');
G = A*Ke / ( rho*A*L*s^2 + b*s +c);

% Plotting the unit step response
step(G);
grid;
ylim([0,14]);
```

The plot obtained executing this code is the same that the one depicted in the problem statement (see Fig. 3.50).

Problem 3.3

A cardiopulmonary bypass pump (extracorporeal pump or heart–lung machine) rotates at an angular speed $\omega(t)$. This speed is regulated by a control voltage $v(t)$. The system is discretized using a sampling time $T_s = 0.2$ and a zero-order hold (ZOH). The discrete-time dynamics of the pump is described by the following difference equation:

$$\begin{aligned} &\omega_d(k+2) - 1.792\omega_d(k+1) + 0.9048\omega_d(k) \\ &\quad = v_d(k+2) - 1.55v_d(k+1) + 0.588v_d(k) \end{aligned}$$

(a) Obtain the transfer function of the discrete-time system, drawing a map of its zeros and poles.
(b) Plot the discrete-time system step response for $0 \le t \le 25$.

Solution 3.3

(a) The system can be described according to Fig. 3.51.

The discrete-time system can be described by its transfer function in the z domain. For this purpose, the difference equation Z transform is obtained

$$\begin{aligned} &\mathcal{Z}[\omega_d(k+2) - 1.792\omega_d(k+1) + 0.9048\omega_d(k)] \\ &\quad = \mathcal{Z}[v_d(k+2) - 1.55v_d(k+1) + 0.588v_d(k)] \end{aligned}$$

$$\begin{aligned} &\mathcal{Z}[\omega_d(k+2)] - \mathcal{Z}[1.792\omega_d(k+1)] + \mathcal{Z}[0.9048\omega_d(k)] \\ &\quad = \mathcal{Z}[v_d(k+2)] - \mathcal{Z}[1.55v_d(k+1)] + \mathcal{Z}[0.588v_d(k)] \end{aligned}$$

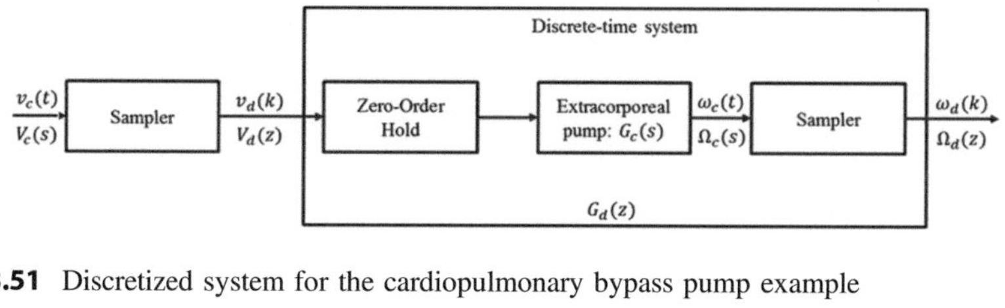

Fig. 3.51 Discretized system for the cardiopulmonary bypass pump example

$$z^2\Omega_d(z) - 1.792z\,\Omega_d(z) + 0.9048\,\Omega_d(z)$$
$$= z^2 V_d(z) - 1.55z\,V_d(z) + 0.588V_d(z)$$

$$\left[z^2 - 1.792z + 0.9048\right]\Omega_d(z) = \left[z^2 - 1.55z + 0.588\right]V_d(z)$$

$$G_d(z) = \frac{\Omega_d(z)}{V_d(z)} = \frac{z^2 - 1.55z + 0.588}{z^2 - 1.792z + 0.9048}$$

The zeros of this function are obtained by solving the equation

$$z^2 - 1.55z + 0.588 = 0$$
$$z = \frac{1.55 \pm \sqrt{1.55^2 - 4 \cdot 1 \cdot 0.588}}{2} = \frac{1.55 \pm 0.2247}{2} = \begin{cases} z = 0.6626 \\ z = 0.8874 \end{cases}$$

Likewise, the poles are obtained by solving the equation

$$z^2 - 1.792z + 0.9048 = 0$$
$$z = \frac{1.792 \pm \sqrt{1.792^2 - 4 \cdot 1 \cdot 0.9048}}{2} = \frac{1.792 \pm 0.6387i}{2}$$
$$\begin{cases} z = 0.8960 + 0.3193i \\ z = 0.8960 - 0.3193i \end{cases}$$

This paragraph can also be solved using the following MATLAB code

```
%% Discrete-time transfer function

% Defining discrete-time transfer function
Ts=0.2; %Sampling time
z=tf('z',Ts);
Gd = (z^2-1.55*z+0.588) / (z^2-1.792*z+0.9048);

% Obtaining the zeros and poles of Gd(z)
zerosGd = zero(Gd);
polesGd = pole(Gd);
display(zerosGd);
display(polesGd);

% Plotting the pole-zero map of Gd(z)
pzmap(Gd);
axis equal;
```

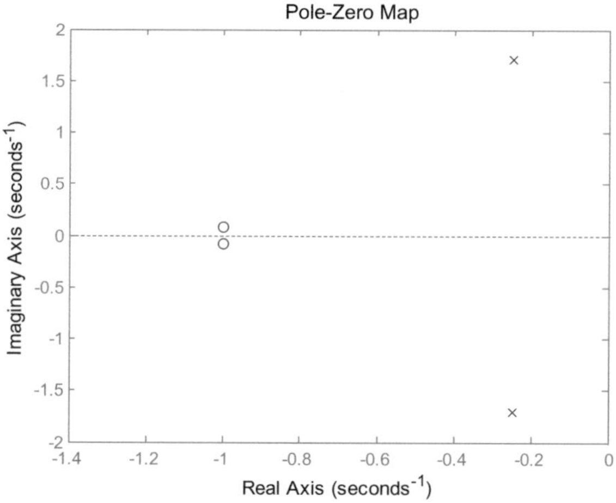

Fig. 3.52 Pole–zero map for the cardiopulmonary bypass pump system

The execution of this code results in the following transfer function, zeros and poles

```
zerosGd =
     0.8874
     0.6626

polesGd =
   0.8960 + 0.3193i
   0.8960 - 0.3193i
```

These results are the same that the previously obtained. The code also generates the pole–zero map shown in Fig. 3.52.

(b) To plot the discrete-time system step response, the following MATLAB code is added

```
% Straightforward way to plot the discrete-time response
figure;
step(Gd,tf);
ylim([-0.4,1.5]);
ylabel('\omega_d');
```

The result obtained is depicted in Fig. 3.53.

Problem 3.4

A bionic arm is a nonlinear system that can be modelled according to the following equation

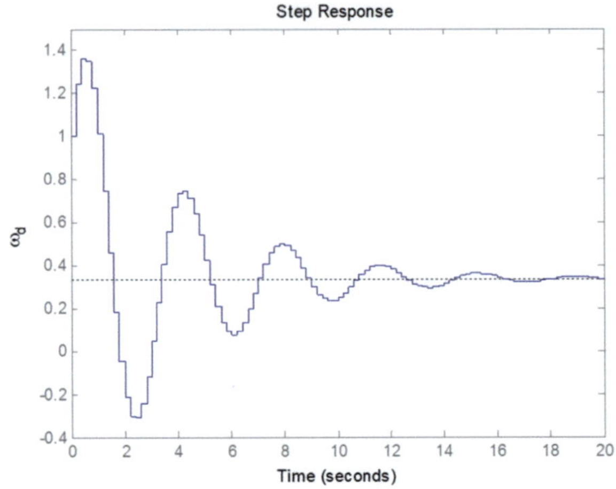

Fig. 3.53 Step response for the discretized system

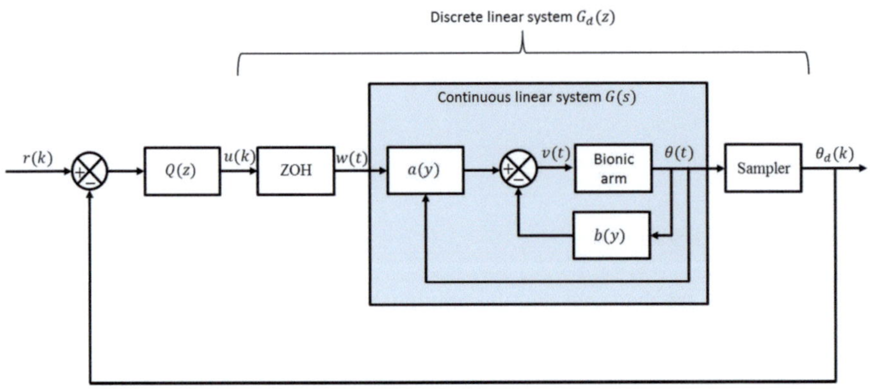

Fig. 3.54 Bionic arm control

$$a[\theta(t)]\frac{\mathrm{d}^2\theta(t)}{\mathrm{d}t^2} = v(t) + b\left[\theta(t), \frac{\mathrm{d}\theta(t)}{\mathrm{d}t}\right],$$

where $v(t)$ is the electric voltage exciting the motors moving the arm, $\theta(t)$ is the angle of the joint defining the arm position, and $a[\cdot]$ and $b[\cdot]$ are two nonlinear functions. The system is linearized using a double feedback loop, as it is depicted in Fig. 3.54. The resulting linear system is defined by the transfer function:

$$G(s) = \frac{1}{s^2}$$

To control the arm, a PID controller is used, which can be described by its transfer function

$$Q(z) = K_c\left(2\alpha + \frac{z-1}{T_s z} + \alpha^2 \frac{T_s z}{z-1}\right),$$

where $K_c = 5$ and $\alpha = 3$. The discretization of the control loop is accomplished using a sampler ($T_s = 0.1$) and a zero-order hold (ZOH). For the closed-loop system:

(a) Obtain the transfer function.
(b) Determine its zeros and poles location. Plot a pole–zero map, indicating whether the system is stable or not.
(c) Identify the dominant poles and obtain their magnitude, natural frequency and damping ratio. Based on these parameters, obtain approximate values for the step response overshoot and time settling.
(d) Obtain the step response overshoot and time settling considering the whole system (not only de dominant poles).
(e) Plot the overshoot and time settling as a function of K_c. Repeat this process to obtain the plot as a function of α.
(f) Plot the overshoot and time settling as a function of both K_c and α.

Solution 3.4

(a) To obtain the closed-loop discrete-time transfer function, the following MATLAB code is used

```
%% Obtaining closed-loop discrete transfer function

% Defining constants
Ts =0.1; %Sampling time
Kc = 5; %PID first constant
alpha = 3; %PID second constant

% Defining open-loop discrete transfer function
s = tf('s'); % "s" is the elementary continuous transfer function
z = tf('z',Ts); %"z" is the elementary discrete transfer function
G = 1/s^2; %Continuous linear system equivalent to bionic arm
Gd = c2d(G,Ts); %Discrete linear system equivalent to bionic arm
Q = Kc * ( 2*alpha + (z-1)/(Ts*z) + alpha^2*(Ts*z)/(z-1) ); %PID
M = feedback(Q*Gd,1); %Closed-loop discrete transfer function
M = minreal(M); %Obtaining minimal realization of the closed-loop tf
```

```
display(M);
```

The obtained result is

```
M =

  0.4225 z^3 - 0.2275 z^2 - 0.4 z + 0.25
  -------------------------------------------
  z^4 - 2.577 z^3 + 2.772 z^2 - 1.4 z + 0.25
```

That is,

$$M(z) = \frac{0.42z^3 - 0.23z^2 - 0.4z + 0.25}{z^4 - 2.58z^3 + 2.77z^2 - 1.4z + 0.25}$$

(b) To determine the zeros and poles in $M(z)$, and to plot the corresponding map, the following MATLAB code is added

```
%% Obtaining zeros, poles & their map

% Obtaining zeros
zer = zero(M); %Zeros
display(zer);

% Obtaining poles
p = pole(M); %Poles
display(p);

% Obtaining poles & zeros map
pzmap(M);
axis equal;
```

obtaining

```
zer =

  -1.0000 + 0.0000i
   0.7692 + 0.0000i
   0.7692 - 0.0000i

p =

   0.6857 + 0.5827i
   0.6857 - 0.5827i
   0.8374 + 0.0000i
   0.3687 + 0.0000i
```

The corresponding pole–zero map is shown in Fig. 3.55.

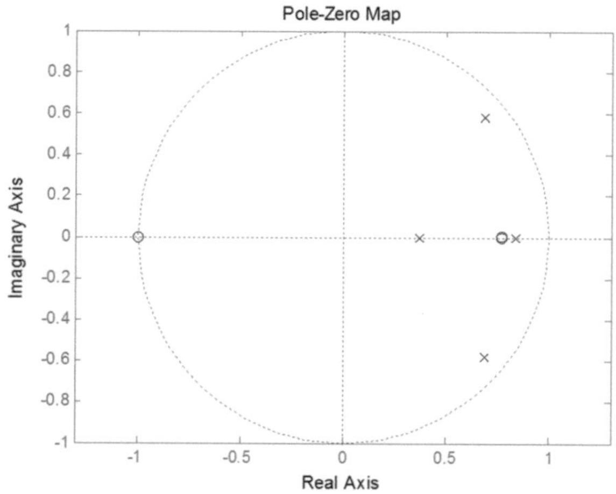

Fig. 3.55 Pole–zero map for the bionic arm system

As it can be seen, the system has a double zero at $z = 0.77$ and another zero at $z = -1$. It also have two real poles at $z = 0.37$ and at $z = 0.84$; as well as pair of complex conjugates poles at $z = 0.69 \pm 0.58j$.

The system is stable as it has all its poles within the unit circle in the z-plane.

(c) The dominant poles are those that are closest to the unit circle or, in other words, those that are farthest from the origin, and therefore, they have greatest magnitude. To determine them, let us first proceed to calculate the poles of $M(z)$, that is, those values of z that make null the denominator, according to the equation

$$z^4 - 2.58z^3 + 2.77z^2 - 1.4z + 0.25 = 0$$

These poles in the z-plane are related to the poles in the s-plane by the definition of the z variable as

$$z \equiv e^{s\,T_s}$$

Therefore, once obtained the poles in z, the corresponding poles in s are determined by

$$s = \frac{\text{Ln}\, z}{T_s}$$

The natural frequency corresponding to every pole is derived as

$$\omega_n = |s| = \frac{|\text{Ln}\, z|}{T_s}$$

On the other hand, the damping ratio can be obtained by

$$\xi = -\cos[\arg(s)] = -\cos\left[\arg\left(\frac{\mathrm{Ln}\, z}{T_s}\right)\right] = -\cos[\arg(\mathrm{Ln}\, z)]$$

Using these results, the step response overshoot can be obtained by

$$O_s = e^{\frac{-\xi\pi}{\sqrt{1-\xi^2}}}$$

Likewise, the settling time can be determined by the equation

$$ts \approx \frac{4}{\xi\omega_n}$$

All this values can be obtained by adding the following MATLAB code:

```
%% Obtaining dominant poles and its approximate response

% Obtaining magnitude, natural freq. & damping ratio for every pole
[wn,xi,p] = damp(M); %Natural frequency, damping ratio & poles
magp = abs(p); %Magnitude

% Obtaining dominant poles
gmp = max(magp); %Greatest magnitude of the poles
pd = p( abs(p)==gmp ); %Dominant poles has the greatest magnitude
wnd = wn( abs(p)==gmp ); %Dominant poles' natural frequency
wnd = unique (wnd); %Obtaining unique values for natural frequency
xid = xi( abs(p)==gmp ); %Dominant poles' damping ratio
xid = unique (xid); %Obtaining unique values for damping ratio
magd = gmp; %Dominant poles' magnitude
display(wnd);
display(xid);
display(magd);

% Obtaining approximate response corresponding to dominant poles
Os = exp( (-xid*pi) / sqrt(1-xid^2 )); %Overshoot
ts = 4 / (xid * wnd); %Settling time
display(Os);
display(ts);
```

The obtained result is:

```
wnd =

    7.1220


xid =

   0.1482
```

```
magd =
       0.8998


Os =
      0.6245


ts =
      3.7899
```

That is, the dominant poles can be defined by a magnitude $|z| = 0.8998$, a natural frequency $\omega_n = 7.1220$ and a damping ratio $\xi = 0.1482$. Applying these values to the performance equations, an approximate value of 62.45% is obtained for the overshoot and 3.7899 s for the settling time.

(d) To obtain the step response overshoot and time settling considering the whole system (not only de dominant poles), the following MATLAB code is added:

```
%% Obtaining overshoot & settling time for the whole system

Sr = stepinfo(M); %Step response
Os = Sr.Overshoot; %Overshoot
ts = Sr.SettlingTime; %Settling time
display(Os);
display(ts);
```

Executing this code, the following result is obtained:

```
Os =
   105.9225


ts =
    4.1000
```

That is, the step response has an overshoot of 105.9225% and a settling time of 4.1 s. The difference with the values obtained in the previous paragraph is due to the influence of the remaining $M(z)$ roots (zeros and non-dominant poles).

(e) First, a graphic containing both the overshoot and the settling time as a function of K_c will be plotted. Latter, a second graphic with, again, the overshoot and the settling time, now as a function of α, will also be plotted. For this purpose, the following MATLAB code is added:

```
%% Obtaining overshoot & settling time as a function of
    % each PID parameter

% Defining constants and variables
n = 100; %Number of different values for PID parameters
Kcv = linspace(0,12,n); %Vector of PID first constant (Kc)
alphav = linspace(0,5,n); %Vector of PID second constant (alpha)
Os = zeros(1,n); %Allocate space for overshoot vector
ts = zeros(1,n); %Allocate space for settling time vector
tsmax = 30; %Maximum value of settling time for plotting

% Computing overshoot & settling time for different values of Kc
alpha = 3; %PID second constant
for i = 1:n %For every value of Kc
    Kc = Kcv(i); %Obtaining Kc value
    Q = Kc * ( 2*alpha + (z-1)/(Ts*z) + alpha^2*(Ts*z)/(z-1) ); %PID
    M = feedback(Q*Gd,1); %Closed-loop discrete tf
    M = minreal(M);  %Minimal realization of the closed-loop tf

    Sr = stepinfo(M); %Step response
    Os(i) = Sr.Overshoot; %Overshoot
    ts(i) = Sr.SettlingTime; %Settling time
end
ts( ts>tsmax ) = nan; %Avoiding too large settling time in plots

% Plotting overshoot & settling time as a function of Kc
figure;
hAx = plotyy(Kcv,Os,Kcv,ts);
title(['Step response for \alpha=',num2str(alpha)]);
xlabel('K_c');
ylabel(hAx(1),'Overshoot (%') %Left y-axis
ylabel(hAx(2),'Settling time') %Right y-axis

% Computing overshoot & settling time for different values of alpha
Kc = 5; %PID first constant
for i = 1:n %For every value of alpha
    alpha = alphav(i); %Obtaining alpha value
    Q = Kc * ( 2*alpha + (z-1)/(Ts*z) + alpha^2*(Ts*z)/(z-1) ); %PID
    M = feedback(Q*Gd,1); %Closed-loop discrete tf
    M = minreal(M);  %Minimal realization of the closed-loop tf

    Sr = stepinfo(M); %Step response
    Os(i) = Sr.Overshoot; %Overshoot
    ts(i) = Sr.SettlingTime; %Settling time
end
ts( ts>tsmax ) = nan; %Avoiding too large settling time in plots

% Plotting overshoot & settling time as a function of Kc
figure;
hAx = plotyy(alphav,Os,alphav,ts);
title(['Step response for K_c=',num2str(Kc)]);
xlabel('\alpha');
ylabel(hAx(1),'Overshoot (%') %Left y-axis
ylabel(hAx(2),'Settling time') %Right y-axis
```

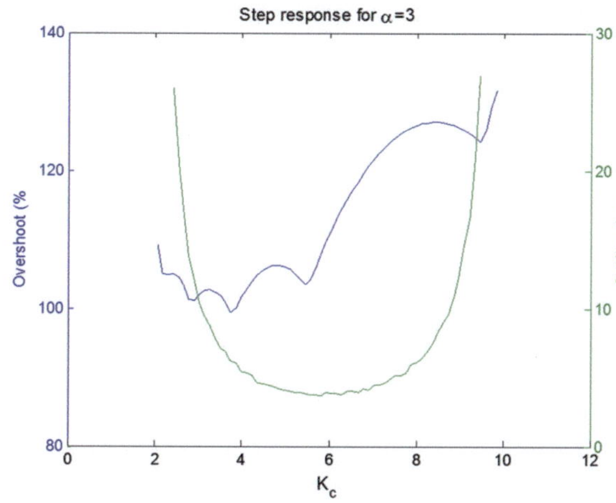

Fig. 3.56 Overshoot and settling time for different values of K_c

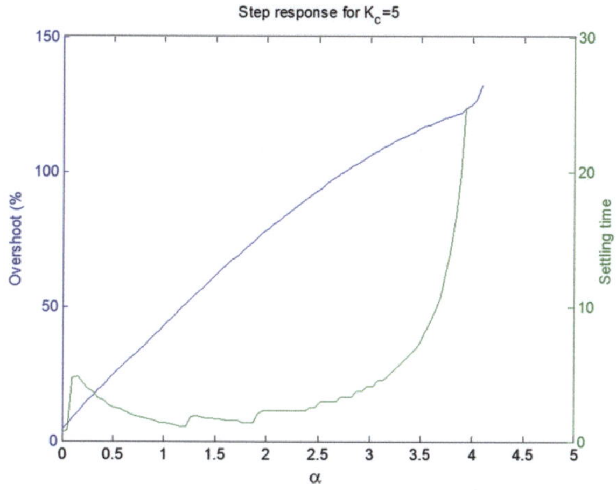

Fig. 3.57 Overshoot and settling time for different values of α

The execution of this code obtains the plots shown in Figs. 3.56 and 3.57. In these plots, there are regions where no value of the overshoot and the settling time is shown. The reason is in these regions, corresponding to certain values of K_c and α, the system is unstable and no overshoot nor settling time can be defined.

(f) A graphic containing the step response overshoot and/or the settling time as a function of two variables, K_c and α, is a three-dimensional graphic. To address this issue, a conventional 3D plot can be used where, for instance, K_c and α are in the X- and Y-axis, and the overshoot (or the settling time) is represented in the Z-axis.

However, a different approach will be used here, drawing a level plot (2D) where the *Z*-axis is substituted for a colour scale. For this goal, the following MATLAB code is added:

```
%% Obtaining overshoot & settling time as a simultaneous
    % function of two PID parameter

% Defining constants and variables
n = 100; %Number of different values for PID parameters
Kcv = linspace(0.01,12,n); %Vector of PID first constant (Kc)
alphav = linspace(0,5,n); %Vector of PID second constant (alpha)
Os = zeros(n,n); %Allocate space for overshoot vector
ts = zeros(n,n); %Allocate space for settling time vector

% Computing overshoot & settling time for different values of PID
for i = 1:n %For every value of Kc
    for j = 1:n %For every value of alpha
        Kc = Kcv(i); %Obtaining Kc value
        alpha = alphav(j); %Obtaining alpha value
        Q = Kc*( 2*alpha + (z-1)/(Ts*z) + alpha^2*(Ts*z)/(z-1) );
%PID
        M = feedback(Q*Gd,1); %Closed-loop discrete tf
        M = minreal(M);  %Minimal realization of the closed-loop tf

        Sr = stepinfo(M); %Step response
        Os(i,j) = Sr.Overshoot; %Overshoot
        ts(i,j) = Sr.SettlingTime; %Settling time
    end
end
ts( ts>tsmax ) = nan; %Avoiding too large settling time in plots

% Plotting overshoot as a function of Kc & alpha
figure;
[KcG,alphaG] = meshgrid(Kcv,alphav);
contourf(KcG,alphaG,Os);
colorbar;
title('Overshoot (%)');
xlabel('K_c');
ylabel('\alpha');

% Plotting settling time as a function of Kc & alpha
figure;
contourf(KcG,alphaG,ts);
colorbar;
title('Settling time');
xlabel('K_c');
ylabel('\alpha');
```

The result obtained executing this code is depicted in Figs. 3.58 and 3.59.

Problem 3.5

Extravascular administration of drugs (oral, rectal, intramuscular, transdermal, subcutaneous) can be modelled by two connected compartments. In the first of these compartments, for example the digestive tract, there is an amount of drug $x(t)$

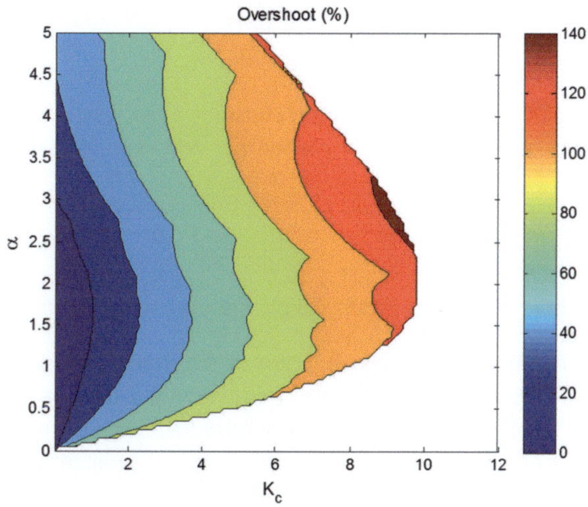

Fig. 3.58 Overshoot with respect to variations in the parameters K_c and α

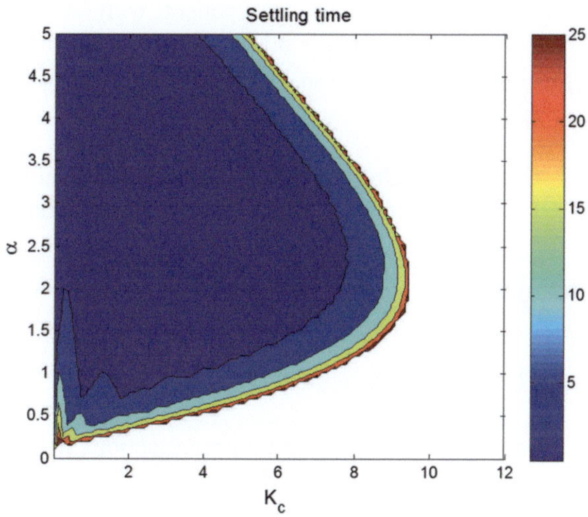

Fig. 3.59 Settling time with respect to variations in the parameters K_c and α

which increases with the administration of the drug and decreases by the absorption of the drug in the blood plasma, which is reflected in the following equation:

$$\frac{\mathrm{d}x(t)}{\mathrm{d}t} = q(t) - K_a x(t),$$

where $q(t)$ is the rate of administered drug, and K_a is the absorption constant. In the second compartment, which represents the blood plasma, there is an amount of drug $p(t)$ which increases with absorption and decreases by elimination of the drug, which is reflected in the following equation

$$\frac{dp(t)}{dt} = K_a x(t) - K_e p(t),$$

where K_e is the elimination constant. Finally, the plasmatic concentration $c(t)$ is the amount of drug in the plasma per unit volume, i.e.

$$c(t) = \frac{p(t)}{V_d},$$

where V_d is a constant called the apparent volume of distribution.

(a) Obtain the block diagram of the system and derive its transfer function.
(b) Find the order, natural frequency and damping ratio of the system. Derive the values of the constants required for the system to be undamped.
(c) Obtain the evolution of plasmatic concentration when an D_0 oral dose is administered.
(d) Derive the peak plasma concentration c_p and the time t_p in which it occurs.

Consider now the case of a patient, with an apparent volume of $V_d = 10\,\text{L}$, to which an oral dose $D_0 = 500\,\text{mg}$ of a certain drug is administered. Under these circumstances, a peak concentration of $c_p = 30\,\frac{\text{mg}}{\text{L}}$ is reached 8 h later of the drug administration.

(e) Obtain the values of the absorption and elimination constants.
(f) Draw the evolution of plasmatic concentration.

Solution 3.5

(a) The system's functional diagram is depicted in Fig. 3.60.

More specifically, using transfer functions in the s domain, the block diagram can be drawn as shown in Fig. 3.61.

The transfer function of the first block can be obtained from its differential equation

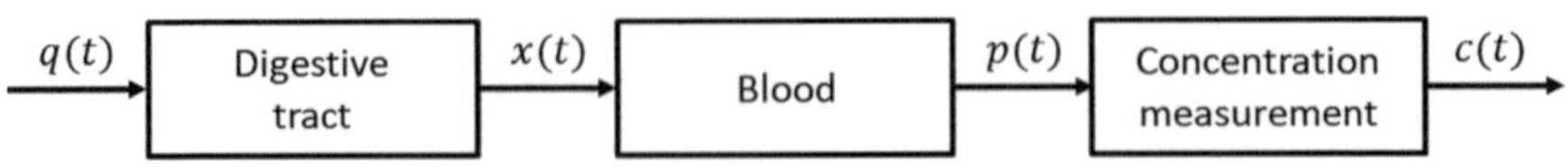

Fig. 3.60 Functional diagram of the extravascular administration of drugs system

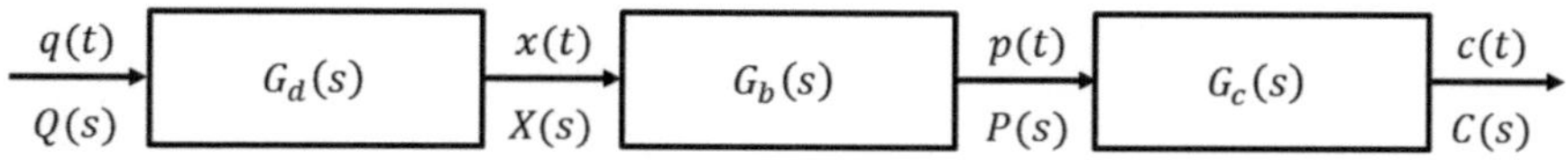

Fig. 3.61 Block diagram in the s domain

$$\begin{aligned}
\frac{\mathrm{d}x(t)}{\mathrm{d}t} &= q(t) - K_a x(t) \\
\mathcal{L}\left[\frac{\mathrm{d}x(t)}{\mathrm{d}t}\right] &= \mathcal{L}[q(t)] - \mathcal{L}[K_a x(t)] \\
sX(s) &= Q(s) - K_a X(s) \\
(s + K_a)X(s) &= Q(s) \\
G_d(s) \equiv \frac{X(s)}{Q(s)} &= \frac{1}{s + K_a}
\end{aligned}$$

For the second block, it can be written as

$$\begin{aligned}
\frac{\mathrm{d}p(t)}{\mathrm{d}t} &= K_a x(t) - K_e p(t) \\
\mathcal{L}\left[\frac{\mathrm{d}p(t)}{\mathrm{d}t}\right] &= \mathcal{L}[K_a x(t)] - \mathcal{L}[K_e p(t)] \\
sP(s) &= K_a X(s) - K_e P(s) \\
(s + K_e)P(s) &= K_a X(s) \\
G_b(s) \equiv \frac{P(s)}{X(s)} &= \frac{K_a}{s + K_e}
\end{aligned}$$

And, finally, in the third block

$$\begin{aligned}
c(t) &= \frac{p(t)}{V_d} \\
\mathcal{L}[c(t)] &= \mathcal{L}\left[\frac{p(t)}{V_d}\right] \\
C(s) &= \frac{P(s)}{V_d} \\
G_c(s) \equiv \frac{C(s)}{P(s)} &= \frac{1}{V_d}
\end{aligned}$$

Therefore, the overall transfer function of the system will be

$$G(s) \equiv \frac{C(s)}{Q(s)} = G_d(s) \cdot G_b(s) \cdot G_c(s)$$

$$G(s) = \frac{1}{s+K_a} \cdot \frac{K_a}{s+K_e} \cdot \frac{1}{V_d}$$

$$G(s) = \frac{K_a}{V_d} \frac{1}{(s+K_a)(s+K_e)}$$

(b) From the obtained transfer function in the previous paragraph, it is clear that it is a second-order system. In its canonical form, it is expressed as

$$G(s) = \frac{\frac{K_a}{V_d}}{s^2 + (K_a + K_e)s + K_a K_e} = \frac{K\omega_n^2}{s^2 + 2\zeta\omega_n s + \omega_n^2}$$

So,

$$\omega_n^2 = K_a K_e; \quad 2\zeta\omega_n = K_a + K_e$$

The natural frequency is

$$\omega_n = \sqrt{K_a K_e}$$

For the damping ratio, it can be written that

$$\zeta = \frac{K_a + K_e}{2\omega_n} = \frac{K_a + K_e}{2\sqrt{K_a K_e}}$$

In order for the system to be underdamped, it must be ensured that the damping ratio is less than 1 ($\zeta < 1$). The value of ζ can be rewritten as

$$\zeta = \frac{\frac{1}{K_a}(K_a + K_e)}{\frac{1}{K_a} 2\sqrt{K_a K_e}} = \frac{1 + \frac{K_e}{K_a}}{2\sqrt{\frac{K_a K_e}{K_a^2}}} = \frac{1 + \frac{K_e}{K_a}}{2\sqrt{\frac{K_e}{K_a}}}$$

Denoting the ratio between the elimination and absorption ratios as $R \equiv K_e/K_a$, it can be derived that

$$\zeta = \frac{1+R}{2\sqrt{R}}$$

In order for the system to be underdamped, it must be fulfilled that

$$\begin{aligned}
&\zeta = \frac{1+R}{2\sqrt{R}} < 1 \\
&1+R < 2\sqrt{R} \\
&(1+R)^2 < \left(2\sqrt{R}\right)^2 \\
&1+2R+R^2 < 4R \\
&1-2R+R^2 < 0 \\
&(R-1)^2 < 0
\end{aligned}$$

But this condition is not fulfilled by any real value of R, so no combination of the constants causes the system to have an underdamped behaviour.

(c) When an oral dose D_0 is administered, the rate of drug dispensed can be modelled as an impulse, such as

$$\begin{aligned}
&q(t) = D_0\delta(t) \\
&Q(s) = \mathcal{L}[q(t)] = \mathcal{L}[D_0\delta(t)] = D_0 \cdot \mathcal{L}[\delta(t)] = D_0 \cdot 1 = D_0
\end{aligned}$$

So, the plasmatic concentration is

$$C(s) = G(s)Q(s) = \frac{K_a}{V_d}\frac{1}{(s+K_a)(s+K_e)}D_0$$

The corresponding time evolution can be derived as

$$c(t) = \mathcal{L}^{-1}[C(s)] = \mathcal{L}^{-1}\left[\frac{K_a}{V_d}\frac{1}{(s+K_a)(s+K_e)}D_0\right]$$

To obtain it, the following MATLAB code can be used:

```
function main

%% Paragraph c)
% Symbolic computation of plasmatic concentration
syms s Ka Ke Vd D0 %Symbolic variables and constants
G = (Ka/Vd) / ( (s+Ka) * (s+Ke) ); %System's transfer function
Q = D0; %System's input in s-domain (oral drug dosis)
C = G * Q; %System's output in s-domain (plasmatic concentration)
c = ilaplace(C); %System's output in t-domain (plasmatic concentr.)
pretty(c);
```

The obtained result is the following:

```
D0 Ka exp(-Ke t)   D0 Ka exp(-Ka t)
---------------- - ----------------
  Vd (Ka - Ke)       Vd (Ka - Ke)
```

that is, the plasmatic concentration evolution is:

$$c(t) = \frac{D_0 K_a}{V_d(K_a - K_e)} \left(\mathrm{e}^{-K_e t} - \mathrm{e}^{-K_a t} \right)$$

(d) At the peak of the plasmatic concentration, its derivative is zero, so at the time of the peak $t = t_p$

$$\left. \frac{\mathrm{d}c(t)}{\mathrm{d}t} \right|_{t=t_p} = 0$$

To solve this equation, the following MATLAB code is added:

```
%% Paragraph d)
% Symbolic computation of peak plasmatic concentration
syms t tp %Symbolic variables and constants
dc_dt = diff(c,t); %Plasmatic concentration time derivative
tp = solve(dc_dt == 0,t); %Peak time
pretty(tp);
cp = subs(c,t,tp); %Peak plasmatic concentration
pretty(cp);
```

The obtained result is the following:

```
   / Ka \
log| -- |
   \ Ke /
---------
 Ka - Ke
```

```
           /          / Ka \ \                   /          / Ka \ \
           |   Ke log| -- | |                   |   Ka log| -- | |
           |          \ Ke / |                   |          \ Ke / |
D0 Ka exp| - ------------ |   D0 Ka exp| - ------------ |
           \       Ka - Ke   /                   \       Ka - Ke   /
-------------------------- - --------------------------
        Vd (Ka - Ke)                          Vd (Ka - Ke)
```

that is, the plasmatic concentration peak value is reached at

$$t_p = \frac{\mathrm{Ln}\left(\frac{K_a}{K_e}\right)}{K_a - K_e},$$

and its value is

$$c_p \equiv c(t_p) = \frac{D_0 K_a}{V_d(K_a - K_e)}\left(\mathrm{e}^{-K_e t_p} - \mathrm{e}^{-K_a t_p}\right)$$

$$c_p = \frac{D_0 K_a}{V_d(K_a - K_e)}\left(\mathrm{e}^{-K_e \frac{\mathrm{Ln}\left(\frac{K_a}{K_e}\right)}{K_a - K_e}} - \mathrm{e}^{-K_a \frac{\mathrm{Ln}\left(\frac{K_a}{K_e}\right)}{K_a - K_e}}\right)$$

(e) Once known the peak plasma concentration and the time when this value is reached, it is possible to determine the constants K_a and K_e by solving the above equations. There is no analytical solution for them, so a numerical result can be obtained by adding the following MATLAB code:

```
%% Paragraph e)
% Numerical computation of the absorption and elimination rates
D0 = 500; %Initial drug dosis (bolus)
Vd = 10; %Volume of distribution
Par.D0 = D0; %Initial drug dosis (bolus)
Par.Vd = Vd; %Volume of distribution
Par.tp = 8; %Peak time (experimental value)
Par.cp = 30; %Peak concentration (experimental value)
X = fsolve(@(X)AbsElimRateEquations(X,Par),[2,1]); %Solve equations
Ka = X(1); %Absortion rate
Ke = X(2); %Elimination rate
display(Ka);
display(Ke)


% Define de equations for the absorption and elimination rates
%   They are computed as a function of the peak response
function F = AbsElimRateEquations(X,Par)
%X: Vector containing absortion and elimination rates
%Par: Structure containing model's parameters
%F: Set of equations' result (should be zero)

%Error in the equation for the final value of recovered persons
Ka = X(1); %Absortion rate
Ke = X(2); %Elimination rate

D0 = Par.D0; %Initial drug dosis (bolus)
Vd = Par.Vd; %Volume of distribution
tp = log(Ka/Ke)/(Ka - Ke); %Peak time (computed value)
A = D0*Ka/(Vd*(Ka - Ke)); %Intermediate constant
cp = A *exp(-Ke*tp) - ...
     A*exp(-Ka*tp); %Peak concentration (computed value)

eq1 = tp - Par.tp; %Equation for peak time (should be zero)
eq2 = cp - Par.cp; %Equation for peak concentration (should be zero)
F = [eq1 eq2]; %Set of equations (should be zero)
```

The obtained result is the following:

```
Ka =

    0.2164


Ke =

    0.0639
```

that is, the absorption and elimination constants are

$$K_a = 0.2164\,h^{-1}; \quad K_e = 0.0639\,h^{-1}$$

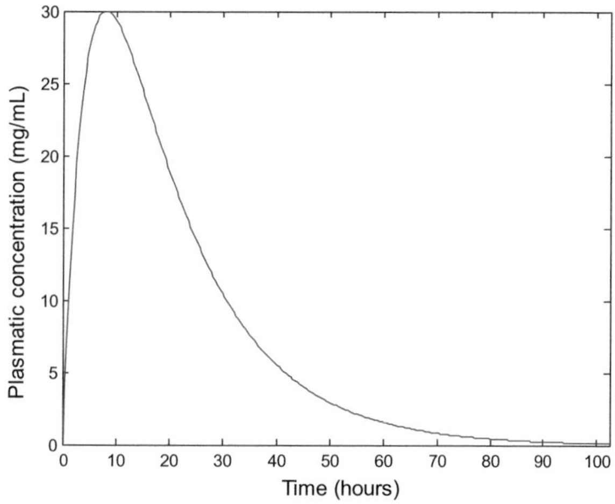

Fig. 3.62 Plasmatic concentration for an oral dose over time

(f) In this case, the following MATLAB code is added to the `function main`:

```
%% Paragraph f)
% Numerical computation of the absortion and elimination rates
s = tf('s');
G = (Ka/Vd) / ( (s+Ka) * (s+Ke) ); %System's transfer function
[c,t] = impulse(G*D0);
plot(t,c);
xlim([0,t(end)]);
xlabel('Time (hours)');
ylabel('Plasmatic concentration (mg/mL)');
```

The obtained result is depicted in Fig. 3.62.

Frequency Domain Analysis of Continuous and Discrete Systems

4

This chapter analyses the response of the system when sinusoidal input signals are applied. While the impulse and step functions are useful test signals to use in the characterization of linear systems as it was described in the preceding chapter, it is difficult to find in practice abrupt steps and impulsive changes, while periodic behaviour is frequent in physiology. As periodic signals can be decomposed into its sinusoidal components, sinusoidal signals represent useful inputs to dynamic systems. Also frequency response method can be used to derive a transfer function starting from input–output measurements when sinusoidal signals are applied as inputs.

In this chapter, we will study analytically the frequency response of linear time-invariant systems and its graphical representation mainly as Bode diagrams, both for continuous- and discrete-time systems by applying the bilinear transformation mapping. Stability both of continuous and of discretized systems will be also stated in terms of stability margins, and relation between time response and frequency response will also be stated.

Frequency Response Analysis of Continuous Systems

While impulse, step and ramp signals are usually applied as test input signals to analyse system's response, in physiology periodic phenomena are a common occurrence instead.

Then, considering a sinusoidal input with amplitude U and frequency ω_0 as

$$u(t) = U\sin(\omega_0 t) \tag{4.1}$$

the steady-state response will be also sinusoidal with same frequency, amplitude Y and phase shift as

J. Fernández de Cañete et al., *Automatic Control Systems in Biomedical Engineering*,
https://doi.org/10.1007/978-3-319-75717-9_4

$$y(t) = Y\sin(\omega_0 t + \phi) \tag{4.2}$$

Considering a linear time-invariant dynamic system given by its transfer function $G(s)$, both Laplace transforms of input and output of the system will be related by

$$Y(s) = G(s)\,U(s) \tag{4.3}$$

and replacing $s = j\omega$ so as to make a steady-state analysis, then

$$Y(j\omega) = G(j\omega)\,U(j\omega) \tag{4.4}$$

Since all terms are complex magnitudes,

$$Y(j\omega) = |Y(j\omega)|\ \angle Y(j\omega) \tag{4.5}$$

$$G(j\omega) = |G(j\omega)|\ \angle G(j\omega) \tag{4.6}$$

$$U(j\omega) = |U(j\omega)|\ \angle U(j\omega) \tag{4.7}$$

The amplitude and phase delay of the output signal $y(t)$ then will be given by

$$\frac{Y}{X} = |G(j\omega_0)| \quad \phi = \angle G(j\omega_0) \tag{4.8}$$

The complex function $G(j\omega)$ predicts how the system will respond to sinusoidal inputs of variable frequency and is called the frequency response of the system, obtained by substituting $s = j\omega$ into the system transfer function $G(s)$. In fact, $|G(j\omega)|$ represents the gain and $\angle G(j\omega)$ the phase difference of the system respectively when a sinusoidal input with ω frequency is applied (Fig. 4.1).

Example 4.1

The use of analogy in biomedical engineering enables modelling physiologic systems by means of electric circuits. A simple model of respiratory dynamics can be assumed as an RLC electric circuit, where R, L and C represent airway resistance, inertance and air storage capacity in the lungs (Fig. 4.2).

Obtain the output response $P_a(t)$ of this system when sinusoidal input signals $P_i(t)$ of unitary amplitude and variable frequencies of $\omega_1 = \pi$, $\omega_2 = 8\pi$ and $\omega_3 = 16\pi$ are applied, assuming $R = 0.3,\ L = 0.01$ and $C = 0.1$ as structural parameters.

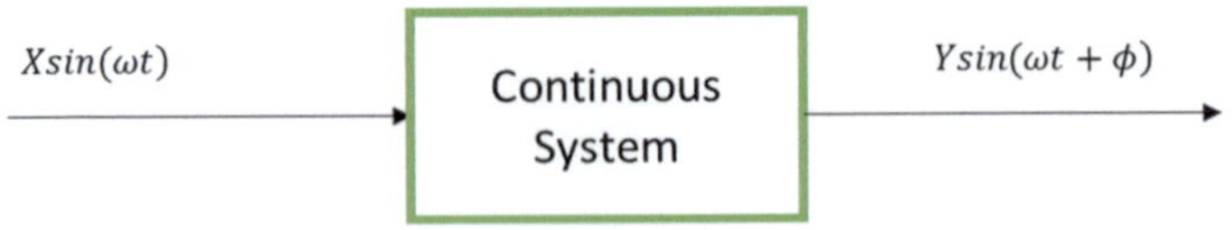

Fig. 4.1 Frequency response of a continuous linear system

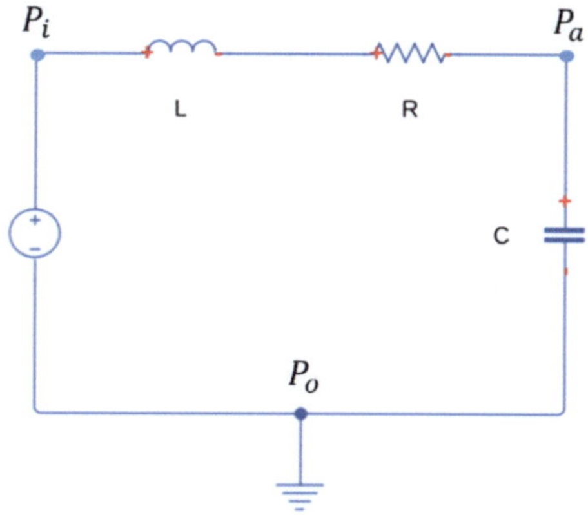

Fig. 4.2 Analog electric model of global respiratory dynamics

Solution

Applying Kirchhoff laws, it results in a dynamic equation relating input pressure at the airway opening $P_i(t)$ with alveolar output pressure $P_a(t)$ as

$$LC\ddot{P}_a(t) + RC\dot{P}_a(t) + P_a(t) = P_i(t) \tag{4.9}$$

being $P_0(t)$ ambient pressure set to zero. The transfer function can be easily developed taking Laplace transform at each side of Eq. (4.10) and grouping them so that

$$G(s) = \frac{P_a(s)}{P_i(s)} = \frac{1}{LCs^2 + RCs + 1} \tag{4.10}$$

The frequency response will be obtained by replacing $s = j\omega$; thus,

$$G(j\omega) = \frac{1}{LC(j\omega)^2 + RCj\omega + 1} \tag{4.11}$$

where

$$|G(j\omega)| = \frac{1}{\sqrt{(1 - LC\omega^2)^2 + R^2C^2\omega^2}} \tag{4.12}$$

$$\angle G(j\omega) = -\text{atan}\left(\frac{RC\omega}{1 - LC\omega^2}\right) \tag{4.13}$$

representing the relationship between $P_a(t)$ and $P_i(t)$ with ω variable frequency in terms of gain and phase shift.

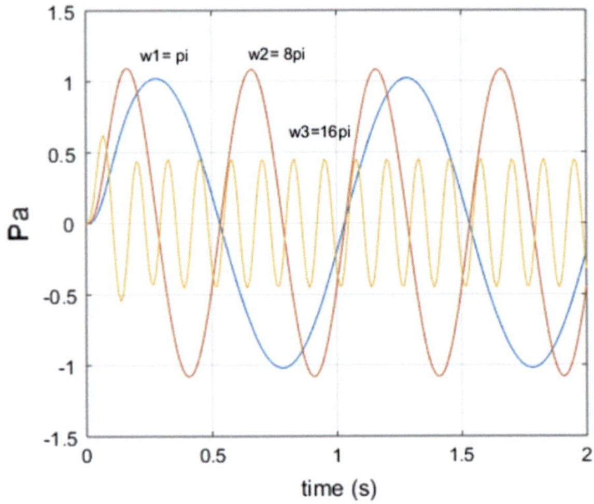

Fig. 4.3 Output response of model of global respiratory dynamics when sinusoidal signal inputs of variable frequency are applied

The output response can be plotted (Fig. 4.3) by using the MATLAB code below:

```
≫s=zpk('s');
≫R=0.3;L=0.01;C=0.1;
≫G=1/(L*C*s^2+R*C*s+1);

≫t=0:0.01:2;
≫wo=2*pi;w1=4*pi;w2=16*pi;

≫u1=sin(wo*t);u2=sin(w1*t); u3=sin(w2*t);
≫y1=lsim(G,u1,t); y2=lsim(G,u2,t); y3=lsim(G,u3,t);
≫plot(t,y1,t,y2,t,y3)
```

In order to derive the frequency response of closed-loop systems, let us consider the single control loop system configuration as depicted in Fig. 4.4.

The closed-loop transfer function is given by

$$M(s) = \frac{Y(s)}{R(s)} = \frac{KK_h G(s)}{1 + KG(s)H(s)} \tag{4.14}$$

so that the frequency response will be computed by substituting $s = j\omega$, in this way

$$M(j\omega) = |M(j\omega)| \angle M(j\omega) = \frac{Y(j\omega)}{R(j\omega)} = \frac{KK_h G(j\omega)}{1 + KG(j\omega)H(j\omega)} \tag{4.15}$$

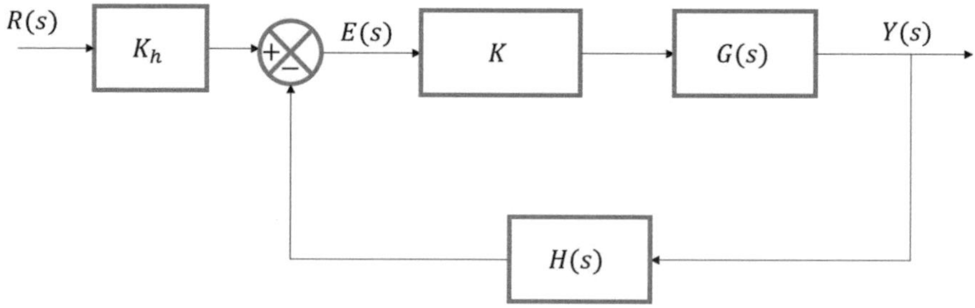

Fig. 4.4 Single closed-loop system configuration

that can be expressed as

$$|M(j\omega)| = \frac{|KK_hG(j\omega)|}{|1 + KG(j\omega)H(j\omega)|} \tag{4.16}$$

$$\angle M(j\omega) = \angle KK_hG(j\omega) - \angle(1 + KG(j\omega)H(j\omega)) \tag{4.17}$$

The frequency response plays an important role in the analysis and control of closed-loop systems through both the gain and the phase characteristics, as the time response does which was already described.

Bode Diagrams of Continuous Systems

Bode plot is one of the most useful graphical methods applied for control analysis and design based on the frequency response, by presenting the gain characteristic in logarithmic form while leaving the phase curve on linear scale.

The gain of the frequency response is thus expressed in decibels (dB) as defined by

$$|G(j\omega)|_{\text{dB}} = 20\log_{10}|G(j\omega)| \tag{4.18}$$

The logarithmic nature of the Bode diagram represents one advantage since it is possible to see the contributions of poles and zeros of a dynamic system to the frequency response analysis and also to the control task when both poles and zeros are added to the open loop.

Thereby, considering a general transfer function described as

$$G(s) = \frac{K(\tau_{c1}s + 1)(\tau_{c2}s + 1)\ldots(\tau_{\text{cm}}s + 1)}{s^N(\tau_{p1}s + 1)(\tau_{p2}s + 1)\ldots(\tau_{\text{pn}}s + 1)} \tag{4.19}$$

then, the general frequency response will be given by

$$G(j\omega)=\frac{K(\tau_{c1}j\omega+1)(\tau_{c2}j\omega+1)\ldots(\tau_{cm}j\omega+1)}{(j\omega)^N\left(\tau_{p1}j\omega+1\right)\left(\tau_{p2}j\omega+1\right)\ldots\left(\tau_{pn}j\omega+1\right)} \tag{4.20}$$

Applying the logarithmic scale, we have the gain characteristic as

$$|G(j\omega)|_{\mathrm{dB}}=20\log_{10}\frac{K\sqrt{\tau_{c1}\omega^2+1}\sqrt{\tau_{c2}\omega^2+1}\ldots\sqrt{\tau_{\mathrm{cm}}\omega^2+1}}{\omega^N\sqrt{\tau_{p1}\omega^2+1}\sqrt{\tau_{p2}\omega^2+1}\ldots\sqrt{\tau_{\mathrm{pn}}\omega^2+1}} \tag{4.21}$$

and following the logarithmic rules, it results

$$\begin{aligned}|G(\omega)|_{\mathrm{dB}}=&\,20\log_{10}K+\sum_{i=1}^{m}20\log_{10}(\tau_{\mathrm{ci}}\omega^2+1)^{1/2}\\&+20\log_{10}\omega^{-N}+\sum_{j=1}^{m}20\log_{10}(\tau_{\mathrm{pj}}\omega^2+1)^{-1/2}\end{aligned} \tag{4.22}$$

Therefore, the logarithmic approach enables considering the contribution of each term be pole or zero in an additive fashion. Similarly, the phase characteristic will be given by

$$\angle G(\omega)=\sum_{i=1}^{m}\mathrm{atan}(\omega\tau_{ci})-N\frac{\pi}{2}-\sum_{j=1}^{n}\mathrm{atan}\left(\omega\tau_{pj}\right) \tag{4.23}$$

The gain and phase contributions of each component can be of the form $(j\omega^{-1})$ in case of integrators, so the gain curve represents a straight line with −20 dB/decade of slope which is infinite at very low frequency and zero at very high frequency instead. Each integrator contributes to the phase curve with $-\pi/2$ rad each (Fig. 4.5).

In case of single zeros as $(\tau_c j\omega+1)$ (poles as $\left(\tau_p j\omega+1\right)^{-1}$), each will contribute separately to the Bode diagram. At very low frequency, zeros' (poles') contribution is null and at very high frequency zeros' (poles') contribution is $\infty(-\infty)$ following a straight line of +20 dB/decade (−20 dB(decade), and at $\omega=1/\tau_c(\omega=1/\tau_p)$ is 3 dB (−3 dB). Each zero (pole) adds a shift of $\mathrm{atan}(\tau_c\omega)$ $(-\mathrm{atan}(\tau_p\omega))$ to the phase characteristic which approaches to $+\pi/2(-\pi/2)$ at high frequency (Fig. 4.6).

Complex zeros and poles contribute with second-order factors as

$$G(s)=\frac{K(\tau_{c1}s+1)(\tau_{c2}s+1)\ldots\left(\left(\frac{s}{\omega_{\mathrm{nc}}}\right)^2+2\xi_c\left(\frac{s}{\omega_{\mathrm{nc}}}\right)+1\right)\ldots}{s^N\left(\tau_{p1}s+1\right)\left(\tau_{p2}s+1\right)\ldots\left(\left(\frac{s}{\omega_{\mathrm{np}}}\right)^2+2\xi_p\left(\frac{s}{\omega_{\mathrm{np}}}\right)+1\right)\ldots} \tag{4.24}$$

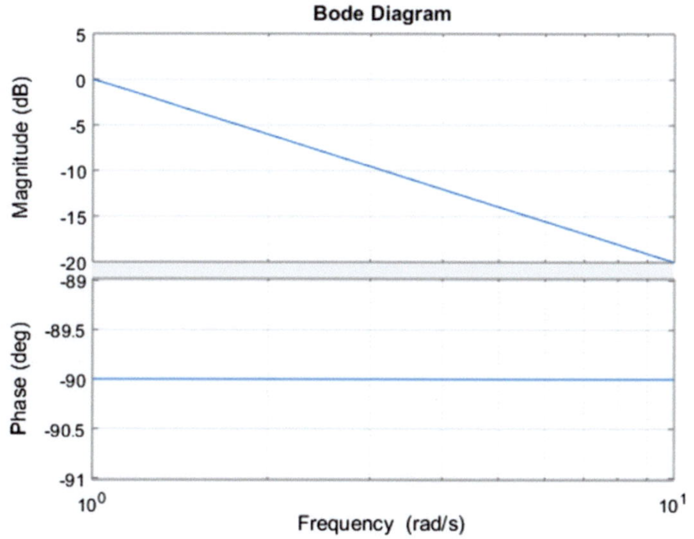

Fig. 4.5 Bode plot of an integrator

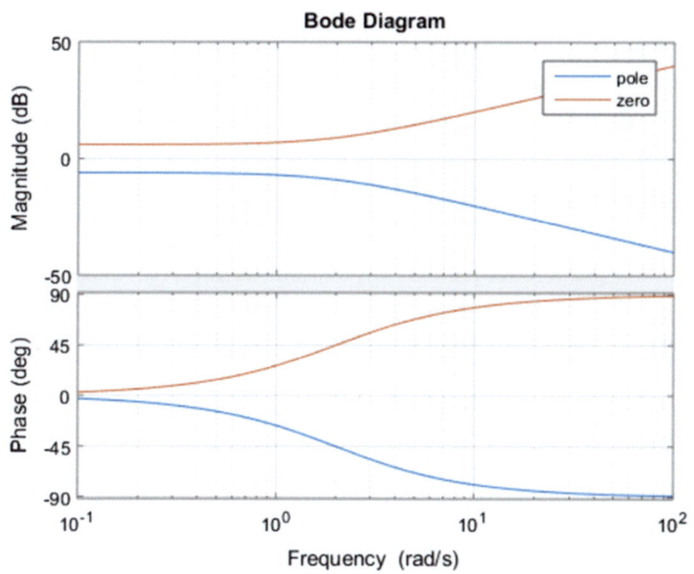

Fig. 4.6 Bode plot of both a single zero and pole at $\tau = 1$

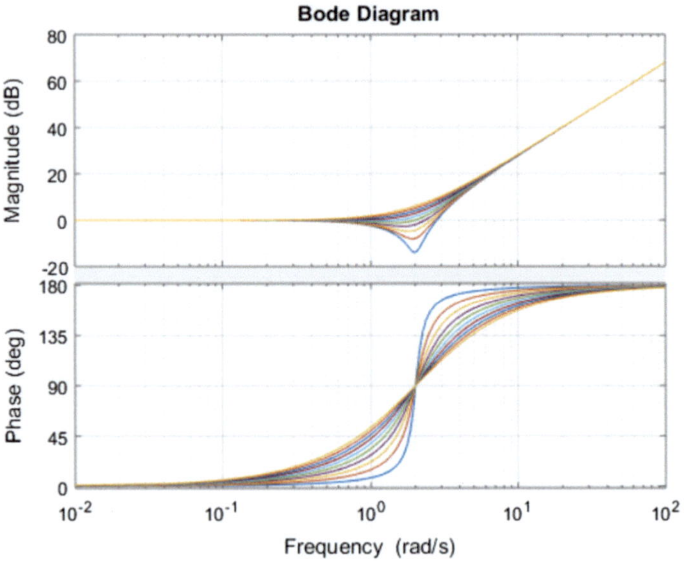

Fig. 4.7 Bode plot of a complex zero at $\omega_n = 1$

then, the general frequency response will be given by

$$G(j\omega) = \frac{K(\tau_{c1}j\omega + 1)(\tau_{c2}j\omega + 1)\ldots\left(\left(\frac{j\omega}{\omega_{\text{nc}}}\right)^2 + 2\xi_c\left(\frac{j\omega}{\omega_{\text{nc}}}\right) + 1\right)\ldots}{(j\omega)^N(\tau_{p1}j\omega + 1)(\tau_{p2}j\omega + 1)\ldots\left(\left(\frac{j\omega}{\omega_{\text{np}}}\right)^2 + 2\xi_p\left(\frac{j\omega}{\omega_{\text{np}}}\right) + 1\right)\ldots} \tag{4.25}$$

The gain and phase contributions depend on the relative values of natural frequency ω_{nc} (ω_{np}) and damping coefficient ξ_c (ξ_p) for each complex zero (pole). In fact, it results in a family of curves for each relative value of ω_n and ξ (Figs. 4.7 and 4.8).

If a time delay term is present into the open loop, the resulting Bode contribution of

$$G(s) = e^{-t_o s} \tag{4.26}$$

will be given by

$$G(j\omega) = e^{-j\omega t_o} = 1\angle - \omega t_o \tag{4.27}$$

Therefore, no contribution is made in gain characteristic, while a linear function of frequency with negative slope is the phase contribution (Fig. 4.9).

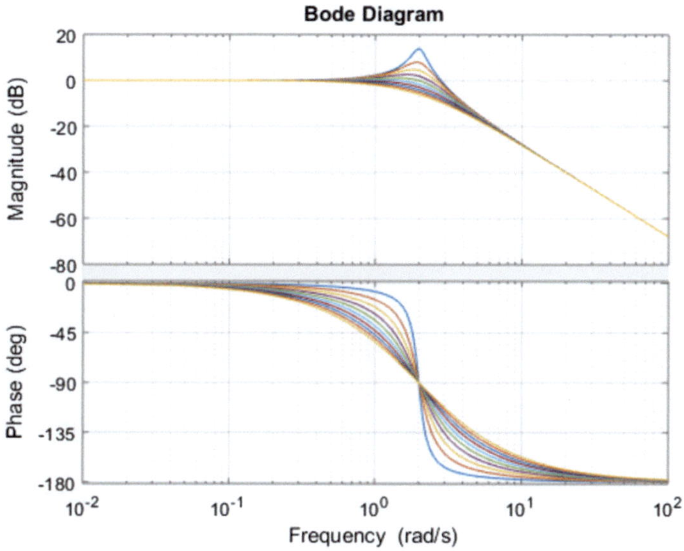

Fig. 4.8 Bode plot of a complex pole at $\omega_n = 1$

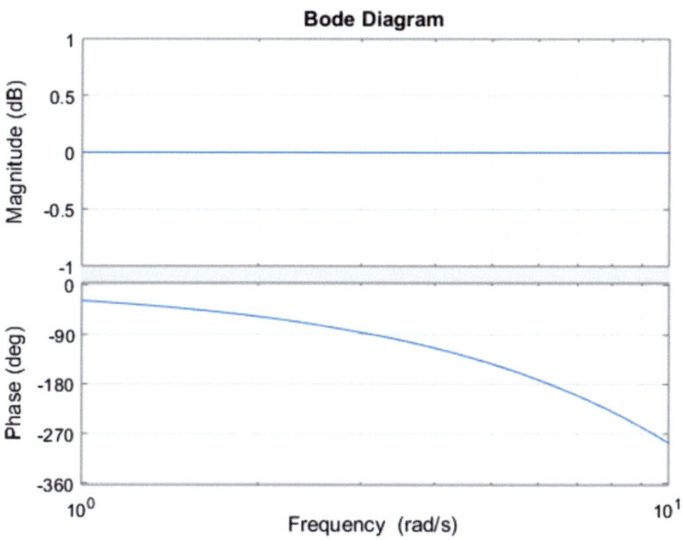

Fig. 4.9 Bode plot of a single delay of $t_o = 0.5$ s

Example 4.2

Draw the Bode plot of the analog electric circuit of the global respiratory dynamics for increasing values of $R = R_1 = 0.01$, $R_2 = 0.1$ and $R_3 = 1$ while maintaining values of $L = 0.01$ and $C = 0.1$.

Solution

As the frequency response of the circuit was defined as

$$G(j\omega) = \frac{1}{LC(j\omega)^2 + RCj\omega + 1} = \frac{1}{\left(\frac{j\omega}{\omega_n}\right)^2 + 2\xi\left(\frac{j\omega}{\omega_n}\right) + 1} \tag{4.28}$$

Then by identifying terms, we have

$$\omega_n = \frac{1}{\sqrt{LC}} = 31.62\ \frac{\text{rad}}{\text{s}} \tag{4.29}$$

$$\xi_1 = \frac{R_1 C\omega_n}{2} = 0.0158 \quad \xi_2 = 0.1581 \quad \xi_3 = 1.5810$$

A list of MATLAB commands is made so as to display the Bode diagram (Fig. 4.10)

```
≫s=zpk('s');
≫R=0.01;L=0.01;C=0.1;

≫for i=1:3;
  G=1/(L*C*s^2+R*C*s+1);
  bode(G)
```

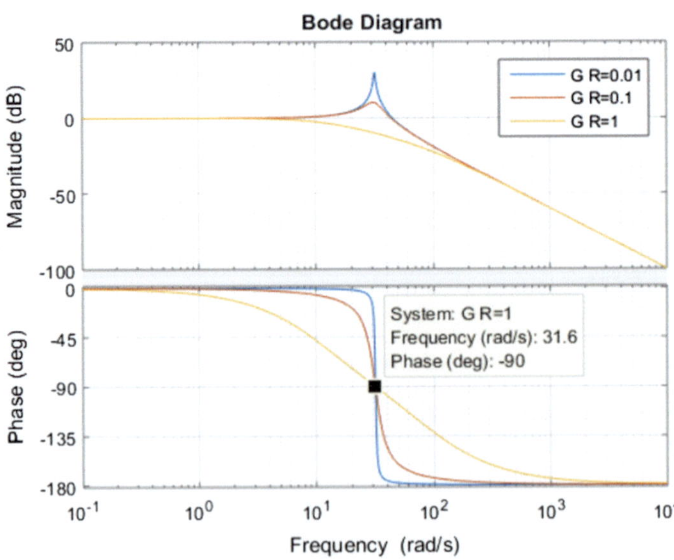

Fig. 4.10 Bode plot of the model of global respiratory dynamics for variable R

```
    R = 10*R
    hold on
≫end
```

It can be seen how the amplitudes of the output pressure $P_a(t)$ decrease as R increases when it applied a sinusoidal input pressure $P_i(t)$ (as it could happen in asthmatic patients) since the system gain is reduced in dB mainly around frequency $\omega_n = \frac{1}{\sqrt{LC}}$ =31.68 and a sudden delay can occurred at this frequency.

Frequency Response Analysis of Discrete-Time Systems

The frequency response of discrete-time systems has the same meaning of that of continuous-time systems. Thus, following the same approach as before, considering a discrete sinusoidal input with amplitude U and frequency ω_0 as

$$u(kT) = U\sin(\omega_0 kT) \tag{4.30}$$

the steady-state response will be also sinusoidal with same frequency, amplitude Y and phase shift as

$$y(kT) = Y\sin(\omega_0 kT + \phi) \tag{4.31}$$

Considering a linear time-invariant dynamic discrete-time system given by its transfer function $G(z)$, both Z transforms of input and output of the system will be related by

$$Y(z) = G(z)\,U\,(z) \tag{4.32}$$

and replacing $z = e^{sT}|_{s=j\omega} = e^{j\omega T}$ so as to make a steady-state analysis, then

$$Y\left(e^{j\omega T}\right) = G\left(e^{j\omega T}\right)\,U\,\left(e^{j\omega T}\right) \tag{4.33}$$

$$G\left(e^{j\omega T}\right) = \left|G\left(e^{j\omega T}\right)\right|\,\angle G\left(e^{j\omega T}\right) \tag{4.34}$$

The amplitude and phase delay of the sinusoidal output signal $y(kT)$ then will be given similarly by the relation

$$\frac{Y}{X} = \left|G\left(e^{j\omega_o T}\right)\right| \quad \phi = \angle G\left(e^{j\omega_o T}\right) \tag{4.35}$$

The complex function $G(e^{j\omega T})$ is the discrete frequency response of the system and predicts how the system will respond to sinusoidal discrete-time inputs of variable frequency at each sampling time kT, obtained by substituting $z = e^{j\omega T}$ into

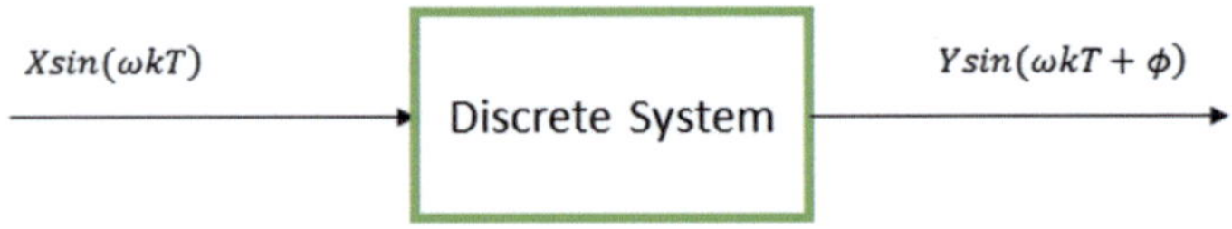

Fig. 4.11 Frequency response of a discrete-time linear system

the system transfer function $G(z)$. In fact, $|G(e^{j\omega T})|$ represents the gain and $\angle G(e^{j\omega T})$ the phase difference of the system when an input sinusoidal discrete-time input with ω frequency is applied (Fig. 4.11).

Bode Diagrams of Discrete-Time Systems

Discrete Bode plots can also be obtained by applying the logarithmic scale as it was done for continuous systems. It should be noted that discrete Bode plots are periodic with period ω_s since

$$G\left(e^{j(\omega+\omega_s)T}\right) = G\left(e^{j\omega T}e^{j2\pi}\right) = G\left(e^{j\omega T}\right) \tag{4.36}$$

so Bode characteristics are not drawn above the Nyquist frequency $\frac{\omega_s}{2}$ because both of symmetry of the frequency response and of lack of compliance of sampling theorem. Indeed, is commonly applied the bilinear transformation is commonly applied by mapping z-plane to w-plane, as it was described formerly as

$$z = \frac{1+\frac{T}{2}w}{1-\frac{T}{2}w} \rightarrow w = \frac{2}{T}\frac{z-1}{z+1} \tag{4.37}$$

which transformed the unit circle in z-plane $z = e^{j\omega T}$ into the imaginary axis in w-plane since

$$w = \frac{2}{T}\frac{e^{j\omega T}-1}{e^{j\omega T}+1} = j\frac{2}{T}\tan\left(\frac{\omega T}{2}\right) = jv \tag{4.38}$$

so Bode diagram of $G(w)|_{w=jv}$ is usually plotted instead of that of $G(z)|_{z=e^{j\omega T}}$. In this way, the gain of the frequency response is thus expressed in decibels (dB) as

$$|G(jv)|_{\text{dB}} = 20\log_{10}|G(jv)| \tag{4.39}$$

while the phase characteristic will be given by $\angle G(jv)$. The relation between the transformed frequency v and the original frequency ω in w-plane and s-plane, respectively (Fig. 4.12), is given by

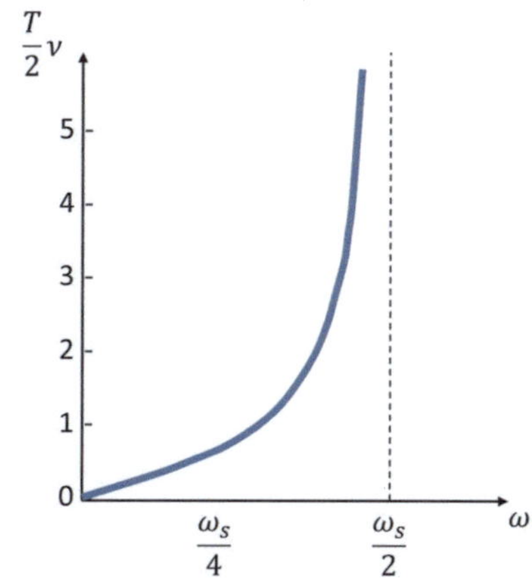

Fig. 4.12 Relation between frequency in w-plane and s-plane

$$v = \frac{2}{T}\tan\left(\frac{\omega T}{2}\right) 0 \leq \omega \leq \frac{\omega_s}{2} \tag{4.40}$$

Therefore, it should be noted that the Bode $G(jv)$ in w-plane presents certain particularities. Firstly, $\lim_{v\to\infty} G(jv)$ is constant and corresponds to $\lim_{\omega\to\omega_s/2} G(j\omega)$ since ω is limited to $0 \leq \omega \leq \frac{\omega_s}{2}$ and then $0 \leq v \leq \infty$. Besides, $G(w)$ is in general a non-minimum phase transfer function, which exhibits zeros with negative phase characteristic. As there is a frequency distortion between s-plane and w-plane, it would be recommended to settle the working domain of $\omega < \frac{\omega_s}{10}$ since in this case

$$v = \frac{2}{T}\tan\left(\frac{\omega T}{2}\right) \cong \omega \tag{4.41}$$

and frequency distortion in w-plane would be diminished.

Example 4.3
Draw the discrete Bode plot in z-plane and w-plane of the discretized model of the analog electric circuit of the global respiratory dynamics for $R = 0.01, L = 0.01$ and $C = 0.1$

Solution
As the transfer function of the circuit was defined as

$$G(s) = \frac{1}{LCs^2 + RCs + 1} = \frac{1}{0.001s^2 + 0.001s + 1} = \frac{1000}{s^2 + s + 1000} \tag{4.42}$$

with $\omega_n = \frac{1}{\sqrt{LC}} = 31.62 \frac{\text{rad}}{\text{s}}$. Therefore, the sampling time T can be selected to satisfy $\omega_n < \frac{\omega_s}{10}$, then $\omega_s > 10w_n$ so $T < \frac{2\pi}{10\omega_n}$, that is $T < 0.02$.

The discretized analog electric circuit with $T = 0.01$ will be given by

$$G(z) = Z\{G_{H0}(s)G(s)\} = \left(1 - z^{-1}\right)Z\left\{\frac{G(s)}{s}\right\} \tag{4.43}$$

$$\begin{aligned} G(z) &= \left(\frac{z-1}{z}\right)Z\left\{\frac{1}{s(0.001s^2 + 0.001s + 1)}\right\} \\ G(z) &= \frac{0.04942z + 0.04926}{z^2 - 1.891z + 0.99} \end{aligned} \tag{4.44}$$

Then by applying the bilinear transformation on $G(z)$, we get

$$G(w) = G(z)|_{z=\frac{1+0.05w}{1-0.05w}} = \frac{0.04942\left(\frac{1+0.05w}{1-0.05w}\right) + 0.04926}{\left(\frac{1+0.05w}{1-0.05w}\right)^2 - 1.891\left(\frac{1+0.05w}{1-0.05w}\right) + 0.99} \tag{4.45}$$

$$G(w) = \frac{-4.251e - 05w^2 - 5.076w + 1017}{w^2 + 1.025w + 1017} \tag{4.46}$$

A list of MATLAB commands is made so as to display the discrete Bode diagram both in s-plane, z-plane, and w-plane for comparison purposes (see Fig. 4.13).

```
≫s=tf('s');
≫R=0.01;L=0.01;C=0.1
```

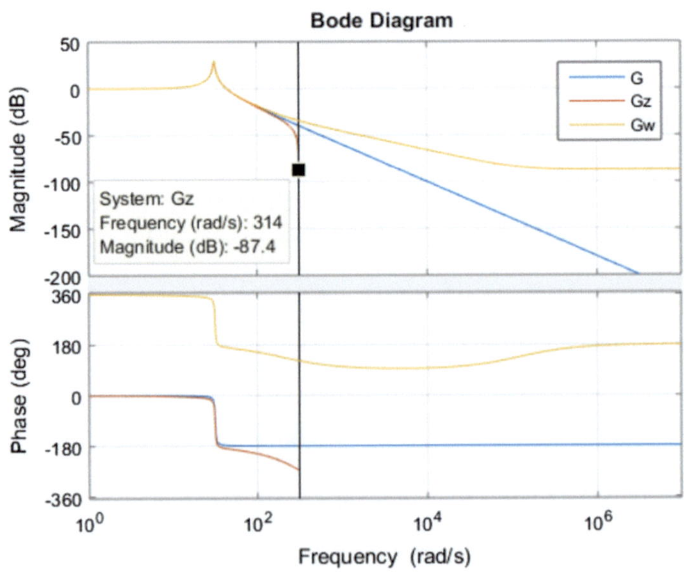

Fig. 4.13 Bode plot of the model of global respiratory dynamics in s-plane, z-plane and w-plane

```
≫G=1/(L*C*s^2+R*C*s+1);

≫T=0.01;
≫Gz=c2d(G,T,'zoh');
≫Gw=d2c(Gz,'tustin');

≫bode(G,Gz,Gw)
```

The comparative Bode plot shows the increasing discrepancies between Bode in s-plane with those of w-plane and z-plane with increasing frequency, this latter limited to the Nyquist frequency $\frac{\omega_s}{2} = \frac{\pi}{T} = 314$ rad/s.

Relation Between Transient and Frequency Response

In the design of control systems using the frequency response analysis, it is necessary to define a set of frequency response features starting from the Bode plots of the dynamic systems. These frequency response features are tightly related to those of transient response already described.

Considering the closed-loop frequency response of a linear and invariant time system as shown in Fig. 4.4 with transfer function

$$M(j\omega) = |M(j\omega)| \angle M(j\omega) = \frac{KG(j\omega)}{1 + KG(j\omega)H(j\omega)} \tag{4.47}$$

and particularizing for a prototype of a second-order system with specific values of damping ratio ξ and undamped natural frequency ω_n as

$$M(j\omega) = \frac{Y(j\omega)}{R(j\omega)} = \frac{\omega_n^2}{s^2 + 2\xi\omega_n s + \omega_n^2} \tag{4.48}$$

the following frequency response features will be next defined:

- **Peak Resonance**. It is defined as the maximum value of $|M(j\omega)|$ and gives an indication of the relative stability of the closed-loop system. This is an easily measurable parameter which can be related to ξ through

$$M_p = \frac{1}{2\xi\sqrt{1-\xi^2}} \qquad \xi \leq 0.707 \tag{4.49}$$

- **Resonant Frequency**. It refers to the frequency at which the peak resonance takes place and is related to ξ and ω_n as

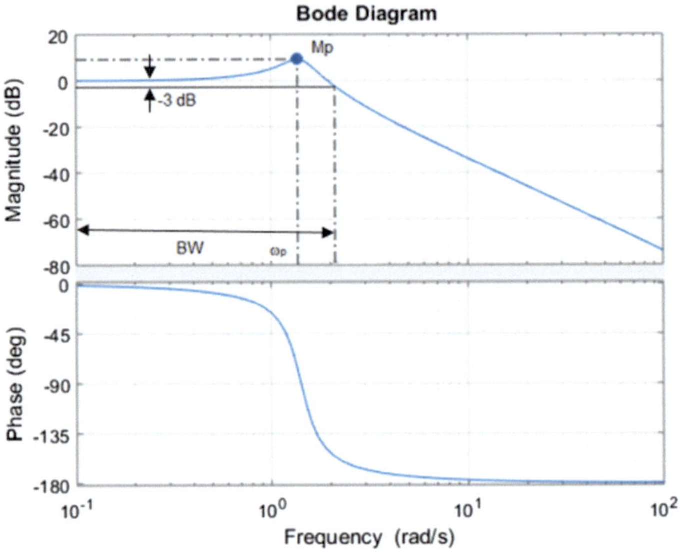

Fig. 4.14 Closed-loop frequency response features

$$\omega_p = \omega_n\sqrt{1-2\xi^2} \quad \xi \leq 0.707 \tag{4.50}$$

- **Bandwidth**. Defined as the frequency at which $|M(j\omega)|$ drops 3 dB down the zero frequency value and gives a measure of the transient response being expressed as

$$\mathrm{BW} = \omega_n\left[\left(1-2\xi^2\right)+\sqrt{4\xi^4-4\xi^2+2}\right]^{1/2} \tag{4.51}$$

A large bandwidth corresponds to a faster transient response, and conversely small bandwidth tends to make the system slower. For a fixed value of natural frequency, the bandwidth decreases as the damping ratio increases.

In Fig. 4.14, it is shown the closed-loop frequency features over the gain characteristic of $M(j\omega)$ Bode.

Example 4.4
In Fig. 4.15, it is revisited a feedback control system designed to regulate the concentration of a drug in the circulatory system $q(t)$ by varying the input dose $i(t)$, by using a sensor to measure concentration with gain K_m and a pump with gain K_p as actuator, as it was similarly described in Example 3.11.

Assuming a transfer function model $T(s) = \frac{0.1}{s^2+0.17s+0.006}$ with $K_p = 0.1$, $K_m = 5$, draw the open-loop Bode plot and closed-loop Bode plots for $K = 1$ and

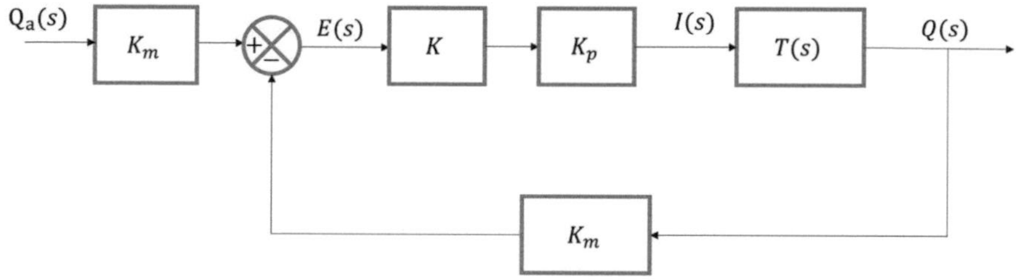

Fig. 4.15 Closed-loop control system for regulation of concentration of drug

$K = 10$ and determine the closed-loop features: peak resonance, resonant frequency and bandwidth.

Solution

In first place, we have to obtain the open-loop transfer function, defined by

$$G(s) = \frac{Q(s)}{E(s)} = KK_pT(s) \tag{4.52}$$

$$G(s) = \frac{0.01K}{s^2 + 0.17s + 0.006} \tag{4.53}$$

while the closed-loop transfer function will be given by

$$\begin{aligned} M(s) &= \frac{Q(s)}{Q_a(s)} = \frac{K_mG(s)}{1 + K_mG(s)} \\ M(s) &= \frac{Q(s)}{Q_a(s)} = \frac{0.05G(s)}{1 + 0.05G(s)} \end{aligned} \tag{4.54}$$

A list of MATLAB commands is made so as to display the Bode diagram both in open loop (Fig. 4.16) and closed loop (Fig. 4.17) for K variable.

```
≫s=tf('s');
≫Km=5;Kp=0.1;
≫K=1;

≫T=0.1/(s^2+0.17*s+0.006);
≫G1=K*Kp*T;

≫K=10;
≫G2=K*Kp*T;

≫bode(G1,G2)
```

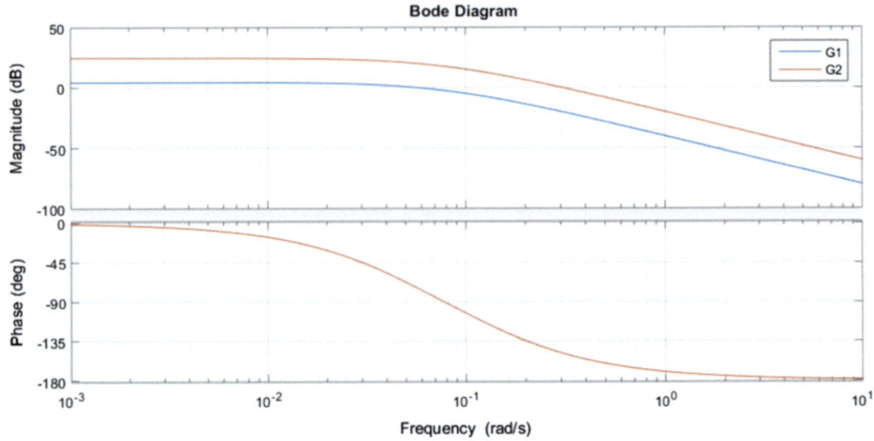

Fig. 4.16 Open-loop frequency response of system for regulation of concentration of drug both for K variable

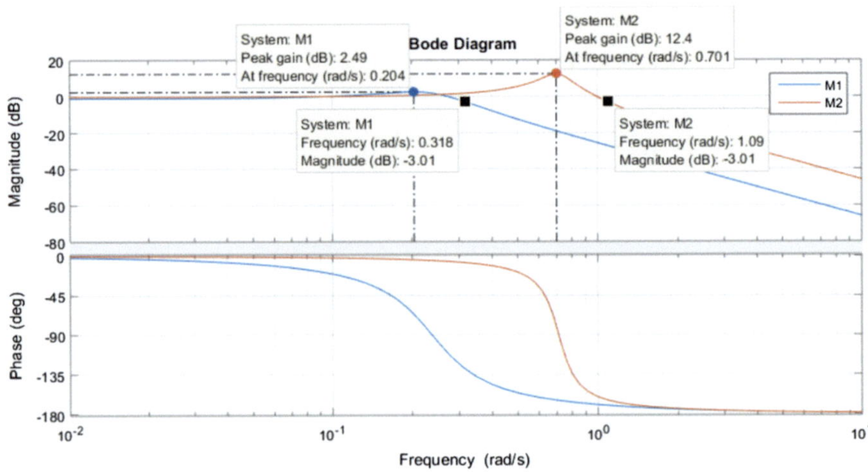

Fig. 4.17 Closed-loop frequency response of system for regulation of concentration of drug both for K variable

```
≫M1=Km*feedback(G1,Km,-1)
≫M2=Km*feedback(G2,Km,-1)

≫bode(M1,M2)
```

The bandwidth of $M_1(j\omega)$ is 0.318 rad/s, while that of $M_2(j\omega)$ is 1.09 rad/s. As far as K increases, the bandwidth of the closed-loop control system increases and therefore its transient response is faster. The peak resonance of $M_1(j\omega)$ is 2.49 dB at resonant frequency 0.204 rad/s, while that of $M_2(j\omega)$ is 12.4 dB at resonant frequency 0.701 rad/s; thus, $M_2(s)$ will exhibit a large maximum overshoot as compared to $M_1(s)$.

Relative Stability of Continuous- and Discrete-Time Systems in the Frequency Domain

When designing a control system, the stability of closed-loop system must be assured. Besides due to the model inaccuracies presented when considering transfer function as a valid representation of the real system, it would be also necessary an adequate relative stability, so it would be necessary assuring not only the closed-loop stability of the system but its degree of stability also.

In the stability analysis that is going to be carried out, the Bode diagram of the open-loop transfer function of the system is going to be used as a means to assess the relative, closed stability both of continuous and of discrete-time systems.

Starting from the continuous closed-loop transfer function of Fig. 4.4 given by

$$M(s) = \frac{Y(s)}{R(s)} = \frac{KK_h G(s)}{1 + KG(s)H(s)} \tag{4.55}$$

the frequency response will be computed as

$$M(j\omega) = |M(j\omega)| \angle M(j\omega) = \frac{Y(j\omega)}{R(j\omega)} = \frac{KK_h G(j\omega)}{1 + KG(j\omega)H(j\omega)} \tag{4.56}$$

so we will work with the Bode diagram of the open-loop transfer function $KG(j\omega)H(j\omega)$ assuming only minimum phase systems. In addition, if unity feedback is considered, the open-loop transfer function coincides with $KG(j\omega)$ being K the controller gain. The critical stability point will be given by

$$\begin{aligned} 1 + KG(j\omega)H(j\omega) &= 0 \rightarrow KG(j\omega)H(j\omega) = -1 \\ -1 &= |1| \angle 180^\circ = 0\ \text{dB} \angle 180^\circ \end{aligned} \tag{4.57}$$

Relative stability of a closed-loop system is usually expressed in terms of phase and gain margins of the open-loop transfer function measured as its relative closeness to the point -1, defined as

1. **Phase Margin**: Is the amount of additional phase lag ϕ in the gain crossover frequency ω_{cg} required to bring the system to the edge of instability, where the gain crossover frequency is the frequency at which magnitude of the open-loop transfer function $KG(j\omega)H(j\omega)$ is unitary (0 dB), that is

$$\begin{aligned} \phi &= 180^\circ + \angle KG(j\omega_{cg})H(j\omega_{cg}) \\ &\left|KG(j\omega_{cg})H(j\omega_{cg})\right|_{dB} = 0 \end{aligned} \tag{4.58}$$

2. **Gain Margin**: Is the reciprocal of the magnitude $|KG(j\omega)H(j\omega)|$ in that frequency denoted as phase crossover frequency ω_{cp} at which the phase angle is -180°, so that the gain margin K_g is defined as

$$
\begin{aligned}
K_g &= \frac{1}{\left|KG(j\omega_{cp})H(j\omega_{cp})\right|} \\
K_{gdB} &= 20\log_{10}K_g = -20\log_{10}\left|KG(j\omega_{cp})H(j\omega_{cp})\right|
\end{aligned}
\tag{4.59}
$$

Both the phase margin and the gain margin can be determined directly form the Bode plot of $KG(j\omega)H(j\omega)$, so that the system will be closed-loop stable when both the phase margin and the gain margin be positive (Fig. 4.18).

Negative stability margins indicate instability; in fact, it is enough that only one is negative to induce instability (Fig. 4.19).

Adequate phase and gain margins ensure against variations in system components and are specified together with defined gain and phase crossover frequency values. The two relative stability margin values define the behaviour of the closed-loop system near the resonant frequency.

To obtain a satisfactory closed-loop behaviour, the phase margin must be between 30° and 60°, and the gain margin must be greater than 6 dB. With these values, a minimum phase system has guaranteed stability, even if the open-loop gain and time constants vary to some degree.

As for relative stability of discrete-time systems, the definitions of phase margin and gain margin of continuous systems apply directly to an open-loop discrete transfer function expressed in the w-plane through the bilinear transformation, as long as the working domain complies to $\omega < \frac{\omega_s}{10}$ to reduce frequency distortion between s-plane and w-plane.

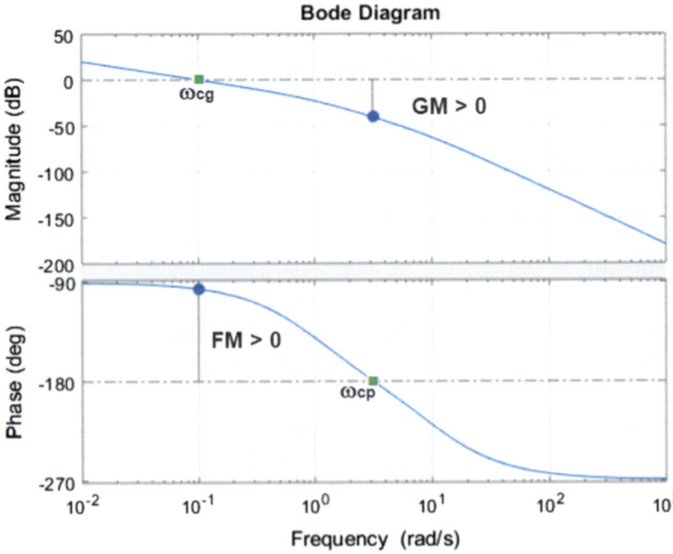

Fig. 4.18 Relative stability margins and corresponding frequencies on a Bode diagram for a stable closed-loop system

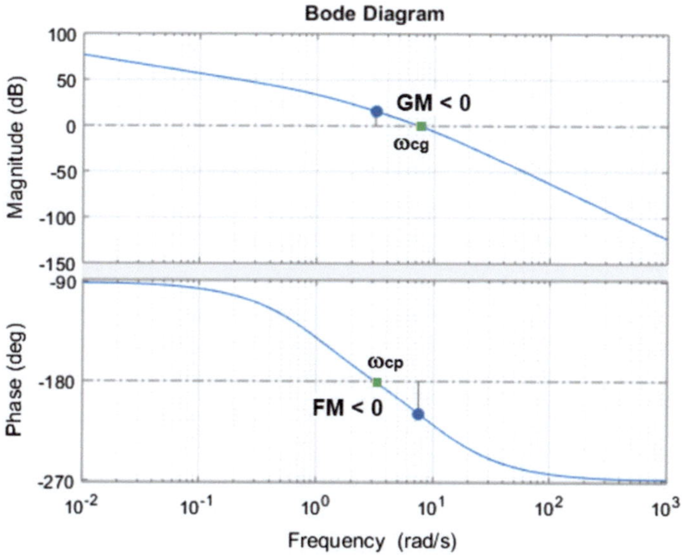

Fig. 4.19 Relative stability margins and corresponding frequencies on a Bode diagram for an unstable closed-loop system

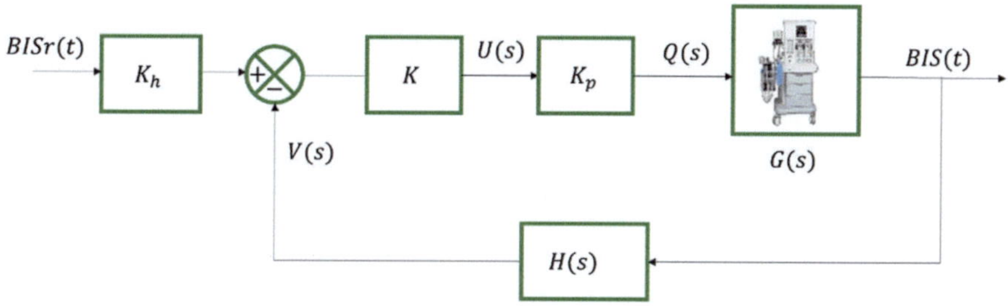

Fig. 4.20 Closed-loop control system for anaesthetic control

Example 4.5

In surgical operations, anaesthesia must be maintained during the entire procedure. The goal is to develop an automatic control system to regulate the depth of anaesthetic measured by the bispectral index (BIS) by using an infusion pump with gain K_p controlled by a gain K to set the infusion rate $Q(s)$ (Fig. 4.20).

The patient dynamics is given by a simplified linear system constituted by a pharmacokinetic system with dynamics

$$G(s) = \frac{BIS(s)}{Q(s)} = \frac{0.1\left(s + \frac{1}{10.6}\right)}{\left(s + \frac{1}{34}\right)\left(s + \frac{1}{3.08}\right)\left(s + \frac{1}{4.8}\right)} \quad (4.60)$$

with sensor dynamics given by first-order dynamics as

$$H(s) = \frac{V(s)}{BIS(s)} = \frac{1}{10s + 1} \tag{4.61}$$

and proportional gains $K_p = 1.2$ and $K_h = 1$, respectively. Obtain the phase and gain margins FM and GM for proportional gain $K = 1$ and derive the critical gain K so that the closed-loop system is marginally stable.

Solution

Firstly, we proceed to derive the open-loop transfer function

$$OL(s) = KK_pG(s)H(s) = \frac{1.2K0.1\left(s + \frac{1}{10.6}\right)}{\left(s + \frac{1}{34}\right)\left(s + \frac{1}{3.08}\right)\left(s + \frac{1}{4.8}\right)} \frac{1}{10s + 1} \tag{4.62}$$

Both phase margin and gain margin will depend on the controller gain K value, starting from a specific value of $K = 1$ and drawing its Bode (Fig. 4.21) and ending searching for $K = K_c$ value which makes the system marginally stable, that is $FM = 0$ and $GM = 0$.

A list of MATLAB commands is shown below to solve this problem.

```
≫s=zpk('s');
≫G=0.1*(s+(1/10.6))/((s+(1/34))*(s+(1/3.08))*(s+(1/4.8)))
G =
```

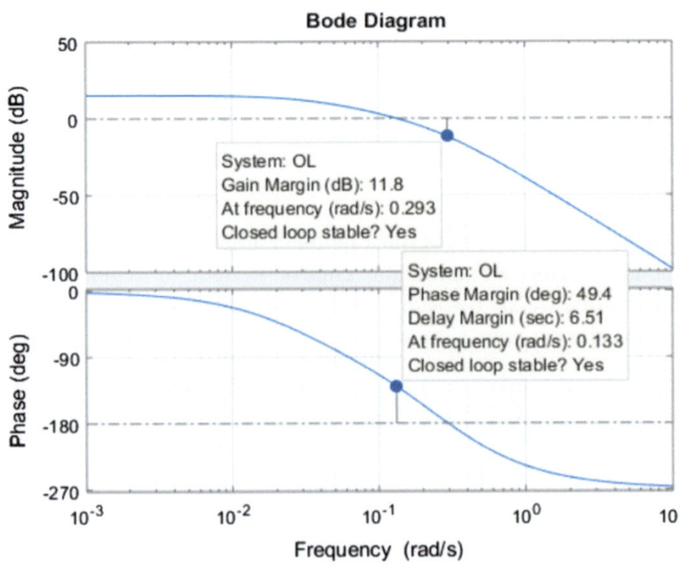

Fig. 4.21 Relative stability margins on Bode diagram for anaesthetic control system

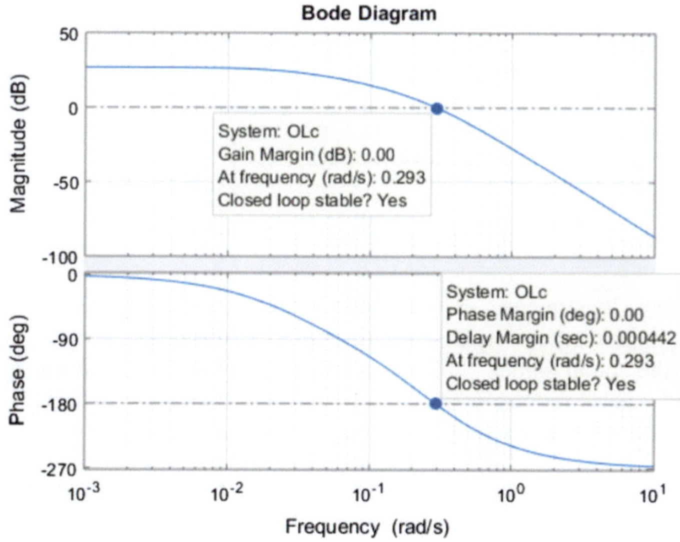

Fig. 4.22 Relative stability margins on Bode diagram for anaesthetic control system under critical stability

```
      0.1 (s+0.09434)
  ---------------------------------------
  (s+0.02941) (s+0.2083) (s+0.3247)
≫H=1/(10*s+1)
≫K=1;Kp=1.2;
≫OL=K*Kp*G*H
≫bode(OL)
```

In order to determine the critical gain K_c, the Bode gain characteristic should be reduced in 11.8 dB, therefore $GM = 11.8\ \text{dB} = 20\log_{10}K$, thus $K = 10^{\frac{1.8}{20}} = 3.8905$ and the corresponding critical Bode can be displayed (Fig. 4.22).

```
≫Kc=3.8905;
≫OLc=Kc*Kp*G*H
≫bode(OLc)
```

Solved Problems

Problem 4.1

A hearing aid is a device that picks up the ambient sound and, after amplification, directs it into the ear canal, trying to correct the possible acoustic impairments of a

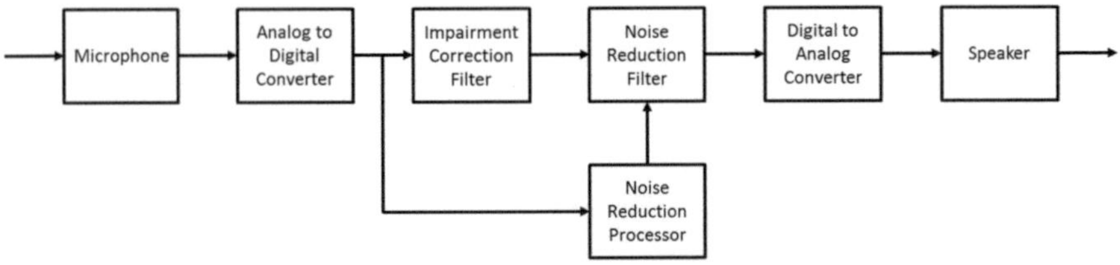

Fig. 4.23 Structure of a hearing aid which is a device

certain patient. Its structure, as it is showed in Fig. 4.23, is made up by a microphone that captures the sound; a digitizer that performs the analog–digital conversion of the sound signal; a filter that compensates the acoustic impairments of the patient; a noise reduction filter (according to the strategy defined by a sound processor); a digital–analog converter; and a speaker.

(a) Design a prototype in your computer to obtain the frequency response of the user's ear. Store it in a file and draw the corresponding result to someone ("patient") with hearing impairments. Repeat the process for another person ("reference") with good hearing whose result will be used to calibrate the system. Compare these results to the threshold level of ISO 226 equal-loudness contours.
(b) Obtain the frequency response of the prototype developed in the previous paragraph using the comparison between the reference response and the standard response (ISO).
(c) Obtain the patient's audiogram. The audiogram depicts the difference between the patient's frequency response and normal (reference) hearing.
(d) Based on the audiogram, obtain the appropriate impairment correction filter for the patient. For a good fit, it is normal to use a large number of coefficients in the filter.
(e) In a given circumstance (conversation in a random noisy environment), the noise reduction algorithm concludes that the corresponding filter must produce attenuation of 5 dB for frequencies below 1 kHz; of 10 dB for frequencies between 1 and 5 kHz; and no attenuation for frequencies above 5 kHz. Obtain the noise reduction filter that suits those circumstances.

Solution 4.1

(a) To obtain the ear's response to a certain frequency, an audio signal of increasing volume is generated, identifying with a mouse-click the moment in which the user detects the sound. To reduce errors, the same sound signal is played repeatedly. This operation is repeated for several frequencies in each of the two ears. The above procedure can be implemented using the following MATLAB code:

```
function main

%% Paragraph a)
TestEarResponse;

% Function to test the ear frequency response
function TestEarResponse

% Defining a volume-increasing sound test signal
SoundLength = 5; %The sound will last 5 seconds
fs = 44100; %Sampling frequency of the sound test signal
t = 0:1/fs:SoundLength; %Time vector for the test signal
nt = length(t); %Number of samples in the test signal
maxSL = 0; %Maximum sound level (in dB)
minSL = -120; 0; %Minimum sound level (in dB)
nfold = 3; %Number of repetitions of each test signal
SL = linspace(minSL,maxSL,nt); %Sound level vector (in dB)
V = db2mag(SL); %Volume of the test signal

% Defining the set of frequencies for the test signals
phon = 0; %Loudness threshold
[~, freq] = iso226(phon); %The set of frequencies defined in ISO226
freq = freq'; %Obtain the frequencies as a column vector
nfreq = length(freq); %Number of frequencies in the test set

% Obtaining the sound level detected for every test signal
Sd = zeros(nfreq,nfold); %Sound level detected matrix
for ear = {'left','right'} %For each ear
    %Draw a graphic with the on-going results
    figure;
    semilogx(freq,ones(nfreq,1),'*');
    ylim([minSL,maxSL]);
    xlim([10,30000]);
    xlabel('Frequency(Hz.)');
    ylabel('Relative Sound Loudness(dB)');
    title(['Ear: ',char(ear)]);
    hold on;

    %Add a button to be pressed when the sound is heard
    btn = uicontrol('Style', 'pushbutton',...
                    'String', 'Press when hear',...
                    'Position', [80 60 100 20]);

    %Test sounds using every frequency
    for iFreq = 1:nfreq
        f = freq(iFreq);
        TestSound = V .* sin (2*pi*f*t); %Test sound
        switch char(ear)
            case 'left'
                channel = 1;
                LeftChannel = TestSound;
                RightChannel = zeros(1,nt);
                color = 'r';
                marker = 'o';
            case 'right'
                channel = 2;
                LeftChannel = zeros(1,nt);
                RightChannel = TestSound;
                color = 'b';
                marker = 'x';
        end
```

```
%Repeat the test several times to reduce errors
for fold = 1:nfold
    %Define the stereo sound to be played
    player = audioplayer([LeftChannel;RightChannel],fs);
    %Define what happen when the "heard" button is pressed
    set(btn,'Callback',  {@PressWhenHear,player});
    %Define what happen if the "heard" button is not pressed
    set(player,'StopFcn',@EndSound);
    play(player); %Start playing the sound

    %Start an infinite loop until the button is pressed
    while 1
        drawnow;
        StopS = get(player,'UserData'); %# of the stop sample
        if StopS %SoptS=0 if the button has not been pressed
            break;
        end
    end %of the infinite loop

    Sd(iFreq,fold) = SL(StopS); %Sound level of stop sample
    %Draw the result for a single test signal
    plot(f,SL(StopS),'Line','none','Marker',marker,...
                    'MarkerEdgeColor',color);
            end %of the repetitions with the same frequency

        end %of the set of frequencies

        %Draw the frequency response for an ear
        semilogx(freq,mean(Sd,2),'Color',color);
        grid on

        %Write a datasheet with the results of a single ear
        xlswrite('audiogram',[freq,Sd],channel);

    end %of each ear


    % Function to define what happen when the button is pressed
    function PressWhenHear(~,~,player)
    pause(player); %Pause the player


    % Function to define what happen when sound is stopped (or paused)
    function EndSound(player, ~)
    StopS = get(player,'CurrentSample');  %Number of the stop sample
    ns = get(player,'TotalSamples'); %Total number of samples
    if StopS == 1
        StopS = ns;
    end
    set(player,'UserData',StopS)
```

In the previous code, to define the frequency set of the sound signals, the values defined in ISO 226 are used. A function that implements this standard can be obtained from the MATLAB® Central File Exchange Web.

The result is an Excel file called "audiogram.xls" which contains two sheets: one for the left ear and one for the right ear. In each sheet, the rows represent the different frequencies; and the columns, the volume of the signal at the time of sound detection in each of its plays (remember that each test sound is repeated several times to reduce errors).

The previous code is executed first time for a normal hearing person, and the obtained file is renamed as "audiogram reference.xls". It is then run second time for a patient with hearing impairments, and the file obtained is renamed as "audiogram patient.xls".

To calibrate the prototype, the frequency response of both ears will be used for the reference person, averaging and smoothing the experimental data (assuming that the prototype has a symmetrical response in the left and right channels).

In order to be able to compare the results with the ISO 226 standard, the best fit of the reference response to that standard is also computed. The procedure has been adding an offset to the normalized response trying its fitting to the reference experimental points. Different offset values are used, choosing the one that minimizes the root mean square error (RMSE) between the ISO response and the experimental reference points. To accomplish these tasks, the following code is added to the `function main`:

```
% Calibrating the prototype
[Fref,Sref,Siso] = ObtainReferenceResponse('audiogram reference');


%Function to calibrate the prototype
function [Fref,Sref,Siso] = ObtainReferenceResponse(RefFileName)
%RefFileName: name of the file with the reference's ear response
%Fref: Frequency vector for the reference
%Sref: Sond Level Reference
%Siso: Sound level reference according to ISO 226

% Reading file with the ears' frequency response data
datasheet = cell(1,2); %Datasheets corresponding to each ear
for channel = 1:2
    datasheet{channel} = xlsread(RefFileName,channel);
end
Fref = datasheet{1}(:,1); %Vector of frequencies
Sd = [datasheet{1}(:,2:end) ,...
      datasheet{2}(:,2:end)]; %Sound level detected (both ears)

% Obtaining the sound level reference response
Sref = mean(Sd,2); %Mean of every value available in both ears
Sref = smooth(Sref,3); %Slight smooth of the reference response

% Obtaining the best fit between the reference and the ISO 226
splISO = iso226(0); %ISO 226 threshold acoustic level
offset = fminsearch(... %Obtains the offset minimizing RMSE
         @(offset)getRMSE(offset,Fref,Sd,splISO)... %Get RMSE
         , 0 ... %Initial value for offset
         );
Siso = splISO' + offset; %Sound level reference according to ISO 226
```

```
% Function to compute the fitting error (RMSE) between
%   - the reference experimental response &
%   - the ISO 226 response for a certain offset
%   Fitting is optimized for a certain range of frequencies
function RMSE = getRMSE(offset,Fref,SRef,Siso)
%offset: Value of the offset to be added to the ISO level
%Fref: Frequency vector for the reference
%Sref: Sond Level Reference
%Siso: Sound level reference according to ISO 226
%RMSE: Root Mean Square Error of the fitting

%Constraint the fitting to a certain range (band) of frequencies
f1 = 100; %Initial frequency of the band
f2 = 5000; %Final frequency of the band
SRef = SRef( Fref>=f1 & Fref<=f2 , :);
Siso = Siso( Fref>=f1 & Fref<=f2 );
Fref = Fref( Fref>=f1 & Fref<=f2);

%Compute the fitting error (RMSE)
nfreq = length(Fref);
nfold = size(SRef,2);
CumError = 0;
for ifreq = 1:nfreq
    for ifold = 1:nfold
        %Fittting error for a single value
        error = SRef(ifreq,ifold) - (Siso(ifreq) + offset);
        CumError = CumError + error^2; %Cummulative cuadratic error
    end
end
N = nfreq * nfold; %Number of fitting errors
RMSE = sqrt(CumError/N); %Root Mean Square Error of the fitting
```

To draw a graph with the frequency response of each ear, both for the reference person and for the patient, the following code can be added to the function main:

```
% Calibrating the prototype
[Fref,Sref,Siso] = ObtainReferenceResponse('audiogram reference');

% Drawing the reference ears' frequency response
DrawEarResponse('audiogram reference',Fref,Sref,Siso);
```

and, additionally, the following MATLAB function is added:

```
% Function to draw the ear frequency response
function DrawEarResponse(FileName,Fref,Sref,Siso)
%FileName: name of the Excel file with the ear response
%Fref: Frequency vector for the reference
%Sref: Sond Level Reference
%Siso: Sound level reference according to ISO 226
```

```
% Reading file with the ears' frequency response data
datasheet = cell(1,2); %Datasheets corresponding to each ear
for channel = 1:2
    datasheet{channel} = xlsread(FileName,channel);
end

% Drawing the sound level detected for each ear
for ear = {'left','right'}
    figure;
    switch char(ear)
        case 'left'
            channel = 1;
            color = 'r';
            marker = 'o';
        case 'right'
            channel = 2;
            color = 'b';
            marker = 'x';
    end

    %Obtaining frequency response data
    freq = datasheet{channel}(:,1); %Vector of frequencies


Sd = datasheet{channel}(:,2:end); %Sound level detected matrix
nfold = size(Sd,2); %Number of repetitions of each test signal

%Draw the frequency response for an ear
semilogx(freq,mean(Sd,2),'Color',color);
hold on;

%Draw the reference frequency response
semilogx(Fref,Sref,'color',[0,0.5,0]);

%Draw the ISO frequency response
semilogx(Fref,Siso,'k');

%Draw the result for every repetition of the test sound
for fold = 1:nfold
     semilogx(freq,Sd(:,fold),'Line','none',...
             'Marker',marker,'MarkerEdgeColor',color);
end %of the repetitions with the same frequency
xlim([10,30000]);
xlabel('Frequency(Hz.)');
ylabel('Relative Sound Loudness(dB)');
title(['Ear: ',char(ear)]);

legend({'Tested ear','Normal ear','ISO threshold'},...
        'Location','North');
end %of each ear
```

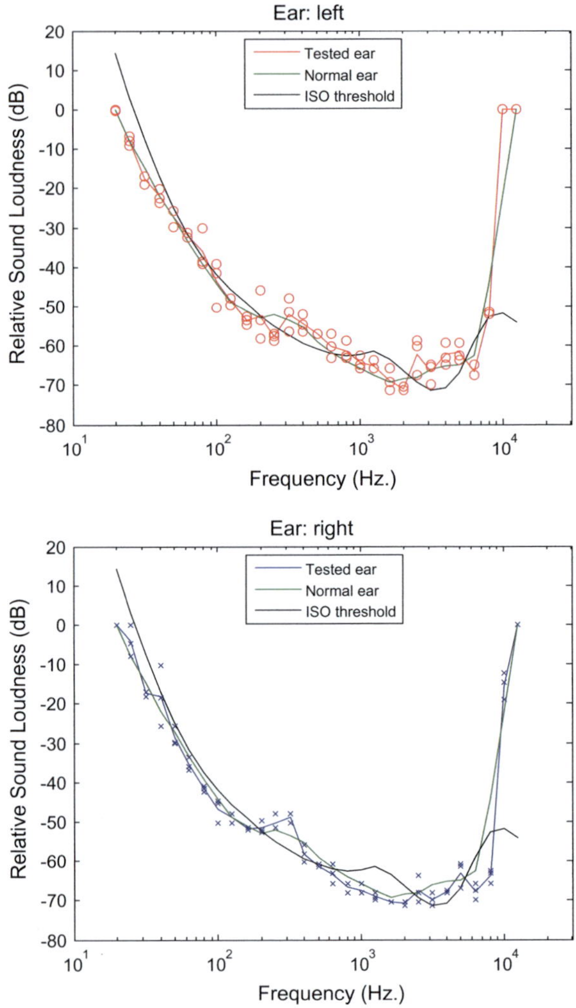

Fig. 4.24 Frequency response of the left ear (top) and right ear (bottom), for the reference person

The obtained result for the reference person is shown Fig. 4.24. The level chosen as the reference and the hearing threshold level according to ISO 226 are also depicted.

For the patient, the obtained results are shown in Fig. 4.25.

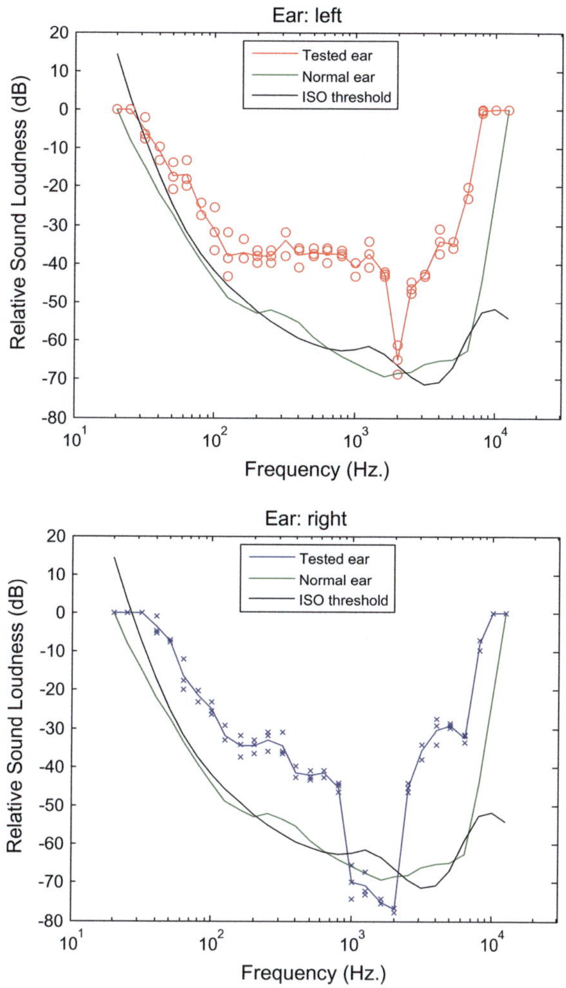

Fig. 4.25 Frequency response of the left ear (top) and right ear (bottom), for the patient

(b) To obtain the frequency response of the system developed in the previous section, the reference person's and the normalized (ISO) frequency response will be compared. For this purpose, the following code is added to the function `main`:

```
%% Paragraph b)
DrawPrototypeResponse(Fref,Sref,Siso);
```

and, additionally, the following MATLAB function is added:

```
% Function to draw the prototype frequency response
function DrawPrototypeResponse(Fref,Sref,Siso)
%Fref: Frequency vector for the reference
%Sref: Sond Level Reference
%Siso: Sound level reference according to ISO 226

figure;
ProtResp = Siso - Sref; %Prototype frequency response
semilogx(Fref,ProtResp,'s-');
xlim([10,30000]);
xlabel('Frequency(Hz.)');
ylabel('Gain(dB)');
title('Prototype frequency response');
```

The obtained result is shown in Fig. 4.26. It can be observed that the used prototype has a good response in almost all the frequency range, although it worsens in the low- and, mainly, in the high-frequency bands. This result is consistent with the most common headphone datasheets.

(c) To obtain the patient's audiogram, the response of their ears is compared with the reference response. To do this, the following code is added to the function main:

```
%% Paragraph c)
audiogram = ObtainAudiogram('audiogram patient',Sref);
DrawAudiogram(Fref,audiogram);
```

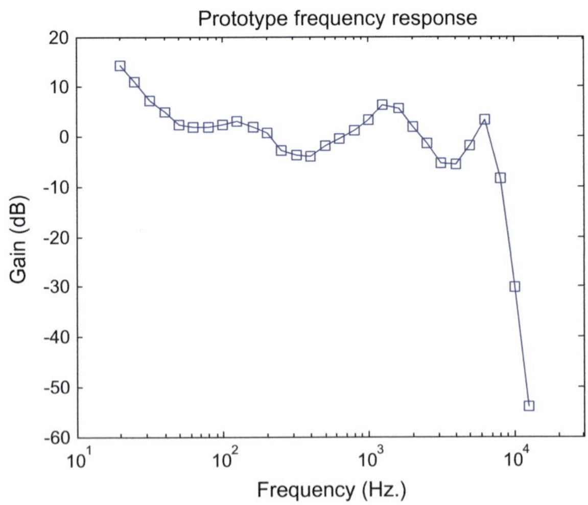

Fig. 4.26 Frequency response of the developed system

and, additionally, the following MATLAB function is added:

```
% Function to obtain the audiogram
function audiogram = ObtainAudiogram(PatientFile,Sref)
%PatienteFile: name of the Excel file with the patient's ear response
%Sref: Sond Level Reference
%audiogram: audiogram data for each ear

% Reading file with the patient's frequency response data
SdPat= cell(1,2); %Sound detected (patient level) for each ear
for channel = 1:2
    datasheet = xlsread(PatientFile,channel);
    Sd = datasheet(:,2:end); %Sound level detected matrix
    SdPat{channel} = mean(Sd,2); %Sound detected (patient level)
end

% Computing the audiogram
audiogram= cell(1,2); %Audiogram for each ear
for channel = 1:2
    audiogram{channel} = Sref - SdPat{channel};
end


% Function to draw the audiogram
function DrawAudiogram(freq,audiogram)
%freq: Vector of frequencies for the audiogram
%audiogram: audiogram data for each ear

figure;
eartxt = {'Left','Right'};
for ear = eartxt
    switch char(ear)
        case 'Left'
            channel = 1;
            color = 'r';
            marker = 'o';
        case 'Right'
            channel = 2;
            color = 'b';
            marker = 'x';
    end
    semilogx(freq,audiogram{channel},'Color',color,...
             'Marker',marker,'MarkerEdgeColor',color);
    hold on;
    xlim([10,30000]);
    xlabel('Frequency(Hz.)');
    ylabel('Hearing Level(dB)');

end
legend({'Left ear','Right ear'},'Location','South');
```

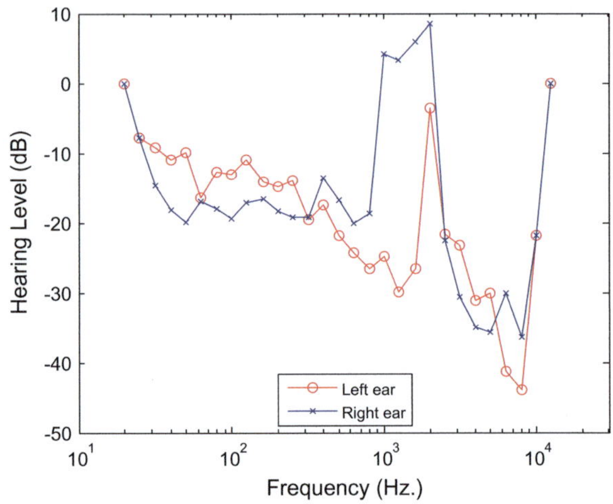

Fig. 4.27 Patient's audiogram

The obtained result is shown in Fig. 4.27.

(d) As the impairment correction filter should be used then it must be chosen just one that compensates the patient's audiogram, that is, its symmetric with respect to the 0 dB axis. Since the shape of the filter is complex, a good design will require a relatively high-order filter. It is common in hearing aids to contain correction filters with several hundred coefficients. The design of the filter can be obtained by adding the following code to the `function main`:

```
%% Paragraph d)
FiltOrd = 300; %Filter order
ObtainCompensationFilter(Fref,audiogram,FiltOrd);
```

and, additionally, the following MATLAB function is added:

```
% Function to obtain the compensation filter
function ObtainCompensationFilter(freq,audiogram,FiltOrd)
%freq: Vector of frequencies for the audiogram
%audiogram: audiogram data for each ear

eartxt = {'Left','Right'};
for ear = eartxt
    %Draw the ideal filter frequency response
    figure;
    switch char(ear)
        case 'Left'
            channel = 1;
            color = 'r';
            marker = 'o';
        case 'Right'
            channel = 2;
            color = 'b';
            marker = 'x';
    end
    H = - audiogram{channel}; %Filter frequency response
    semilogx(freq,H,'Color',color,'Marker',marker,...
                'MarkerEdgeColor',color);
    hold on;
    xlim([10,30000]);
    xlabel('Frequency(Hz.)');
    ylabel('Filter Gain (dB)');
    title(['Ear: ',char(ear)]);

    %Obtain a digital filter to fit the ideal filter
    freqN = [0; freq / max(freq)]; %Normalized frequency [0,1]
    gainN = [0;db2mag(H)]; %Filter gain
    FilSpec = fdesign.arbmag...
        ('N,F,A',FiltOrd,freqN,gainN); %Filter specifications
    Hd = design(FilSpec); %Obtain the digital filter

    %Draw the designed filter frequency response
    [h,w] =freqz(Hd); %Frequency response of the designed filter
    ff = w/pi*max(freq); %Denormalized frequency vector
    hf = mag2db(abs(h)); %Designed filter gain (dB)
    ff = ff( ff>=min(freq) ); %Discard non desired frequencies
    hf = hf( ff>=min(freq) ); %Discard non desired gains
    semilogx(ff,hf,'color','k');
    legend({'Required','Designed'},'Location','Best');

end
```

The obtained result is shown, for each ear, in Fig. 4.28.

(e) To obtain the noise reduction filter that satisfies the proposed specifications, the following code is added to the `function main`:

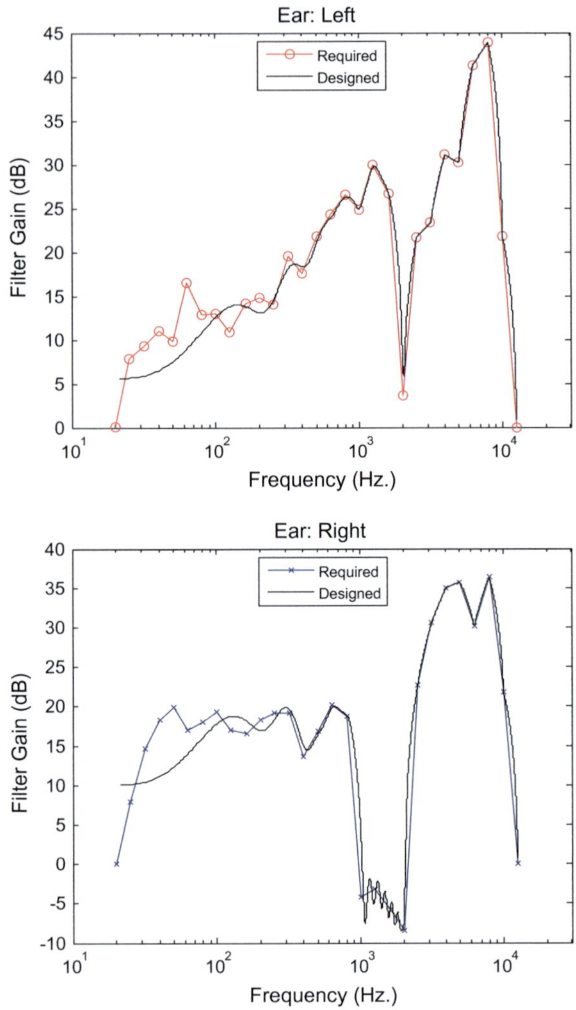

Fig. 4.28 Obtained results for each ear, using the implemented filter

```
%% Paragraph e)
FiltOrd = 300; %Filter order
freq = [20,1000,1001,5000,5001,12500]'; %Frequency vector
NoiseReduction = [5,5,10,10,0,0]'; %Noise reduction (dB)
ObtainNoiseFilter(freq,NoiseReduction,FiltOrd);
```

and, additionally, the following MATLAB function is added:

```
% Function to obtain the noise filter
function ObtainNoiseFilter(freq,NoiseReduction,FiltOrd)
%freq: Vector of frequencies for the audiogram
%NoiseReduction: noise reduction gain values

%Draw the ideal filter frequency response
figure;
H = - NoiseReduction; %Filter frequency response
semilogx(freq,H,'m');
hold on;
xlim([10,30000]);
xlabel('Frequency(Hz.)');
ylabel('Filter Gain (dB)');

%Obtain a digital filter to fit the ideal filter
freqN = [0; freq / max(freq)]; %Normalized frequency [0,1]
gainN = [0;db2mag(H)]; %Filter gain
FilSpec = fdesign.arbmag...
    ('N,F,A',FiltOrd,freqN,gainN); %Filter specifications
Hd = design(FilSpec); %Obtain the digital filter

%Draw the designed filter frequency response
[h,w] =freqz(Hd); %Frequency response of the designed filter
ff = w/pi*max(freq); %Denormalized frequency vector
hf = mag2db(abs(h)); %Designed filter gain (dB)
ff = ff( ff>=min(freq) ); %Discard non desired frequencies
hf = hf( ff>=min(freq) ); %Discard non desired gains
semilogx(ff,hf,'color','k');
legend({'Required','Designed'},'Location','Best');
```

The obtained result is shown in Fig. 4.29.

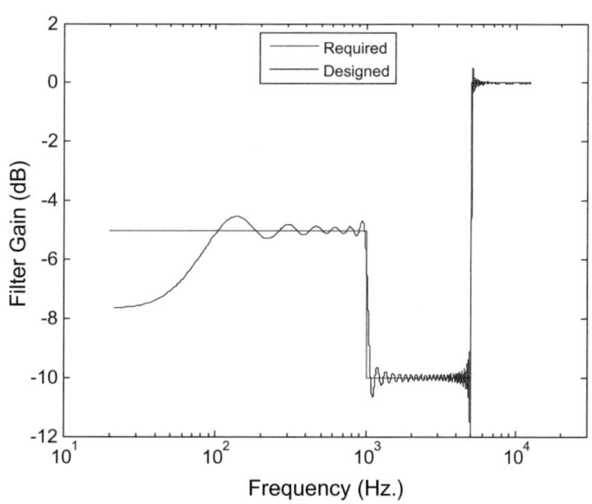

Fig. 4.29 Comparison of the results using the designed and required filters

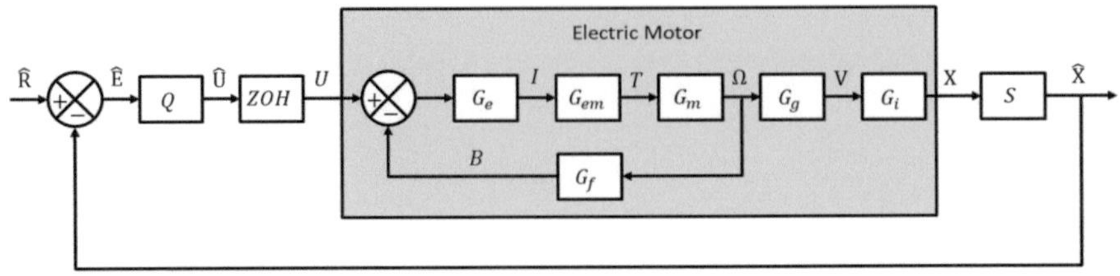

Fig. 4.30 Structure of the positioning control of the axes of a X-ray machine

Problem 4.2

A certain X-ray machine used for radiology has a 3-axis Cartesian movement control, operated by an electric motor in each axis. Machine's movements take place each time on a single axis. The positioning of one of the axes corresponds to the block diagram of Fig. 4.30.

The electric motor, including the load, is defined by an electric block (G_e) which transforms the electric voltage at the motor input (U), jointly with the back-electromotive force generated by the rotation of the motor (B), in an electric current (I); an electromechanical block (G_{em}) that transforms the current into a mechanical torque (T); a mechanical block (G_m) that converts the torque into a rotational speed (Ω); a feedback block (G_f) which transforms the rotational speed into a back-electromotive force (B); a gear block (G_g) which transforms the rotational speed into a linear velocity (V); and an integrator block (G_i) converting the linear speed into a position of the machine axis (X). The transfer functions are

$$G_e = \frac{1}{Ls+R}; G_{em} = K_T; G_m = \frac{1}{Js+B}; G_f = K_e; G_g = E; G_i = \frac{1}{s}$$

$$J = J_m + E^2 M_L; B = B_m + E^2 B_L$$

The motor is digitally controlled by a sampler (S); a zero-order hold (ZOH); and a proportional controller ($Q = K_v$). Obtain:

(a) The continuous transfer function of the motor $G(s)$, including the load.
(b) The discrete transfer function of the motor once discretized $\hat{G}(z)$, when the sampling time $T = 0.1$ s.

(c) The bilinear transform $\tilde{G}(w)$ of the motor's discrete transfer function.

(d) The frequency response of the motor, considering the transfer functions of the three previous paragraphs.

(e) The frequency response of the complete system.

(f) The step response and the frequency response of the complete system when the sampling time is $T = 0.1, 0.2, 0.5$ s.

Data:
Motor electrical inductance: $L = 0.87$ mH.
Motor electrical resistance: $R = 3.05\,\Omega$.
Motor torque constant: $K_T = 0.0622\,\text{N} \cdot \text{m/A}$.
Motor voltage constant: $K_e = 0.062\,\text{V} \cdot \text{s/rad}$.
Motor rotational inertia: $J_m = 6.79 \cdot 10^{-6}\,\text{N} \cdot \text{m} \cdot \text{s}^2/\text{rad}$.
Motor viscous friction: $B_m = 6.306 \cdot 10^{-6}\,\text{N} \cdot \text{m} \cdot \text{s/rad}$.
Gear ratio: $E = 1/300$ m/rad.
Load mass: $M_L = 100$ kg.
Load viscous friction: $B_L = 0.5\,\text{N} \cdot \text{s/m}$.
Controller proportional constant: $K_v = 48$ V/m.

Solution 4.2

(a) The continuous transfer function of the motor, including the load, is

$$G(s) = \frac{X(s)}{U(s)} = \frac{X(s)}{\Omega(s)} \cdot \frac{\Omega(s)}{U(s)} = \frac{G_1(s)}{1 + G_1(s)G_b(s)} \cdot \left[G_g(s)G_i(s)\right],$$

where $G_1(s) = G_e(s)G_{em}(s)G_m(s)$. To derive this transfer function, the following MATLAB code can be used:

```
function main

%% Paragraph a)
%Obtaining electric motor continuous transfer function

%Motor electrical constants
L = 0.87E-3; %Electrical inductance (H)
R = 3.05; %Electrical resistance (ohms)
Kt = 62.2E-3; %Torque constant (N·m/A)
Ke = 0.062; %Voltage constant (V·s/rad)
%Motor mechanical constants
Jm = 6.79E-6; %Motor rotational inertia (N·m·s^2/rad)
Bm = 6.306E-6; %Motor viscous friction (N·m·s/rad)
E = 1/300; %Gear ratio (m/rad)
%Load contants
ML = 100; %Load mass (Kg)
BL = 0.5; %Load viscous friction (N·s/m)
%Motor-Load joint mechanical constants
J = Jm + E^2*ML; %Rotational inertia
B = Bm + E^2*BL; %Viscous friction
%Controller constants
Kv = 48; %Proportional constant (V/m);

%Defining transfer functions of every block
s = tf('s');
Ge = 1/ (L*s+R); %Electrical section transfer function
Gem = Kt; %Electromechanical transfer function
Gm = 1 / (J*s+B); %Mechanical transfer function
Gf = Ke; %Back electromotive force transfer function
Gg = E; %Gear transfer function
Gi = 1/s; %Integrator transfer function
Q = Kv; %Controller transfer function

%Obtaining motor transfer function
G = feedback(Ge*Gem*Gm , Gf) * Gg * Gi; %Motor transfer function
G = minreal (G);
display(G);
```

The obtained result is the following

```
G =

      213.2
  ---------------------------
  s^3 + 3506 s^2 + 4002 s
```

that is

$$G(s) = \frac{213.2}{s^3 + 3506s^2 + 4002s}$$

(b) The discrete transfer function of the motor can be defined as

$$\hat{G}(z) = \frac{\hat{X}(z)}{\hat{U}(z)}$$

To obtain it, the following MATLAB code can be added

```
%% Paragraph b)
%Obtaining electric motor discrete transfer function
T = 0.1; %Sampling time
Gh = c2d(G,T); %Discretize the motor transfer function
display(Gh);
```

The obtained result is the following

```
Gh =
 0.0002913 z^2 + 0.0002836 z + 4.419e-09
 ---------------------------------------------
 z^3 - 1.892 z^2 + 0.8921 z - 4.836e-20
```

that is

$$\hat{G}(z) = \frac{2.91 \times 10^{-4} z^2 + 2.84 \times 10^{-4} z + 4.42 \times 10^{-9}}{z^3 - 1.89 z^2 + 0.89 z - 4.84 \times 10^{-20}}$$

(c) Bilinear transform obtains an approximate value of a discrete system transfer function, using approximations of the variable z, such as

$$z \equiv e^{sT} = \frac{e^{s\frac{T}{2}}}{e^{-s\frac{T}{2}}}$$

Recalling that the series expansion of the exponential function is

$$e^x = 1 + x + \frac{x^2}{2!} + \frac{x^3}{3!} + \cdots \approx 1 + x,$$

it can be written that

$$z = \frac{e^{s\frac{T}{2}}}{e^{-s\frac{T}{2}}} \approx \frac{1 + \frac{sT}{2}}{1 - \frac{sT}{2}} \equiv \tilde{z}(s)$$

Bilinear transform of a discrete system transfer function $X(z)$ is defined as

$$\tilde{X}(w) = \mathcal{B}[X(z)] \equiv X(z)|_{z=\tilde{z}(w)} = X(z)|_{z=\frac{1+\frac{wT}{2}}{1-\frac{wT}{2}}}$$

In this case,

$$\tilde{G}(w) = \mathcal{B}\left[\hat{G}(z)\right] \equiv \hat{G}(z)\Big|_{z=\tilde{z}(w)} = \hat{G}(z)\Big|_{z=\frac{1+\frac{wT}{2}}{1-\frac{wT}{2}}}$$

$$\tilde{G}(w) = \frac{2.91 \times 10^{-4}\left(\frac{1+\frac{wT}{2}}{1-\frac{wT}{2}}\right)^2 + 2.84 \times 10^{-4}\left(\frac{1+\frac{wT}{2}}{1-\frac{wT}{2}}\right) + 4.42 \times 10^{-9}}{\left(\frac{1+\frac{wT}{2}}{1-\frac{wT}{2}}\right)^3 - 1.89\left(\frac{1+\frac{wT}{2}}{1-\frac{wT}{2}}\right)^2 + 0.89\left(\frac{1+\frac{wT}{2}}{1-\frac{wT}{2}}\right) - 4.84 \times 10^{-20}}$$

To obtain this transfer function, the following MATLAB code is added

```
%% Paragraph c)
%Obtaining bilinear transform of electric motor transfer function
Gt = BilinearTransform(Gh);
display(Gt);

%Obtain the bilinear transform of a transfer function
function Y = BilinearTransform(X)
%X: Input transfer function X(z)
%Y: Output transfer function Y(W)

%Symbolic computation of Gt(W)
syms z W
[numX,denX] = tfdata(X,'v'); %Numerator & denominator of X(z)
T = X.Ts;
XSym = poly2sym(numX,z) / poly2sym(denX,z); %Symbolic X(z)
YSym = subs(XSym,z, (1+T/2*W) / (1-T/2*W) ); %Symbolic Y(W)
[numYSym,denYSym]  = numden(YSym); %Num. & denom. of Y(z)

%Transfer function form of Y(W)
numY = sym2poly(numYSym); %Numerator of Y(W)
denY = sym2poly(denYSym); %Denominator of Y(W)
Y = tf(numY,denY,T); %Y(W) (the "z" should be read as "W")
Y = minreal(Y);
```

The obtained result is the following

```
Gt =
  -2.028e-06 z^3 - 0.003038 z^2 + 0.0008095 z + 1.215
  ---------------------------------------------------
            z^3 + 21.14 z^2 + 22.82 z
```

that is

$$\tilde{G}(w) = \frac{-2.03 \times 10^{-6} w^3 - 3.04 \times 10^{-3} w^2 + 8.10 \times 10^{-4} w + 1.22}{w^3 + 21.14 w^2 + 22.82 w}$$

(d) Recalling that $s = \sigma + j\omega$, the motor frequency response can be obtained as

$$G(\omega) \equiv G(s)|_{\sigma=0} = G(s)|_{s=j\omega}.$$

In this case

$$G(\omega) = G(s)|_{s=j\omega} = \left. \frac{213.2}{s^3 + 3506 s^2 + 4002 s} \right|_{s=j\omega}.$$

$$G(\omega) = \frac{213.2}{(j\omega)^3 + 3506 (j\omega)^2 + 4002 (j\omega)}$$

$$G(\omega) = \frac{213.2}{-j\omega^3 - 3506\omega^2 + 4002 j\omega}$$

$$G(\omega) = \frac{213.2}{-3506\omega^2 + j(-\omega^3 + 4002\,\omega)}$$

So, the frequency response $G(\omega)$ is a complex function of the frequency and, therefore, two elements versus ω can be plotted: magnitude and phase.

The same procedure can be performed for the discrete system. Recalling that $z = e^{sT}$, its frequency response is obtained as

$$\hat{G}(\omega) \equiv \hat{G}(z)\big|_{s=j\omega} = \hat{G}(z)\big|_{z=e^{j\omega}}$$

For the electric motor

$$\hat{G}(\omega) = \hat{G}(z)\big|_{z=e^{j\omega}} = \frac{2.91\times 10^{-4}z^2 + 2.84\times 10^{-4}z + 4.42\times 10^{-9}}{z^3 - 1.89z^2 + 0.89z - 4.84\times 10^{-20}}\Bigg|_{z=e^{j\omega}}$$

$$\hat{G}(\omega) = \frac{2.91\times 10^{-4}(e^{j\omega})^2 + 2.84\times 10^{-4}(e^{j\omega}) + 4.42\times 10^{-9}}{(e^{j\omega})^3 - 1.89(e^{j\omega})^2 + 0.89(e^{j\omega}) - 4.84\times 10^{-20}}$$

Also in this case, the frequency response $\hat{G}(\omega)$ is a complex function of the frequency, from which two representations versus ω can be obtained: magnitude and phase.

Finally, it is possible to obtain an approximate frequency response for the discretized motor using the bilinear transform. Calling $w = u + jv$,

$$\tilde{G}(\omega) \equiv \tilde{G}(w)\big|_{u=0} = \tilde{G}(w)\big|_{w=j\omega}$$

In this case

$$\tilde{G}(\omega) = \frac{-2.03\times 10^{-6}w^3 - 3.04\times 10^{-3}w^2 + 8.10\times 10^{-4}w + 1.22}{w^3 + 21.14w^2 + 22.82w}\Bigg|_{w=j\omega}$$

$$\tilde{G}(\omega) = \frac{-2.03\times 10^{-6}(j\omega)^3 - 3.04\times 10^{-3}(j\omega)^2 + 8.10\times 10^{-4}(j\omega) + 1.22}{(j\omega)^3 + 21.14(j\omega)^2 + 22.82(j\omega)}$$

Once again, the frequency response $\tilde{G}(\omega)$ is a complex function of the frequency, from which two representations versus ω can be obtained: magnitude and phase.

The frequency responses corresponding to the above three transfer functions can be obtained by adding the following MATLAB code:

```
%% Paragraph d)
%Defines s for sigma=0
sigma = 0;
omega = linspace(0.01,200,1000);
f = omega/(2*pi); %Frequency in Hz.
s = sigma + 1j*omega;
%Defines z for sigma=0
z = exp(s*T);
%Defines w for u=0
u = 0;
v = omega;
w = u +1j*v;

%Obtain frequency responses
[ModG,ArgG] = FrequencyResponse(G,s);
[ModGh,ArgGh] = FrequencyResponse(Gh,z);
[ModGt,ArgGt] = FrequencyResponse(Gt,w);

%Draw magnitude frequency response
plot(f,ModG,f,ModGh,f,ModGt); %Magnitude frequency response
xlim([0,f(end)]);
xlabel('Frequency (Hz.)');
ylabel('Magnitude (dB)');
legend({'Continuous','Discrete','Bilinear'},'Location','NorthEast');

%Draw phase frequency response
figure;
plot(f,ArgG,f,ArgGh,f,ArgGt); %Phase frequency response
xlim([0,f(end)]);
xlabel('Frequency (Hz.)');
ylabel('Phase (degrees)');
legend({'Continuous','Discrete','Bilinear'},'Location','NorthEast');

%Draw magnitude frequency response (logarithmic scale)
figure;
semilogx(f,ModG,f,ModGh,f,ModGt); %Magnitude frequency response
xlim([0,(1/T)/2]);
xlabel('Frequency (Hz.)');
ylabel('Magnitude (dB)');
legend({'Continuous','Discrete','Bilinear'},'Location','NorthEast');


%Obtain the frequency response of a transfer function
function [ModX,ArgX] = FrequencyResponse(X,p)
%X: Input transfer function: X(s), X(z) or X(w)
%p: Independent variable of the transfer function: s, z or w
%ModX : Magnitude of the frequency response
%ArgX : Phase of the frequency response

[numX,denX] = tfdata(X,'v'); %Numerator & denominator of X
Xw = polyval(numX,p) ./ polyval(denX,p); %Evaluate X for p
ModX = abs(Xw); %Magnitude of X
ModX = mag2db(ModX); %Magnitude of X in dB
ArgX = angle(Xw); %Phase of X
ArgX = unwrap(ArgX) * 180/pi; %Phase of X (unwrapped)
```

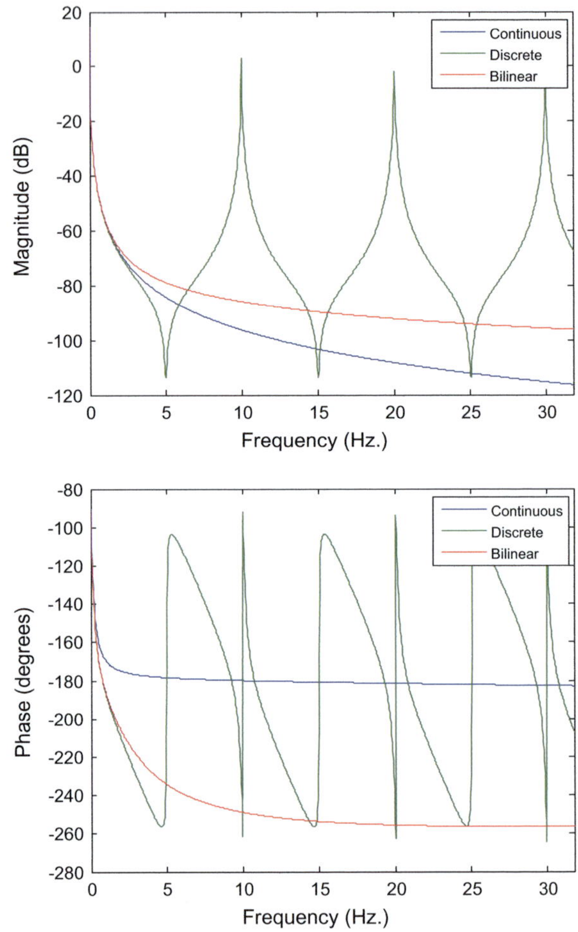

Fig. 4.31 Magnitude (top) and phase (bottom) of the frequency responses

The magnitude and phase of the frequency responses are depicted in Fig. 4.31. It can be verified that the discrete system has a periodic frequency response and that the bilinear transform offers a good approximation to the continuous system, mainly for low frequencies.

It is usual to draw both responses with logarithmic scale frequencies and only up to half of the sampling frequency. For this purpose, the following MATLAB code is added:

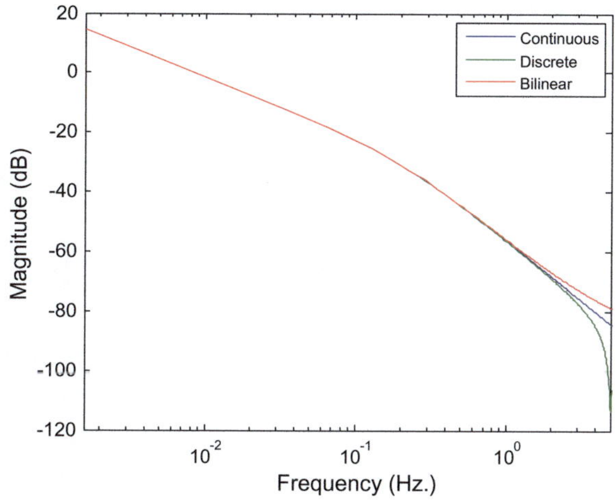

Fig. 4.32 System response with logarithmic scale up to half of the sampling frequency

```
%Draw magnitude frequency response (logarithmic scale)
figure;
semilogx(f,ModG,f,ModGh,f,ModGt); %Magnitude frequency response
xlim([0,(1/T)/2]);
xlabel('Frequency (Hz.)');
ylabel('Magnitude (dB)');
legend({'Continuous','Discrete','Bilinear'},'Location','NorthEast');
```

The obtained result is shown in Fig. 4.32.

(e) The transfer function of the complete system can be obtained from the following expression:

$$\hat{M}(z) = \frac{\hat{X}(z)}{\hat{R}(z)} = \frac{Q(z)\hat{G}(z)}{1 + Q(z)\hat{G}(z)}$$

The corresponding frequency response is

$$\hat{M}(\omega) \equiv \hat{M}(z)\big|_{s=j\omega} = \hat{M}(z)\big|_{z=e^{j\omega}}$$

On the other hand, the bilinear transform corresponding to the complete system is

$$\tilde{M}(w) = \mathcal{B}\left[\hat{M}(z)\right] \equiv \hat{M}(z)\big|_{z=\tilde{z}(w)} = \hat{M}(z)\big|_{z=\frac{1+\frac{wT}{2}}{1-\frac{wT}{2}}}$$

And the frequency response corresponding to this bilinear transform can be written as

$$\tilde{M}(\omega) \equiv \tilde{M}(w)\big|_{u=0} = \tilde{M}(w)\big|_{w=j\omega}$$

Finally, it is possible to consider the complete system without discretizing which is described by its transfer function

$$M(s) = \frac{X(s)}{R(s)} = \frac{Q(s)G(s)}{1+Q(s)G(s)}$$

The corresponding frequency response is

$$M(\omega) \equiv M(s)|_{\sigma=0} = M(s)|_{s=j\omega}$$

The frequency response corresponding to the above three transfer functions can be obtained by adding the following MATLAB code:

```
%% Paragraph e)
Mh = feedback(Q*Gh , 1); %Complete system discrete transfer function
Mt = BilinearTransform(Mh); %Complete system bilinear transform
M = feedback(Q*G , 1); %Complete system continuous transfer function

%Obtain frequency responses
[ModM,ArgM] = FrequencyResponse(M,s);
[ModMh,ArgMh] = FrequencyResponse(Mh,z);
[ModMt,ArgMt] = FrequencyResponse(Mt,w);

%Draw magnitude frequency responses
figure;
semilogx(f,ModM,f,ModMh,f,ModMt); %Magnitude frequency response
xlim([0,(1/T)/2]);
xlabel('Frequency (Hz.)');
ylabel('Magnitude (dB)');
legend({'Continuous','Discrete','Bilinear'},'Location','NorthEast');

%Draw phase frequency responses
figure;
semilogx(f,ArgM,f,ArgMh,f,ArgMt); %Magnitude frequency response
xlim([0,(1/T)/2]);
xlabel('Frequency (Hz.)');
ylabel('Phase (degrees)');
legend({'Continuous','Discrete','Bilinear'},'Location','NorthEast');
```

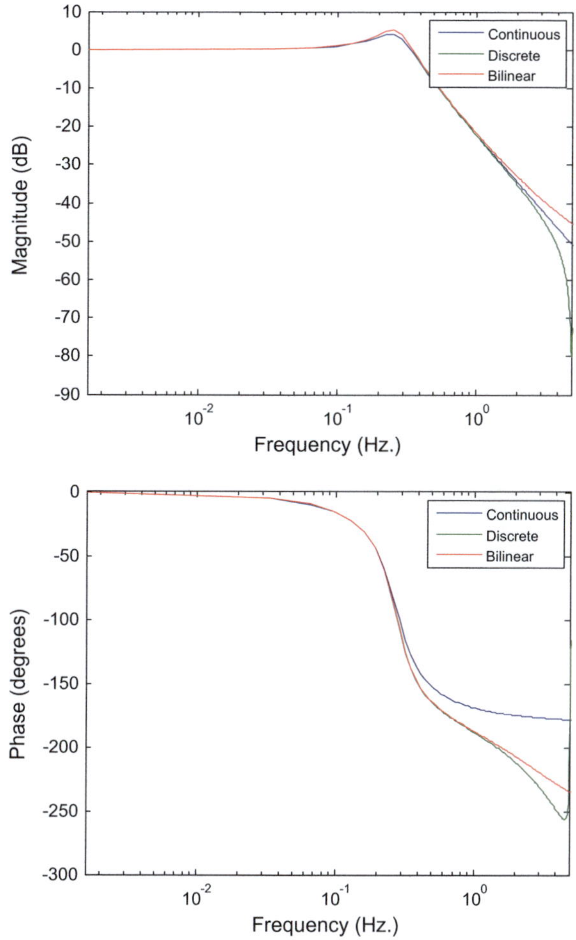

Fig. 4.33 Magnitude (top) and phase (bottom) of the obtained frequency responses

The obtained results are shown in Fig. 4.33. It can be verified that all of them exhibit very similar values, mainly in the low-frequency band.

(f) To obtain the step response and the frequency response of the complete system, using different sampling times, the following MATLAB code is added

```
%% Paragraph f)
%Draw complete system's step response for various sampling times
figure;
Tv=[0.1,0.2,0.5]; %Sampling time vector
nT = length(Tv);
LegText = cell(1,nT);
for i = 1:nT
    T = Tv(i); %Sampling time
    LegText{i} = ['T = ',num2str(T)];
    Gh = c2d(G,T); %Discretize the motor transfer function
    Mh = feedback(Q*Gh , 1); %Complete system transfer function
    step(Mh); %Complete system's step response
    hold all;
end
legend(LegText,'Location','NorthEast');

%Draw complete system's frequency response for various sampling times
figure;
for i = 1:nT
    T = Tv(i); %Sampling time
    Gh = c2d(G,T); %Discretize the motor transfer function
    Mh = feedback(Q*Gh , 1); %Complete system transfer function
    f = linspace(0.01,(1/T)/2,1000); %Frequency in Hz.
    s = 0 + 1j*2*pi*f; %Defines s for sigma=0
    z = exp(s*T); %Defines z for sigma=0
    ModMh = FrequencyResponse(Mh,z); %Magnitude frequency response
    semilogx(f,ModMh); %Draw magnitude frequency response
    hold all;
end
xlabel('Frequency (Hz.)');
ylabel('Magnitude (dB)');
legend(LegText,'Location','NorthEast');
```

The obtained results are shown in Figs. 4.34 and 4.35. It can be seen that increasing sampling time (decreasing sampling frequency), the overshoot and the

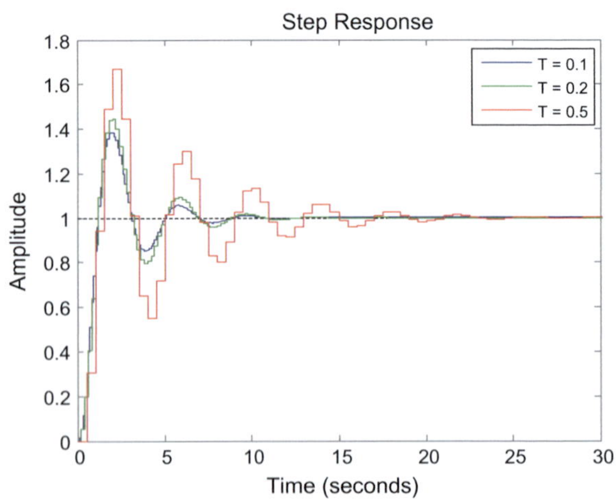

Fig. 4.34 Step response of the complete system

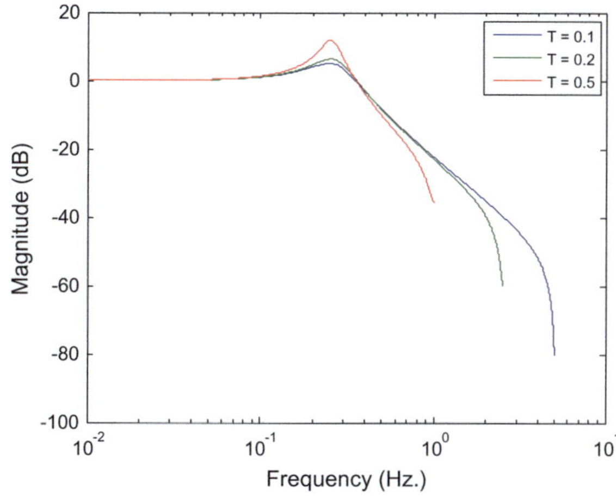

Fig. 4.35 Frequency response of the complete system

settling time are increased, while the resonance peak in the frequency response also increases.

Problem 4.3

Continuous positive airway pressure, CPAP, is a technique developed by Dr. George Gregory and his colleagues in the neonatal intensive care unit at the University of California, San Francisco, in 1971. This technique is specifically designed for non-invasive ventilation, suitable for home treatments of chronic conditions such as sleep apnoea or COPD (chronic obstructive pulmonary disease).

Figure 4.36 shows a CPAP machine. Usually, this device generates a pressurized airflow at the blower output. This pressure prevents the collapse of the airways in patient's throat. The blower consists of the pneumatic part, which includes the impeller. The pressure control of the air directed to the patient airways has to be very accurate in order to have a successful treatment. Furthermore, the controller must have enough stability margins and enough bandwidth to respond to the patient reaction.

The motor for CPAP blower is used during the sleep. The speed control can be designed as a closed loop with unitary feedback with the following plant:

$$G(s) = \frac{e^{-0.1s}}{s}$$

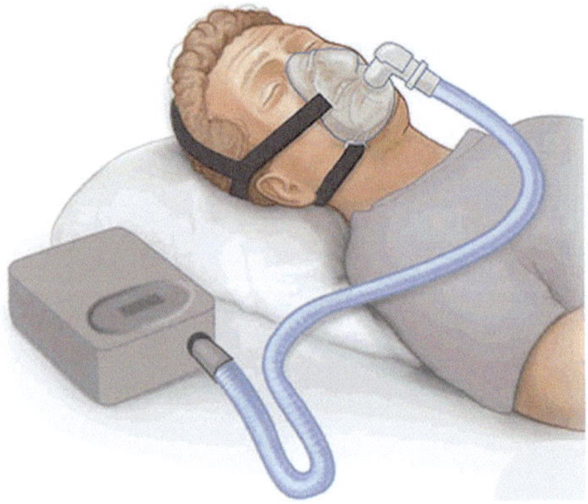

Fig. 4.36 Ilustration of a continuous positive airway pressure (CPAP) machine

(a) Find which values of K make the system stable.
(b) Considering $K = 0.5$ obtain: the phase and gain margins.
(c) Design a PI controller (with $T_i = 0.7\,\text{s}$), that fulfils the following requirements:

- Phase margin greater than 45°.
- Gain margin greater than 17 dB.

For the controlled system, obtain the bandwidth, the resonance frequency and the resonance peak.

Solution 4.3

(a) For the closed-loop system, the transfer function can be obtained according the following procedure

$$G_c(s) = \frac{G(s)}{1+G(s)} = \frac{\frac{Ke^{-2s}}{s}}{1+\frac{Ke^{-2s}}{s}} = \frac{Ke^{-2s}}{s+Ke^{-2s}}$$

Since the transfer function does not have a polynomial expression, it is not possible to use directly the tools of the MATLAB control toolbox. However, it is possible to build a plot of the poles of the function using the following code

Fig. 4.37 Root loci for the CPAP system

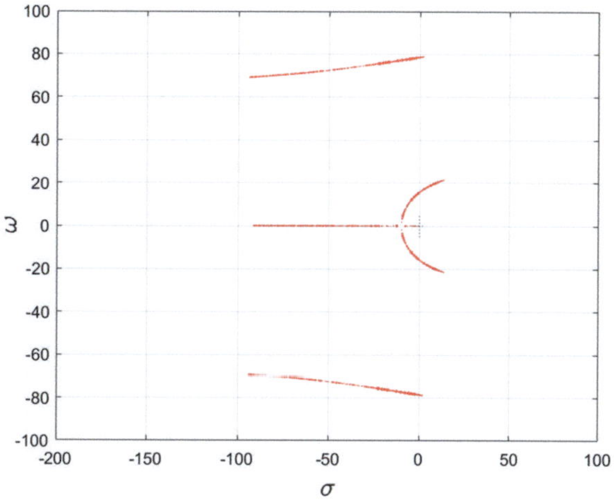


```
% Paragraph a

[sigma,w]=meshgrid(-200:0.1:100,-100:0.1:100);
s=sigma+1j*w;
for K=logspace(-2,2,200)
    Gc=K*exp(-0.1*s)./(s+K*exp(-0.1*s));
    Gcmod=abs(Gc);
    contour(sigma,w,Gcmod,'r');
    line([0,0],[-5,5],'Color','k','LineStyle',':');
    line([-4,3],[0,0],'Color','k','LineStyle',':');
    xlabel('\sigma');
    ylabel('\omega');
    title(['K= ',num2str(K,'%5.2f')]);
    hold all;
    drawnow;
end
```

This is an indirect way to get the root loci. In the dynamic execution of the code, the evolution of the graph can be observed as the value of K increases. From a certain value of K, the poles begin to be in the positive real half-plane and, therefore, hence the system becomes unstable. The graphical result that can be obtained with this code is shown in Fig. 4.37.

As can be seen, the system becomes unstable (the poles start to be in the positive real half-plane) from a value of $K \approx 88$.

This result can be obtained analytically. In order to obtain a stable closed-loop system, the poles of its transfer function, or what is the same, the zeros of its characteristic equation, must have a real negative part. The poles are then calculated by

$$q(s) = 0$$

$$s + Ke^{-0.1s} = 0$$

$$(\sigma + j\omega) + Ke^{-0.1(\sigma + j\omega)} = 0$$

The value of K that makes the system critically stable is therefore that which locates the poles just above the imaginary axis. Hence, its real part σ is zero. Therefore, for this value of the constant, which we will call K_c, we have

$$(0 + j\omega_c) + K_c e^{-0.1(0 + j\omega_c)} = 0$$

$$j\omega_c + K_c e^{-j0.1\omega_c} = 0$$

$$K_c e^{-j0.1\omega_c} = -j\omega_c$$

This is a complex equation that we can divide into two. For the equation corresponding to the argument, we have

$$\arg\left[K_c e^{-j0.1\omega_c}\right] = \arg[-j\omega_c]$$

$$\arg[K_c] + \arg\left[e^{-j0.1\omega_c}\right] = \arg[-j\omega_c]$$

$$0 - 0.1\omega_c = -\frac{\pi}{2}$$

$$\omega_c = \frac{\pi}{0.2}$$

For the equation corresponding to the module, we have

$$\left|K_c e^{-j0.1\omega_c}\right| = |-j\omega_c|$$

$$K_c \left|e^{-j0.1\omega_c}\right| = \omega_c$$

$$K_c = \omega_c = \frac{\pi}{0.2} = 15.7$$

To determine if stability is obtained for values greater or less than the critical value of K_c, we can look at the construction of the previous graph and analyse what is the effect of letting K increase or decrease. We have seen that the expression that allows us to calculate the position of the poles is

$$(\sigma + j\omega) + Ke^{-0.1(\sigma + j\omega)} = 0$$

$$\sigma + j\omega + Ke^{-0.1\sigma} e^{-j0.1\omega} = 0$$

$$\sigma + j\omega + Ke^{-0.1\sigma}(\cos 0.1\omega - j\sin 0.1\omega) = 0$$

$$\sigma + j\omega + Ke^{-0.1\sigma}\cos 0.1\omega - jKe^{-0.1\sigma}\sin 0.1\omega = 0$$

The equation for the real part is

$$\sigma + Ke^{-0.1\sigma}\cos 0.1\omega = 0$$

For this equation to be fulfilled, either $\sigma < 0$, or $\cos 2\omega < 0$. For large values of K, like that values of $\cos 2\omega$ that are between ± 1, the unique way to reduce the value of $Ke^{-2\sigma}\cos 2\omega$ is to let $e^{-2\sigma}$ be small. Hence, $\sigma > 0$. As a conclusion, for large values of K the system is unstable, so the stability condition is

$$K < \frac{\pi}{0.2} \approx 15.7$$

Another way to obtain this result is using the `rlocus` tool of MATLAB. Using the Padé approximation, the exponential function becomes polynomial, so MATLAB can be used to calculate the place of the roots. This is done by the following code

```
s=tf('s')
G=exp(-0.1*s)/s;
Gp=pade(G,4);
rlocus(Gp);
```

The graphical result is obtained in Fig. 4.38.

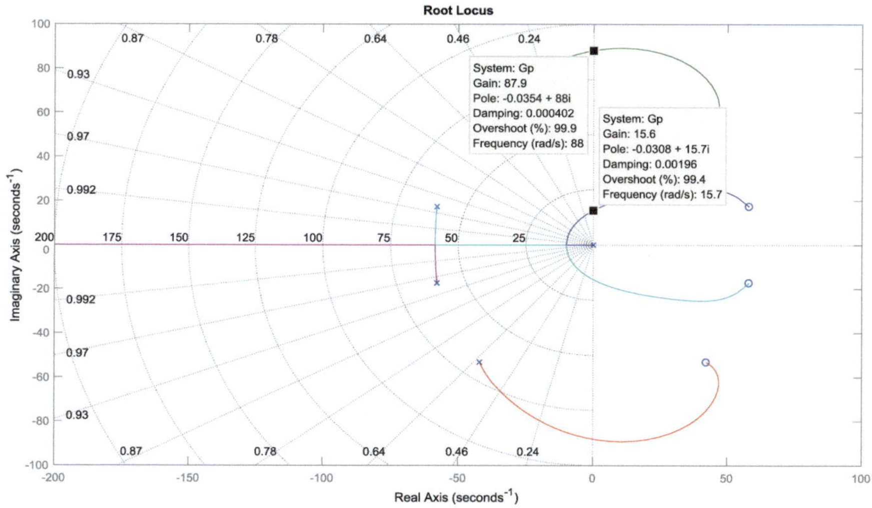

Fig. 4.38 Root locus using rltool, including a Pade approximation

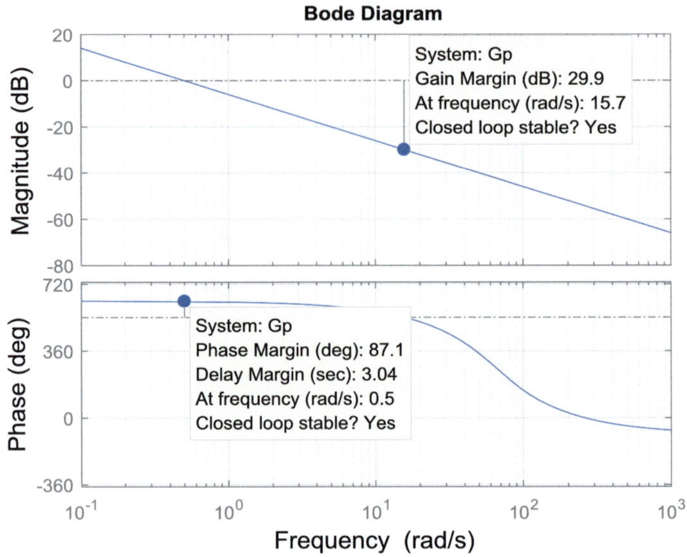

Fig. 4.39 Bode diagram of the open-loop system

The crossing of the imaginary axis by the graph of the root loci can be obtained directly in MATLAB. This way, we can conclude that, approximately, the system is stable for

$$K < 15.7$$

value that coincides with the previously calculated.

(b) The Bode diagram of the open-loop system is used to calculate the gain and phase margins. It can be resolved using the following MATLAB code

```
% Paragraph b
s=tf('s')
K=0.5;
G=K*exp(-0.1*s)/s;
Gp=pade(G,4);
bode(Gp);
```

The result of executing this code is shown in Fig. 4.39.

We can see that the phase margin is 87.1° and the gain margin is 29.9 dB. Hence the system is stable, as corresponds to a value of K less than the critical value.

(c) A PI (C) controller could be described in the following way

$$C = K_p\left(1 + \frac{1}{T_i s}\right) = K_p\frac{s + 1/T_i}{s}$$

For this problem, let us assume $T_i = 0.7s$, and an initial value of 0.5 for K. Hence, the controller could be expressed as:

$$C = 0.5\frac{s + 0.1429}{s}$$

The design of the specified PI controller can be done using the `rltool`. The following code runs the graphical interface of `rltool`, in which it is possible to change manually K_p by moving the poles at the root locus while observing the effect of the phase and gain margins in the open-loop Bode.

```
%Paragragh c
% Closed-loop
s=tf('s')
Kp=0.5;
Ti=0.7;
C=Kp*(s+1/Ti)/s;
G=exp(-0.1*s)/s;
Gp=pade(G,4);
rltool(Gp,C)
```

Figure 4.40 depicts the graphical interface of `rltool` that permits to tune in the best K_p that fulfils the requirements.

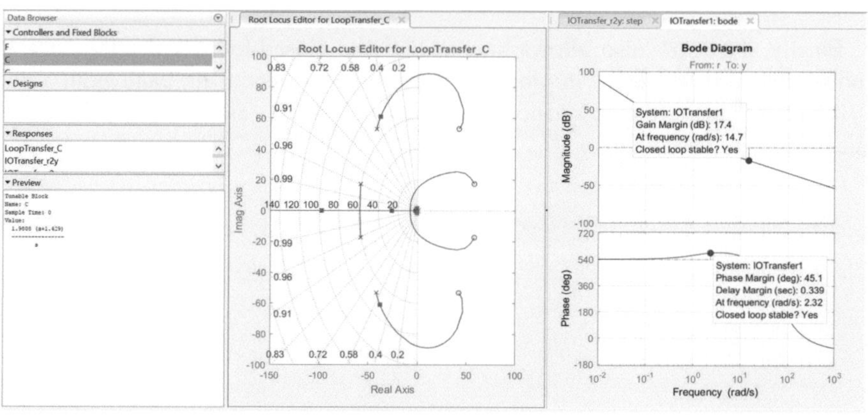

Fig. 4.40 Graphical interface of rltool showing both, the root locus and Bode diagrams for the considered system

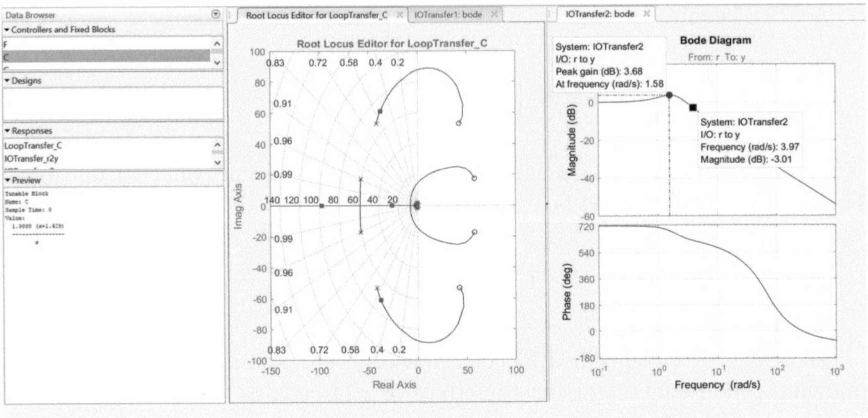

Fig. 4.41 Using rltool for adjusting the bandwidth value, resonance and frequency peaks

With the selected $K_p = 1.9808$, the closed-loop system is stable, with a phase margin of 45.1° and a gain margin of 17.4 dB.

Once the process of tuning has finished, `rtool` allows you to export the designed controller:

```
C =

  1.9808 (s+1.429)
  ----------------
        s
```

Finally, `rltool` also allows you to plot the closed-loop Bode. Figure 4.41 shows this representation, in which it is possible to obtain the bandwidth value ($BW = 3.97$ rad/s), the resonance peak ($M_r = 3.68$ dB) and the resonance frequency ($\omega_r = 1.58$ rad/s).

Computational Tools for the Analysis and Design of Control Systems

5

This chapter focusses on the analysis and design of control systems in order to obtain the desired response to a given input signal (*reference*). That is, for a particular system, e.g. an insulin pump, the problem is to design a proper controller targeting at fulfilling specified requirements according to either that of transient response (overshoot, rise time, etc.) or that of frequency response (phase margin, gain margin, etc.) as well as according to the steady-state response characteristics (steady error).

Different approaches can be adopted in this design, albeit the simplest and intuitive way is to leverage the study of the contribution of the poles of the system in the response. Thus, starting from the system at hand we will need to envisage how the addition of poles/zeros, i.e. adding another system called *controller*, can provide the desired response to the controlled system.

This chapter covers the root locus analysis of systems, a convenient tool for designing control systems based on the position of poles and zeros. After the mathematical justification of the tool, it will be applied to both continuous and discrete systems using the *rltool toolbox* from MATLAB, which largely helps in the analysis and design of control systems both for transient and for frequency response characteristics fulfilment which will be treated in the next chapter.

The Root Locus Analysis of Continuous Systems

The nature and position of the poles of a system directly affect its response. In previous chapters, we have studied the effect of poles according to their nature; i.e. complex poles contribute to a sinusoidal response, while real ones provide a smooth approaching to the steady value, while its position affects to the duration of their effect in the response. Thus, when we want to obtain a desired behaviour of the system at hand, we can apply such knowledge to choose the location of the poles of the system and/or adding additional poles/zeros. For that, we rely on the simplest

J. Fernández de Cañete et al., *Automatic Control Systems in Biomedical Engineering*,
https://doi.org/10.1007/978-3-319-75717-9_5

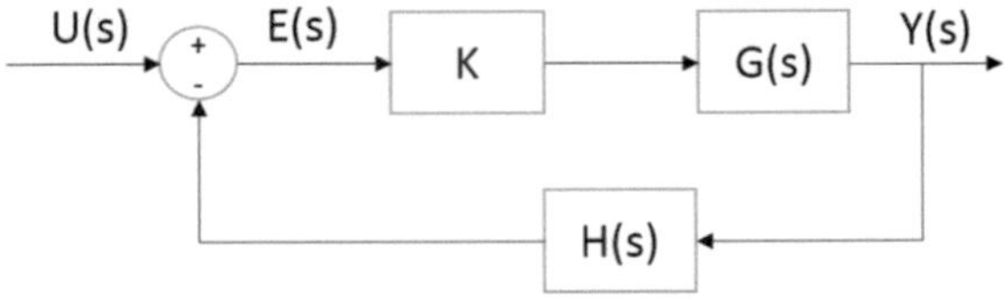

Fig. 5.1 Closed-loop configuration. Through the root locus, all the possible locations of the poles of the closed-loop system can be calculated in function of the parameter K

control configuration, i.e. a closed-loop configuration including a parameter, K, which will alter the original system. The study of the position of the closed-loop poles according to such parameter K is known as *root locus technique*.

The root locus technique was proposed by Evans in 1948 as a graphical representation of the position of the poles of the closed-loop system as long as the parameter K varies. The general configuration is shown in Fig. 5.1, where the output of the system $G(s)$, sensed by the $H(s)$ system, is compared to the input reference $U(s)$. Such difference, called error $E(s)$, is modulated by the parameter K, being $E(s) \cdot K$ the input of $G(s)$.

This is actually a typical *proportional control* where $G(s)$ is the system to be controlled, $H(s)$ represents a sensorial system that measures the output, and K is the proportional gain. When error is zero, in case it happens, it means that $Y(s) = U(s)$, and the input to the system $G(s)$ is also zero; therefore, no actuation is carried out.

Let us instantiate this system with the following example. We are interested in keeping a pre-defined glucose level of a patient by using an insulin pump. Let us assume that the transfer functions of the involved systems are[1]:

$$G_{\text{pump}}(s) = \frac{1}{s+2} \quad G_{\text{pacient}}(s) = \frac{1}{s+1} \quad H_{\text{sensor}}(s) = \frac{1}{s+5}$$

That is, in this case the configuration is shown in Fig. 5.2 top, where the pump delivers insulin to the patient proportionally to the difference between the desired level and the sensed level of glucose in blood.

Note that the gain of the sensor must be considered as a filter in the input. The sensor in this example has a gain of 1/5; i.e. it measures 1/5 of the actual value, and thus, the input value must be properly adequate, adding $K_s = 1/5$. The response of the closed-loop system for $K = 1$ and input reference $U = 1$ are shown in Fig. 5.2 right.

Inspecting Fig. 5.2 bottom, one can envisage that the system is not working as expected, since the patient did not reach the desired glucose level, i.e. the input reference. The solution is to *control* the behaviour of the pump, assuming we cannot modify the sensor, and of course, the patient. Control system will be addressed in Chap. 6, but here we will study the root locus of the system, which is, in fact, the simplest control system: the proportional control system, and our focus here is to select the best K for our purposes, by drawing the root locus.

[1]Transfer functions have kept intentionally simple for the pedagogical sake.

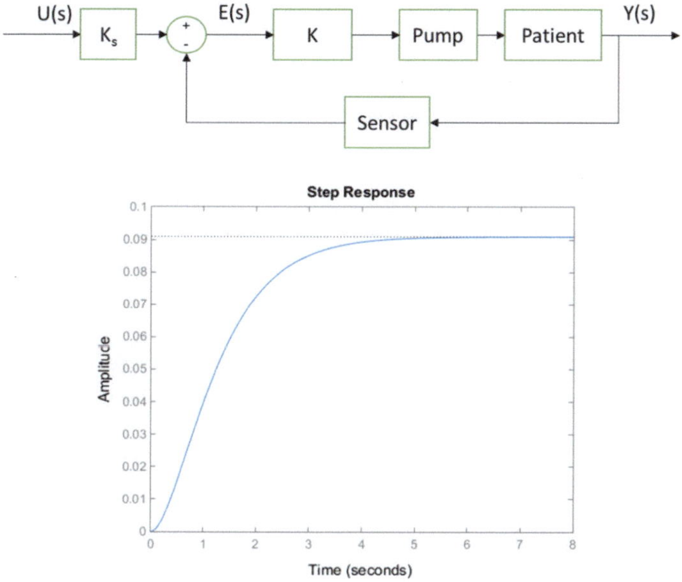

Fig. 5.2 Glucose control system. Top, the closed-loop configuration. Note that in this case the dynamic of the patient must be also modelled. The term K_s models the gain of the sensor. Bottom, system response for $K = 1$

As commented before, the behaviour of the whole system will be defined by the position of poles, and the root locus technique comes to depict their possible locations. Recalling previous chapters, the equivalent transfer function of the closed-loop configuration of Fig. 5.2 is:

$$G_{\text{tot}}(s) = \frac{KG_{\text{pump}} \cdot G_{\text{patient}}}{1 + K \cdot G_{\text{pump}} \cdot G_{\text{patient}} \cdot H_{\text{sensor}}} \tag{5.1}$$

For clarity, we will refer to $G_{\text{pump}}\, G_{\text{patient}}$ as simply G, and H_{sensor} as H, and thus, the transfer function becomes:

$$1 + K \cdot G(s) \cdot H(s) = 0 \tag{5.2}$$

Variations in the parameter K will obviously modify the roots of this equation and therefore the position of the poles. Let us consider the previous example to obtain the closed loop and the position of the poles according to different values of K. From the transfer functions:

$$G(s) = \frac{1}{(s+1)(s+2)} \quad H(s) = \frac{1}{s+5}$$

The closed-loop transfer function when including the parameter K is:

$$G_{\text{tot}}(s) = \frac{K(s+5)}{s^3 + 8s^2 + 17s + 11 + K} \tag{5.3}$$

from which we are interested in its characteristic function, that is:

$$s^3 + 8s^2 + 17s + 10 + K = 0 \tag{5.4}$$

Varying the value of K, we will obtain the different positions of the closed-loop poles within the s-plane. Note that when $K = 0$, the closed-loop configuration actually disappears, and thus, the poles coincide with the poles of the open-loop system, that is, $G(s) \cdot H(s)$.

The root locus technique simply consists of solving the characteristic equation for values of K in the range $[0, \infty)$.[2] Next table shows some examples:

K	P1	P2	P3
0	−5	−2	−1
0.5	−5.04	−1.81	−1.15
0.87941	−5.07	−1.46	−1.46
1	−5.08	−1.46 + 0.18i	−1.46 − 0.18i
5	−5.344	−1.33 + 1.02i	−1.33 + 1.02i
10	−5.6	−1.2 + 1.46i	−1.2 − 1.46i

From this, we can deduce that if $0 \leq K \leq 0.88$ the system is overdamped; that is, it has different and real poles, and for $K \cong 0.88$ the system is critically damped; i.e. it has two real poles located at the same position, and finally if $K \geq 0.88$, the system is underdamped, exhibiting an oscillating behaviour due to the presence of complex conjugated poles. If we depict the position of the poles as long as the value of K is increasing, we obtain the picture of Fig. 5.3.

That is, we can rapidly observe the position of each pole, and therefore, we can infer the behaviour of the final output for a given value of K. However, relying on a systematic calculation of the roots varying the parameter K is not practical at all, and therefore, a set of recipes are available for sketching it.

Root Locus Recipe

There exists a set of simple steps for sketching the root locus of a given system, which are mathematically derived from the complex nature of the characteristic equation. That is, given

$$1 + KG(s)H(s) = 0 \Rightarrow KG(s)H(s) = -1 \tag{5.5}$$

[2]Negative values of K can be also explored, which would impose a positive feedback. This situation occurs, among others, when dealing with delayed systems.

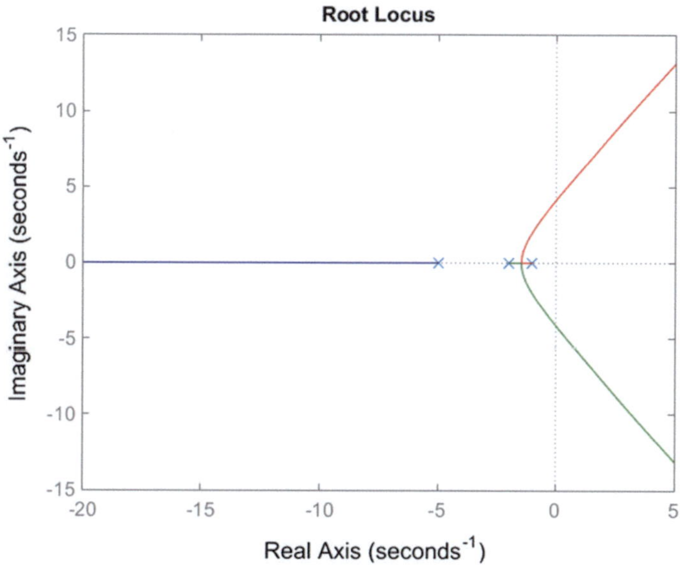

Fig. 5.3 Root locus example. Crosses indicate the poles of the systems G and H. As long as the parameter K increases, the closed-loop poles *move* following the trajectories depicted

we can distinguish the real and complex part of the equation and rewrite it as

$$|KG(s)H(s)|\angle KG(s)H(s) = -1 + j0 \tag{5.6}$$

That is,

$$|KG(s)H(s)| = 1 \tag{5.7}$$

$$\angle KG(s)H(s) = \angle - 1 = \pi(2k+1), \quad k = 0, \ \pm 1, \ \pm 2, \ \ldots \tag{5.8}$$

Equation (5.7) is known as the *magnitude condition* and can be used to calculate the value of the parameter K for a particular position of the poles. Equation (5.8) is called the *angle condition* and permits us to determine whether a particular point of the s-plane belongs or not to the root locus, i.e. if there exists a value for K for which one of the poles of the system are located at that particular point. These two equations are the basis for the root locus method.

Let us apply equations, magnitude and angle conditions to our example in order to test whether the point $s_1 = -7$ belongs to the root locus, as well as computing the value of K. Starting from the open-loop system:

$$G(s)H(s) = \frac{1}{(s+1)(s+2)(s+5)}$$

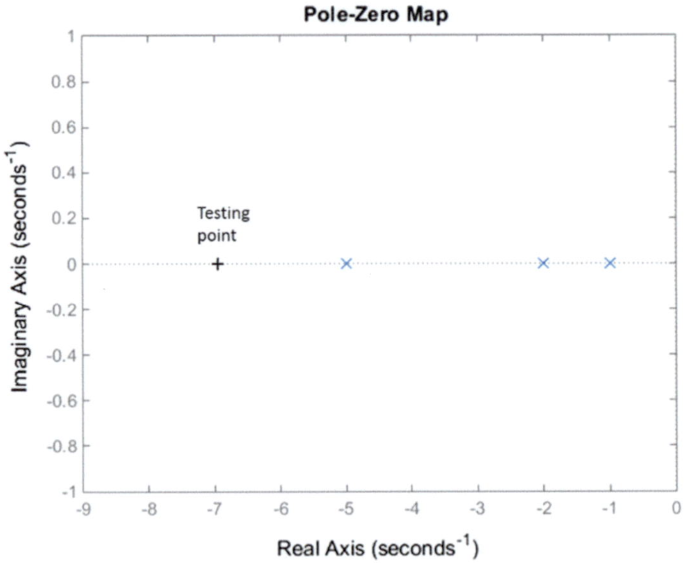

Fig. 5.4 Determining the root locus over the real axis. A test point, in this case $s = -7$, belongs to the root locus if and only if the angle condition holds

For the test point (see Fig. 5.4), angle condition is computed as

$$\angle \frac{K}{(s+1)(s+2)(s+5)}\bigg|_{s=s_1} = \angle(s_1+1) + \angle(s_1+2) + \angle(s_1+5) = 3\pi$$

And therefore it holds for this test point. Actually, Eq. (5.8) holds for any $s \in [-5, -\infty)$

Once we have confirmed that $s_1 = -7$ belongs to the root locus, we want to figure out the value of K that moves one of the poles to this position. For that, we rely on the magnitude condition:

$$\left|\frac{K}{(s+1)(s+2)(s+5)}\right|\bigg|_{s=s_1} = 1 \Rightarrow \frac{K}{|s_1+1||s_1+2||s_1+5|} = 1 \Rightarrow$$
$$K = |s_1+1||s_1+2||s_1+5| = 6 \,*\, 5 \,*\, 2 = 60$$

Actually, going back to Eq. (5.3) and substituting K for 60 yield 3 solutions, $s_1 = -7,\ s_{1,2} = -0.5 \pm 3.1225i$.

In our example, we only consider three poles in the open-loop function. For a general, n-order system, the characteristic equation is given by

$$F(s) = \frac{K(s+z_1)(s+z_2)\ldots(s+z_m)}{(s+p_1)(s+p_2)\ldots(s+p_n)}$$

and the magnitude and angle conditions are computed as

$$\angle F(s) = \sum_{1}^{m} \angle(s+z_i) - \sum_{1}^{n} \angle(s+p_i) = \pi(2k+1), \quad k = 0,\ 1,\ 2,\ \ldots \tag{5.9}$$

$$|F(s)| = \frac{\prod_{1}^{m} |s+z_i|}{\prod_{1}^{n} |s+p_i|} = 1 \tag{5.10}$$

The application of these conditions to every point in the s-plane in order to get the root locus is unfeasible. In practice, a recipe, based on such conditions, is followed to sketch the root locus.

Step 1. Placement of the poles and zeros of $G(s)H(s)$ on the s-plane
Firstly, starting from the characteristic equation

$$1 + K\frac{\prod_{1}^{m}(s+z_i)}{\prod_{1}^{n}(s+p_i)} = 0 \Rightarrow \prod_{1}^{n}(s+p_i) + K\prod_{1}^{m}(s+z_i) = 0 \tag{5.11}$$

mark the poles p_i with a cross "x" and the zeros z_i with a small circle "o" on the s-plane. Note that when $K = 0$, Eq. (5.11) is reduced to:

$$\prod_{1}^{n}(s+p_i) = 0$$

And thus, the poles of the closed-loop system are identical to the poles in the open-loop configuration, that is $G(s)H(s)$. Also, notice that as long as K tends to ∞, the roots of Eq. (5.11) coincide with the zeros of $G(s)H(s)$:

$$\frac{1}{K}\prod_{1}^{n}(s+p_i) + \prod_{1}^{m}(s+z_i) = 0$$

$$\lim_{K\to\infty}\left(\frac{1}{K}\prod_{1}^{n}(s+p_i) + \prod_{1}^{m}(s+z_i)\right) = \prod_{1}^{m}(s+z_i)$$

Thus, in this first step we obtain significant information about the root locus sketch: Root locus starts at the poles of the open-loop system $G(s)H(s)$, when $K = 0$, and tends to the zeros of $G(s)H(s)$ as long as $K\to\infty$. Therefore, the root locus has n strokes, each one starting at a pole of $G(s)H(s)$, and some of them, precisely m, ending at the zeros of $G(s)H(s)$.

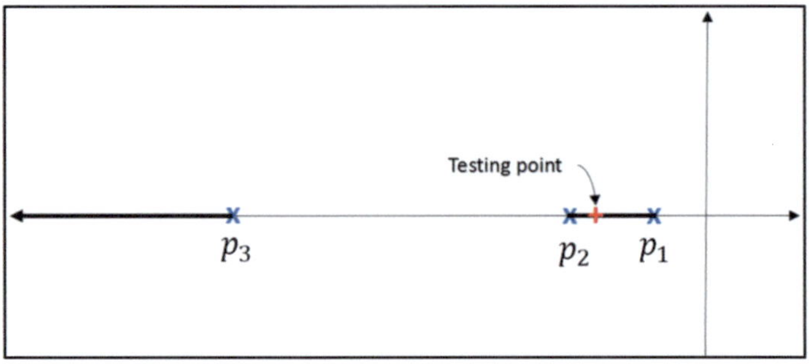

Fig. 5.5 Root locus on the real axis. Based on the angle condition, a point on the real axis will belong to the root locus if the number of poles (and zeros) placed on its right is odd. In this case, the testing point located between p_1 and p_2 belongs to the root locus

Step 2. Root locus on the real axis
In this step, we focus on what parts of the real axis will be part of the root locus. That is, we are going to apply the angle condition to points $s \in \Re$, i.e. $s \in (-\infty, \infty)$. A practical rule to remember about the angle condition on the real axis is: "A given point s^* lying on the real axis belongs to the root locus if and only if the sum of the real poles and real zeros placed on its right is an odd number". This fanciful rule is based on the angle condition and the need of yielding an odd multiple of π.

Figure 5.5 sketches the root locus for our example, showing the open-loop poles at $s = -1$, $s = -2$ and $s = -5$, and testing a point s between p_1 and p_2. In the absence of zeros, the angle condition is computed, in this example, as

$$0 - \sum_{1}^{3} \angle(s + p_i) = \pi + 0 + 0 = \pi = \pi(2k+1),\ k = 0$$

Note that each pole on the left of testing point s contributes to an angle of 0, while the pole on the right contributes to an angle of π. Thus, it is obvious that s is part of the root locus, as well as any point in the interval $(-2, -1)$. Following the same argument, $\forall s \in (-\infty, -10]$, s is part of the root locus.

In this example, observe that a certain direction is given to one segment of the root locus to indicate the path followed by the pole p_3 as long as the parameter K increases. Thus, p_3 tends to $-\infty$ and it can be said that the root locus entails a segment that starts at p_3 and ends at $-\infty$.

Step 3. Assessing asymptotes
In step 1, we stated that the segments of the root locus begin at the poles of $G(s)H(s)$ when $K = 0$ and ends at the zeros when $K \rightarrow \infty$. Normally, in causal systems, the number of zeros, m, is lower than the number of poles, n, and thus only m segments will end at a zero. The rest of segments, concretely m-n, will tend asymptotically to infinity. In the example of Fig. 5.5, $m = 0$ and $n = 3$; i.e. there are

no zeros, and thus, the three segments that start at each pole will asymptotically go to infinity. The angle of the asymptotes with respect to the real axis can be determined by Eq. (5.12).

$$\phi = \frac{\pm\pi\,(2k+1)}{n-m} \quad k = 0,\ 1,\ 2\ldots \tag{5.12}$$

This expression comes from the application of the angle condition to a distant point (ideally at the infinity). At that distant point, the angles from poles and zeros can be assumed as equal and thus they are cancelled, turning the angle condition into

$$\phi = \frac{\pi}{n-m}$$

The point where the asymptotes cross the real axis is determined by the mass centre of the poles and zeros as:

$$s = \sigma_a = \frac{\sum_{i=1}^{n}(-p_i) - \sum_{i=1}^{m}(-z_i)}{n-m} \tag{5.13}$$

In the example, there are three asymptotes, with $\phi = \pm\frac{\pi}{3}$, located at $\sigma_a = \frac{-1-2-5}{3} = -2.333$.

Step 4. Identification of rupture points

In some cases, two or more segments of the root locus collide at a particular point of the s-plane, which is called a *rupture point*. That is, a rupture point occurs when, given a certain value of K, the characteristic equation has roots with a multiplicity greater than 1. Identifying these points will largely help to sketch the root locus further away from the real axis.

Rupture points coincide with the maximum values of K so, in practice, they are computed by deriving K and solving the equation $dK/ds = 0$. The solutions of this equation, s^*, fulfilling that $K(s^*) > 0$ are rupture points.

Let us calculate the rupture points for our example system. The characteristic function is (see Eq. (5.4)):

$$s^3 + 8s^2 + 17s + 10 + K = 0$$

from which

$$K = -s^3 - 8s^2 - 17s - 10$$

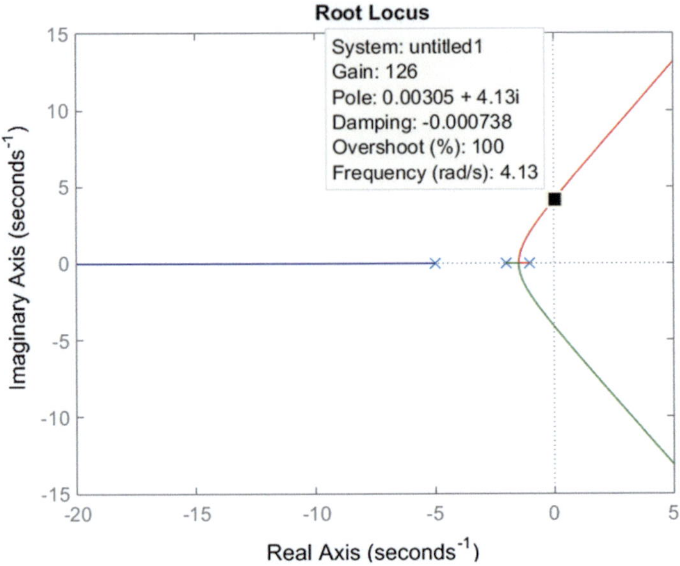

Fig. 5.6 Root locus for the insulin pump example. The critical value of the parameter K and the location of the closed-loop poles have been highlighted

Its derivate is

$$\frac{\mathrm{d}K}{\mathrm{d}s} = -3s^2 - 16s - 17$$

And solving

$$\frac{dK}{ds} = 0 \Rightarrow -3s^2 - 16s - 17 = 0 \Rightarrow s_1 = -3.8685, \; s_2 = -1.4648$$

Checking that it is really a maximum of the curve K, we have that

$$\begin{aligned} K|_{s1} &= -s_1^3 - 8s_1^2 - 17s_1 - 10 < 0 \\ K|_{s2} &= -s_2^3 - 8s_2^2 - 17s_2 - 10 > 0 \end{aligned}$$

and thus $s_2 = -1.4648$ is a rupture point as shown in Fig. 5.6.

Step 5. Determination of starting/ending angles

This step is a straightforward application of the angle condition around the poles and zeros of the open-loop system. It provides us valuable information about the direction of the segments, starting at the poles and ending at the zeros:

$$\varphi_{\text{pole}} = \pi - \sum_{\text{rest}} \varphi_{\text{poles}} + \sum_{\text{rest}} \varphi_{\text{zeroes}}$$
$$\varphi_{\text{zero}} = \pi - \sum_{\text{rest}} \varphi_{\text{zeroes}} + \sum_{\text{rest}} \varphi_{\text{poles}}$$

Step 6. Crossing points of the root locus with the complex axis

This is the last step for sketching a root locus, and it is related to the stability of the closed-loop system. In fact, in this step we calculate the value of the parameter K that turns unstable the closed-loop system, i.e. at least one pole of the will reaches the complex axis.

This step is not always applicable since there are cases where the system is always stable regardless the value of K.[3]

There are different options for calculating the crossing points of the root locus with the complex axis. For our insulin pump example, we will opt for evaluating the characteristic equation at the position of poles in the critical situation, that is $s = 0 \pm j\omega$. Thus,

$$\begin{aligned}
&s^3 + 8s^2 + 17s + 10 + K = 0|_{s=0\pm j\omega} \\
&(j\omega)^3 + 8(j\omega)^2 + 17j\omega + 10 + K = 0 \\
&-j\omega^3 - 8\omega^2 + 17j\omega + 10 + K = 0 \Rightarrow \\
&\left.\begin{array}{l} -8\omega^2 + 10 + K = 0 \\ -j\omega^3 + 17j\omega = 0 \end{array}\right\} \Rightarrow \omega = \sqrt{17}, K = 126
\end{aligned}$$

Thus, in this way we can compute the exact point where the root locus escapes from the left-hand side of the s-plane, i.e. at $0 \pm \sqrt{17}j$, as well as the value of K that moves the poles to such place, i.e. $K = 126$.

The result root locus for the example is shown in Fig. 5.6, which has been obtained by the *rlocus* command of MATLAB, typing:

```
>> rlocus(G*H)
```

Time-Delayed Systems

Time-delayed systems are those exhibiting a reaction time, which is the most common and realistic situation. Focusing on biomedical systems, apart from the small reaction time of devices, like pumps or sensors, which is normally neglected, we have to deal with models for biological processes, for instance the insulin metabolism, which may have a significant delay.

[3]When K is defined in the range $[0, \infty)$

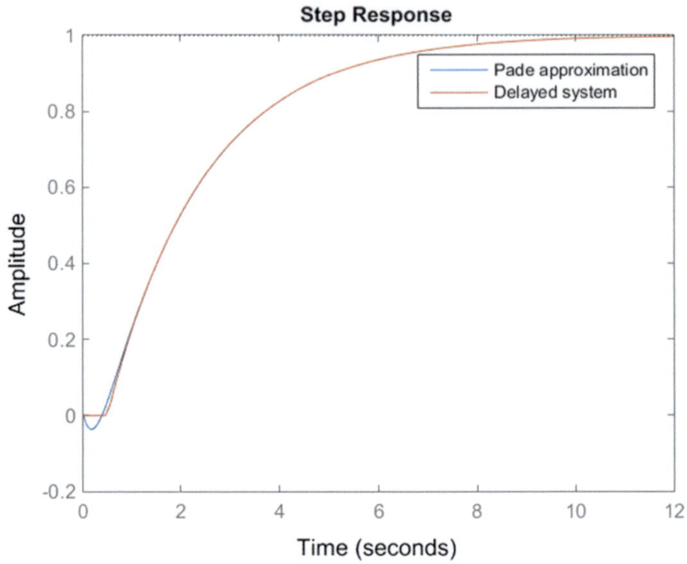

Fig. 5.7 First-order, time-delayed system. In red the ideal delayed system, in blue, the pade approximation of order 1

In general, examples of a time-delayed model are the reaction of a patient to a particular substance or drugs, which can be expressed as a delayed first-order system:

$$G(s) = \frac{K_g e^{-Ts}}{\tau s + 1}$$

where K_g is the gain of the system, T is the delay, and τ is the time constant of the system. For parameters $K_g = 1$, $T = 0.5$, and $\tau = 2$, the response of the system for a unitary step input is shown in Fig. 5.7.

Considering delayed systems for sketching the root locus is not trivial, since their inclusion turns the characteristic equation of the closed-loop system into a non-polynomial equation, and thus, the techniques aforementioned are not applicable. For example, modifying the insulin pump example including a given delay to the model of insulin metabolism will produce an open-loop system G^*H like:

$$G(s)H(s) = \frac{e^{-0.5\,s}}{(s+1)(s+2)(s+5)}$$

and the characteristic equation turns into

$$s^3 + 8s^2 + 17s + 10 + Ke^{-0.5\,s} = 0$$

In these cases, the practical solution is resort to Pade approximations of the delay, turning it into rational models of a given order. Using MATLAB, one can generate such approximation through the command *pade*.

```
>>[num,den]=pade(T,N)
```

where T is the time delay and N is the order of the resultant system which is returned in numerator/denominator format. For example, approximating a time delay of 0.5 s through a first-order system is achieved by:

```
>>[num,den]=pade(5,1)
```

yielding num = [−1 4], den = [1 4], that is:

$$G_{\text{delay}} = \frac{-s+4}{s+4}$$

The ideal first-order delayed system $G(s) = \frac{\mathrm{e}^{-0.5\,\mathrm{s}}}{2s+1}$ and its approximation through the pade function, i.e.$G(s) = \frac{-s+4}{s+4}\frac{1}{2s+1}$, are shown in Fig. 5.7.

From a practical point of view when drawing the root locus of delayed systems, we can directly rely on the *rltool* toolbox of MATLAB, described at the end of this chapter, which automatically takes care of the delay approximation. For instance, these two commands

```
> > G = exp(−0.5 * s)/(2 * s + 1)

G =
                    1
  exp(−0.5 * s)* − − − − − − −−
                  2s + 1
Continuous − time transfer function.

> > rltool(G)
```

will open the *rltool graphical interface* with a warning message (see Fig. 5.8) indicating that a delayed system is being approximated. By default, the second-order Pade approximation is considered, but the designer can modify it. Changing the order to 1, the resultant root locus is shown in Fig. 5.9.

Note that considering the Pade approximation adds, in this case, a zero in s = *4*, that is in the left-hand side of the s-plane. Recalling that the closed-loop poles start from the open-loop poles and end at the zeros (or at infinity), there will be values of

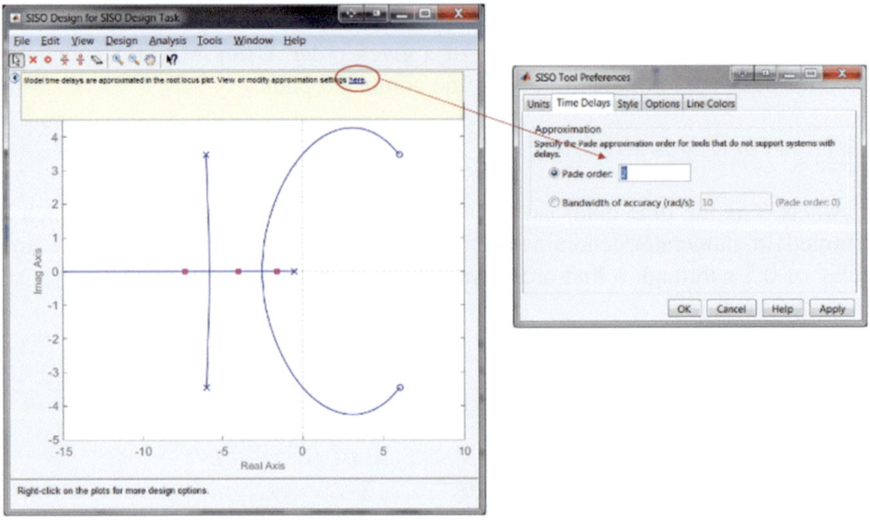

Fig. 5.8 Rltool graphical interface managing time-delayed systems. Rltool triggers a warning message for delayed systems reporting the default approximation, i.e. order 2. The designer can modify such settings through the tool preferences windows

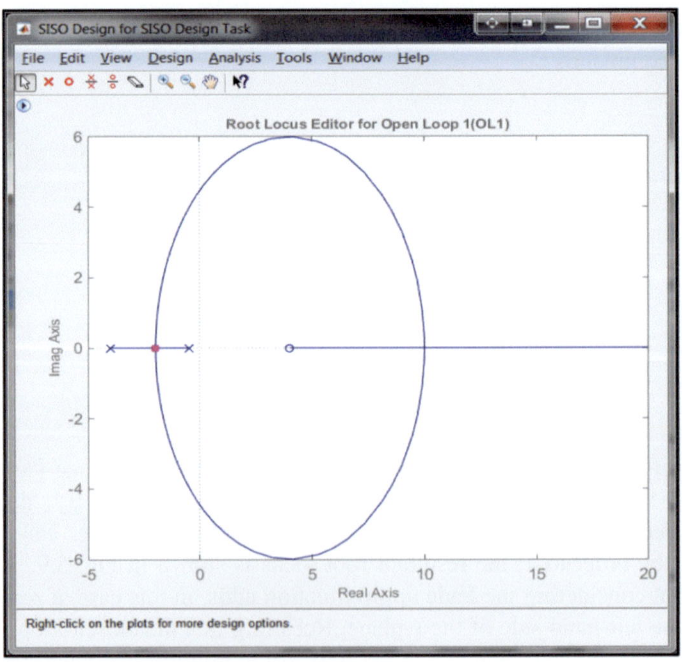

Fig. 5.9 Root locus for a delayed system with a Pade approximation of order 1. Note that, considering a first-order approximation, the closed-loop system includes a zero at the right-hand side of the s-plane

the parameter K that will turn the system unstable. This is the normal situation when considering delays; i.e. they have the effect of reducing the stability of the system at hand.

The Root Locus Analysis of Discrete Systems

The sketch of the root locus for discrete systems follows the same principles and methodology of the case of the continuous systems, albeit some issues must be considered. On the one hand, the discrete plane, z, is different to the continuous plane s. The relation between both planes was discussed in Chap. 3; here, we recall an important difference, the stability criterion: For continuous systems, all poles must be placed in the negative semi-plane, while for discrete system all the poles must be located in the z-plane within a circle centred in (0, 0) and radius 1. On the other, the sample time will affect the discrete system responses, as it affects the position of their open-loop poles, and therefore, different sample times will lead to different root loci.

Following the insulin pump example, let us assume the control is being carried out by any computational device, e.g. a microcontroller, and thus, we need to discretize the models for this discrete analysis. The continuous transfer functions:

$$G(s) = \frac{1}{(s+1)(s+2)} \quad H(s) = \frac{1}{s+5}$$

can be discretized in MATLAB through:

```
>>Gz=c2d(G,0.1);
>>Hz=c2d(H,0.1,'matched')
```

Yielding the following discrete systems:

$$Gz(z) = \frac{(z+0.9048)}{(z-0.9094)(z-0.8146)} \quad Hz(z) = \frac{0.07869}{z-0.6065}$$

That is, the open-loop system entails three poles $p_1 = 0.9094$, $p_2 = 0.8146$, $p_3 = 0.6065$ and a zero at $z = -0.9048$. The characteristic equation of the closed-loop discrete system is, as in the case of continuous systems:

$$1 + KGH = 0$$

In this case:

$$\begin{aligned} &1 + \frac{K \cdot 0.07869(z+0.9048)}{(z-0.9094)(z-0.8146)(z-0.6065)} = 0 \Rightarrow \\ &(z-0.9094)(z-0.8146)(z-0.6065) + K \cdot 0.07869(z+0.9048) = 0 \end{aligned} \tag{5.14}$$

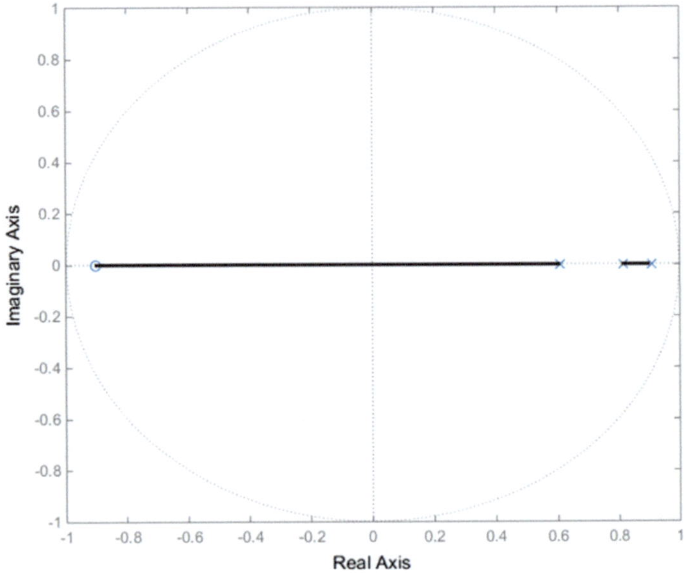

Fig. 5.10 Discrete root locus on the real axis

Solving this equation for values of K in the interval $[0, \infty)$ will generate the root locus in the z-plane.

Similarly to the continuous case, the same recipes can be followed in order to sketch the root locus in the z-plane, since the same angular and magnitude conditions are applicable. For example, the root locus segments laying on the real axis, which is a direct application of the magnitude condition, for our discrete example, see Fig. 5.10, will result in the interval $[-0.9048, 0.6065]$, i.e. from the zero to the most left placed pole, and in the interval between the other poles, i.e. $[0.8146, 0.9094]$

The number of poles (n) and zeros (m) reveals the existence of asymptotes. As in the continuous case, each branch of the root locus will start in an open-loop pole and will end in an open-loop zero, or at infinity implies that the number of branches that ends at the infinity will be $n - m$, and therefore, there will be $n - m$ asymptotes with an orientation:

$$\phi = \frac{\pi}{n - m}$$

Thus in this case, $\phi = \frac{\pi}{2}$, and the position of such two asymptotes in the real axis is computed as:

$$\sigma_a = \frac{\sum_{i=1}^{n}(-p_i) - \sum_{i=1}^{m}(-z_i)}{n - m}$$

which in this case yields $\sigma_a = 1.6176$.

The existence of rupture points is, in this case, clear since there is an interval of the real axis connecting two poles that belongs to the root locus. The exact rupture point is reckoned according to the first derivate of the parameter K. Having the characteristic Eq. (5.14):

$$(z-0.9094)(z-0.8146)(z-0.6065)+K\cdot 0.07869(z+0.9048)=0$$

$$K=\frac{-z^3+2.3305z^2-1.7864z+0.4493}{0.07869(z+0.9048)}$$

Deriving this expression and equating it to zero result in:

$$\frac{\mathrm{d}K}{\mathrm{d}z}=-0.1574z^3-0.0302z^2+0.3319z-0.1625=0$$

With solutions: $z_1=-1.7424,\ z_2=0.6847,\ z_3=0.8657$. These are candidates to be a rupture point. A first analysis about the validity of these results can be done since rupture points actually belong to the root locus, and thus, as they are real values, an actual solution should belong to the intervals already computed for the real axis. In this case, z_1 and z_2 do not belong and they will not valid solutions. This first reasoning will lead to z_3 as the rupture point. The confirmation of this can be done by checking $\frac{\mathrm{d}K}{\mathrm{d}z}>0$, which is only valid for z_3.

The sketch of the root locus for our example is shown in Fig. 5.11, where the asymptote located at $\sigma_a=1.6176$ has been depicted with a dashed line. Note the rupture point and the selection of the compensator $K=71.3$ which moves two of the closed-loop poles to $z=0.952\pm 0.306j$. This is the limit K for assuring the stability of the system since $\sqrt{0.952^2+0.306^2}=1$, and higher values of K will move these poles out of the unit circle.

The MATLAB rltool Toolbox

The *rltool (Root Locus Tool)* is a MATLAB guided user interface (GUI) used to perform the root locus analysis of linear and invariant single-input single-output, continuous and discrete systems.

This application provides a useful tool to carry out both the design and the testing of controllers by using the root locus drawn. In this way, it is possible to change the proportional parameter, *K*, and/or to add poles/zeros and see directly the results by inspecting the system response when closed-loop poles are moved along its root locus. Besides, this user-friendly tool enables the simultaneous analysis and design of controllers by using not only transient response specifications but that of frequency response also.

The rltool can be invoked by typing at the MATLAB command window:

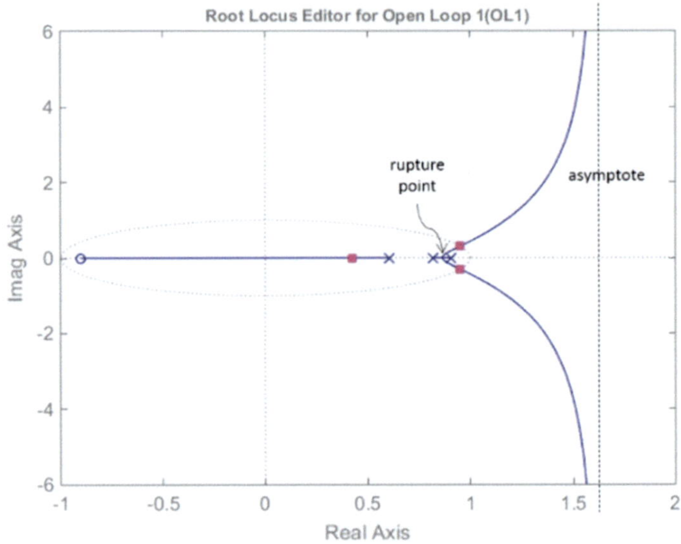

Fig. 5.11 Root locus for a discrete system. The drawing methodology resembles the following for continuous system, albeit the position of the poles contributes differently given the *s*–*z*-plane mapping

```
>> rltool
```

or else by:

```
>> rltool(G)
```

In case, we have previously created the transfer function *G*, that is the open-loop plant. In both cases, the *Control System Designer* window is activated, showing the closed-loop configuration considered, where *F*, *C*,[4] *G* and *H* stand for filter, controller, plant and sensor, respectively. Transfer functions for these elements can be imported from the MATLAB command window, by selecting the *Edit Architecture* option (Fig. 5.12).

For instance, considering the continuous example of the insulin pump, the root locus can be created by:

```
>> s=tf('s');
>> G=(1/(s+2))*(1/(s+1))
>> rltool(G)
```

[4]The rltool interface refers to the variable parameter *K* as *C* which stands for *compensator*.

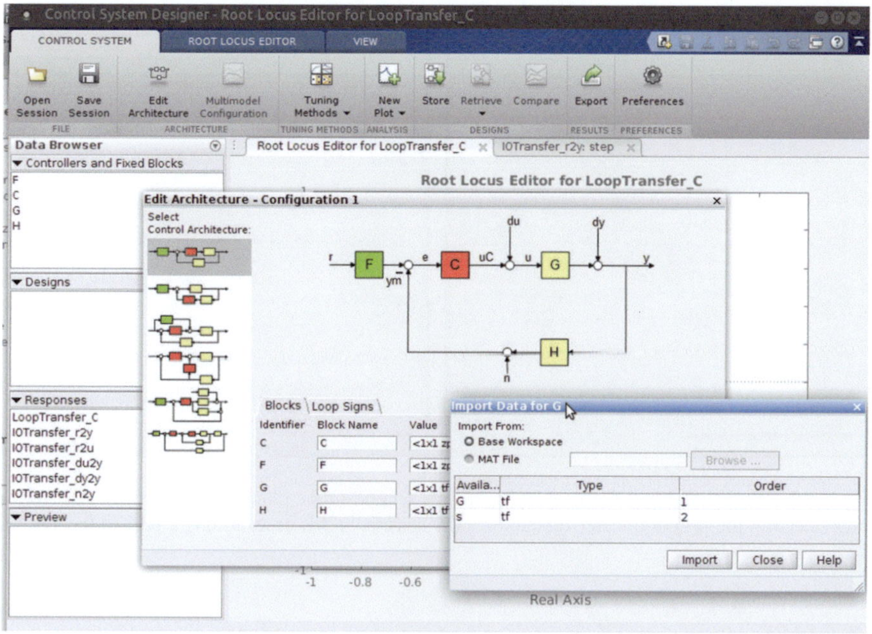

Fig. 5.12 Selection of closed-loop components from the control and estimation tool manager

The *rltool* GUI is opened showing the root locus for G for the case of controller compensator $C = 1$ and $H = 1$. Figure 5.13 left corresponds to the closed-loop system considering an ideal sensor, that is $H = 1$. In order to get the root locus of the complete system, the dynamics of the sensor must be included in the system by importing its transfer function through the *system data* import option. After doing that, we obtain the root locus of the system as shown in Fig. 5.13 right.

Note that the closed-loop pole locations are represented by pink squares. Such markers can be manually dragged along the root locus which will change the value of the gain parameter, i.e. C. That is, we can graphically move the closed-loop poles to a desired location, while the *rltool* application yields the resultant gain value. Contrarily, we may wonder where will be located the poles for a given value of the gain, and thus, the application let us to specify the parameter C in the *Current Compensator* window (Fig. 5.14), automatically updating the position of the closed-loop poles.

The *Compensation Editor* panel allows us designing a more convenient controller rather than the simple proportional one. Thus, the designer can add/modify/delete poles and zeros by selecting the *Pole/Zero* option. This operation can also be made by right-clicking and accessing to the *Add Pole/Zero* or *Delete Pole/Zero* option. Note that the poles and zeros from the original system (in blue colour) cannot be modified.

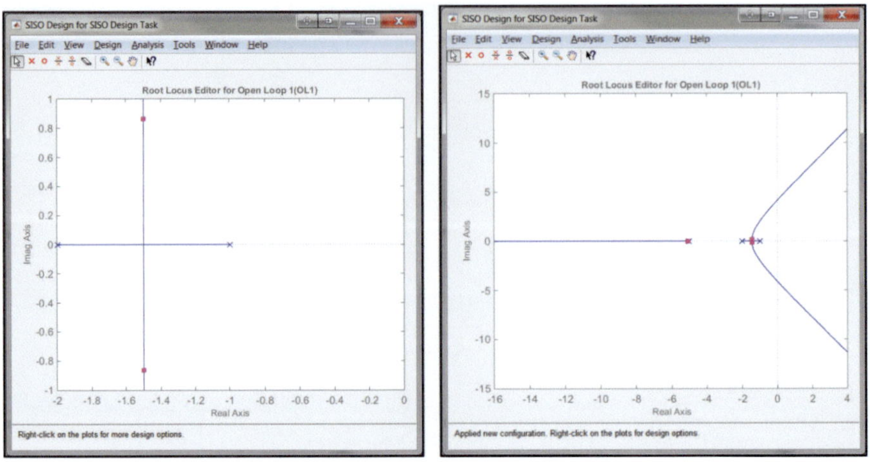

Fig. 5.13 Rltool interface. Left, opening with a second-order plant, considering $H = 1$. Right, after selecting the proper transfer function for the sensor; in this case, adding a pole to the system.

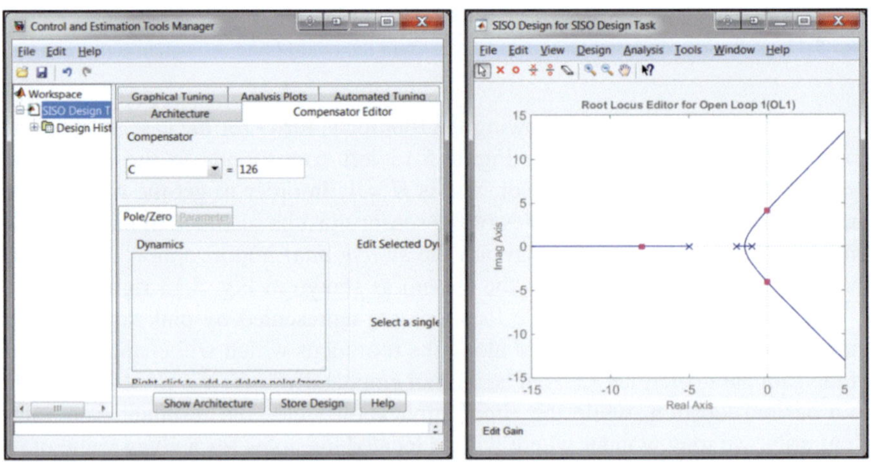

Fig. 5.14 Compensator Editor Window of rltool. In this example, the critical value of the parameter C has been selected, which moves the poles to the complex axis

Let us include a compensator to the controlled system, for instance a *PI controller*. This type of controller adds an integrator, i.e. a pole at $s = 0$ and a zero that will be conveniently placed by the designer, as will be explained in the next chapter. The resultant root locus after adding the integrator and a zero at $s = -5.52$ is shown in Fig. 5.15.

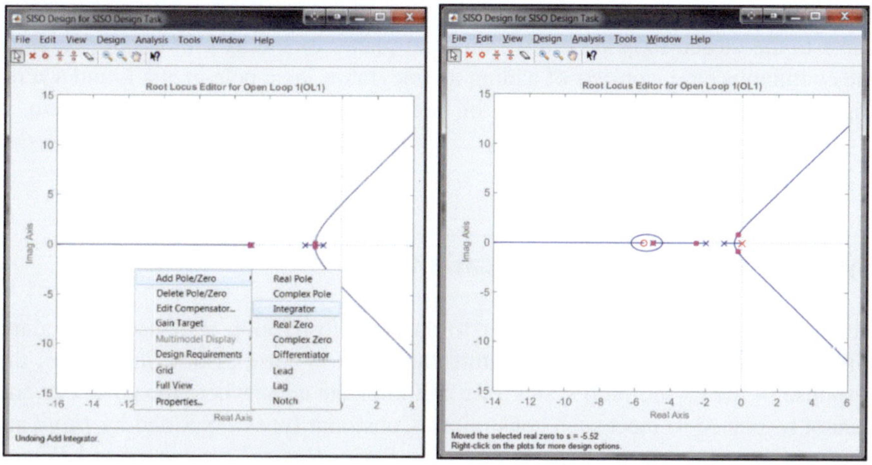

Fig. 5.15 Adding poles and zeros in the rltool. Left, the operation can be carried out through the contextual menu by right-clicking on the *s*-plane. Right, the resultant root locus after the addition of an integrator and a zero. Note that the additional poles/zeros, i.e. those that correspond to the designed controller, are depicted in red

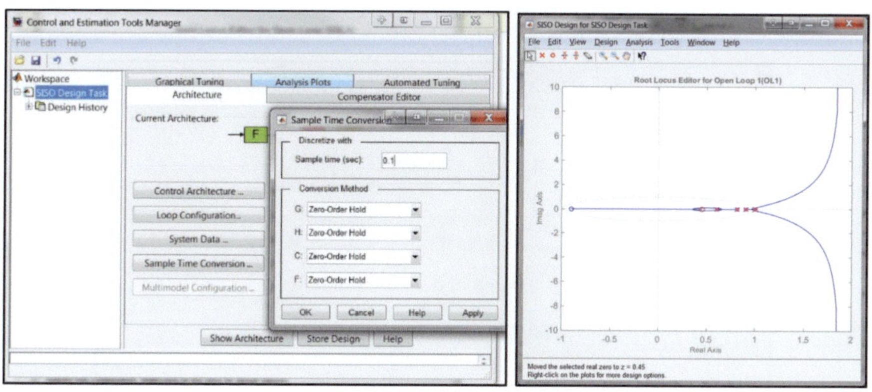

Fig. 5.16 Root locus of a discrete system. The rltool GUI offers the option to discretize the systems involved in the closed loop (left), accordingly producing the root locus of the discrete system (right)

In the case of discrete systems, we can proceed similarly, that is, creating the discrete transfer functions an invoking rltool accordingly. An alternative is to convert the continuous system into discrete directly from the *Control and Estimation Tools Manager*, by selecting the option *Sample Time Conversion*. The tool enables us to select the sample time and the conversion method for each component of the loop. Thus, selecting a sample time of, for example, 0.1 s, we will obtain the discrete root locus shown in Fig. 5.16.

In the same way, the designer can modify the controller by adding poles and zeros to the closed-loop system within the z-plane. Including a PI controller, as in the continuous case, consists of adding an integrator, i.e. a pole in $z = 1$, and a zero. Figure 5.16 right shows the result of adding such an integrator and a zero at $z = 0.45$ to the discrete system.

System Performance Specification

The *rltool* also enables the designer to specify closed-loop response constraints. Given that the location of poles within the plane (s- or z-planes) determines the response of the closed-loop system, we can infer the areas where the poles can and cannot be placed to meet such constraints. Constraint types supported by *rltool* are *Settling Time*, *Percent Overshoot*, *Damping Ratio*, *Natural Frequency* and *Region Constraint* which can be selected by right-clicking on the main windows (see Fig. 5.17).

According to the specified type of restriction, the resultant area varies. Figure 5.18a, b shows the areas with respect to the natural frequency and damping ratio. In this case, figures show the areas for $\omega_n > 1.67$ rad/s and $\xi > 0.29$, where

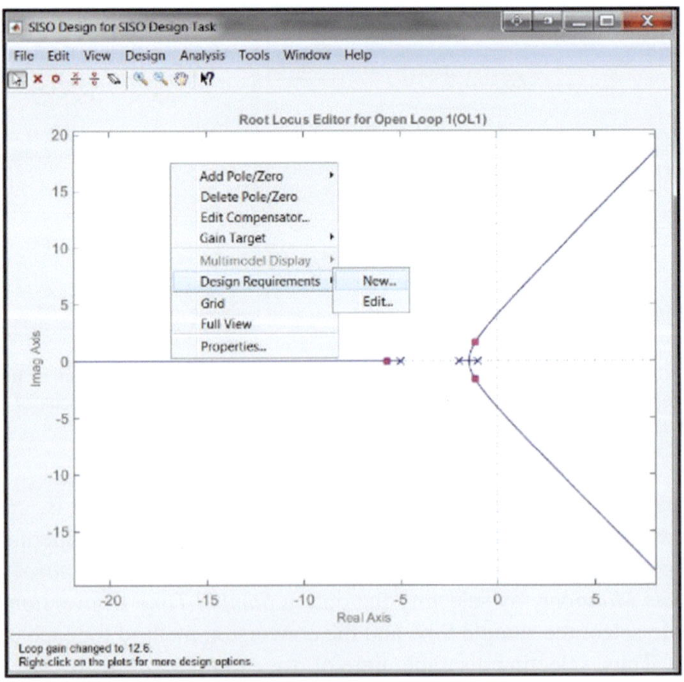

Fig. 5.17 Design requirements setting of *rltool*

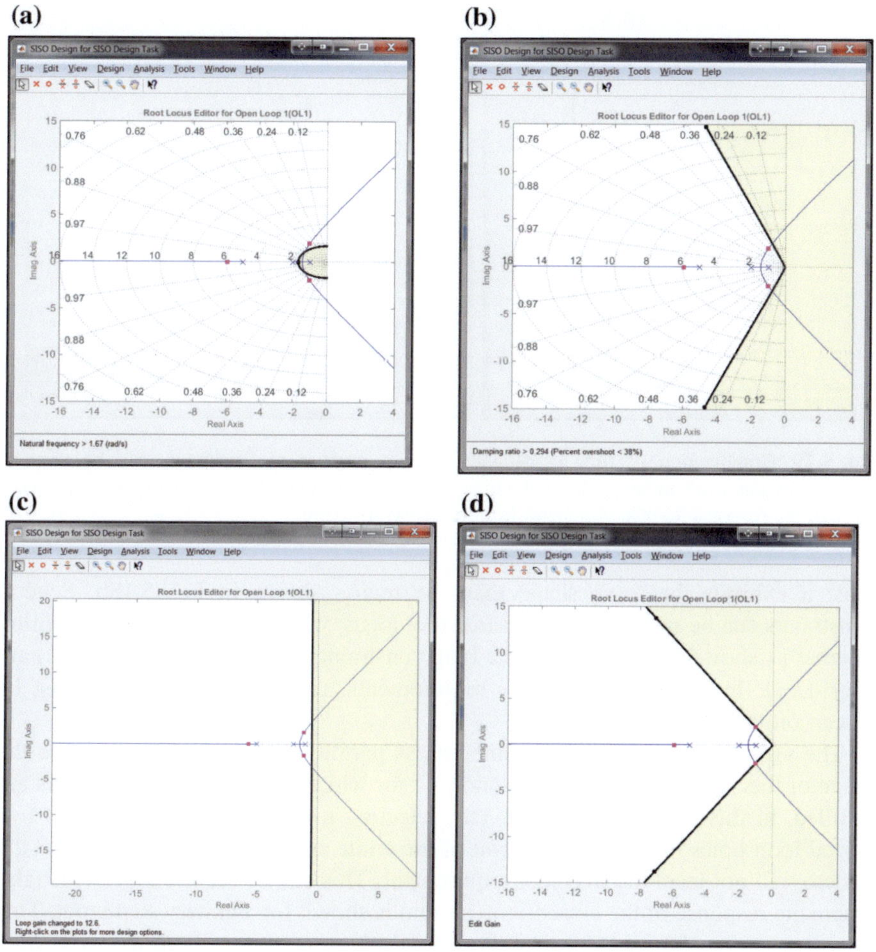

Fig. 5.18 Specification of closed-loop response constraints. rltool determines the areas where closed-loop poles must be located in order to achieve a given set of constraints. **a** and **b** illustrate, respectively, constraints related to the natural frequency and damping ratio. **c** Specify a restriction on the settling time and **d** related to the maximum overshoot

the yellow zones indicate the portion of the plane where the poles must not be placed in order to meet the requirement. For these constraints, the shape of the areas coincides with the grid that establishes the points with the same natural frequency and damping ratio. The grid can be activated by right-clicking on the window and selecting it from the pop-up menu.

Figure 5.18c, d illustrates the areas for a given constraint with respect to the settling time, e.g. $t_{set} < 8$ s and a percent overshoot $M_p < 20(\%)$. Note that in the s-plane the settling time only depends on the real part of the poles, and thus, the area becomes a vertical line, while the overshoot is closely related to the damping ratio;

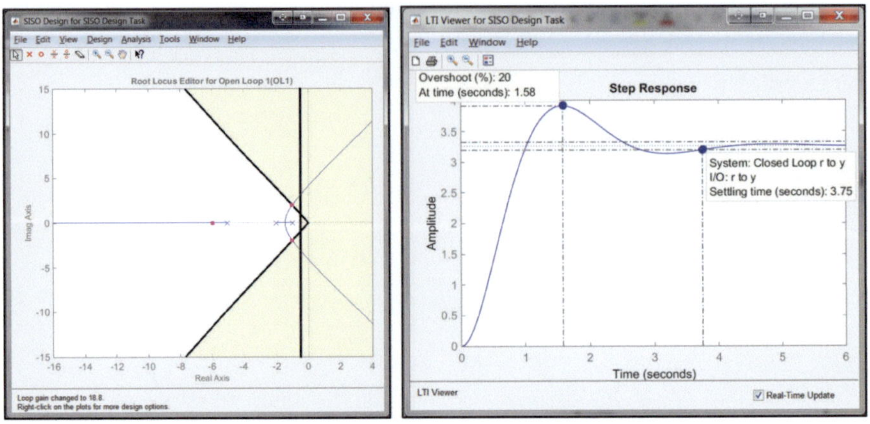

Fig. 5.19 Combination of time constraints. Left, a number of constraints can be defined, overlapping the resultant areas. Right, the time response of the closed-loop system can be observed through the response to step command from the analysis menu

thus, it exhibits the same shape than the region shown in Fig. 5.18b. Several constraints can be specified for the same root locus, which will result in the addition of areas as shown in Fig. 5.19 where both constraints $t_{set} < 8$ s and $M_p < 20(\%)$ are considered. In order to fulfil these requirements, the closed-loop poles must be placed on the white portion.

The visualization of the constrained areas permits the designer to compute the range of the values for the compensator, C, for which the desired requirements are fulfilled. In this case, see Fig. 5.19 left, selecting a value of $C = 18.8$ makes two closed-loop poles moving till the limit of the white area, and thus, for $C > 18.8$ the system will not meet the overshot requirement. This can be seen in Fig. 5.18 right, where the response of the closed-loop system is shown for a unitary step input. This response can be obtained from the rltool *Analysis* menu, selecting *Response to Step command*. Note that when the *Real-Time Update* option is marked (bottom left corner of the image), the output of the system will be updated, while the designer moves the closed-loop poles (changes the compensator). This option largely helps in assessing the effect of the position of the poles in the system response.

It is important to remark that the areas shown by rltool when specifying response requirements are only valid for systems with two dominant poles in the absence of near zeros, as is the case shown in the example. In general, this option serves as an initial guess about the position of the poles for achieving a desired behaviour of the system, albeit visualizing the actual response is highly recommended to confirm the result.

The same system performance design can be done for discrete systems. Given the differences between the s–z-planes presented in Chap. 3, the constraints areas exhibit different shapes to those of the continuous case. Figure 5.20a, b shows the areas for specifying a given natural frequency and damping ratio. Concretely, Fig. 5.20a illustrates the case for $\omega_n > 1.68$ rad/s and $\xi > 0.5$, with the grid option activated. The same constraints regarding the settling time and percent overshoot

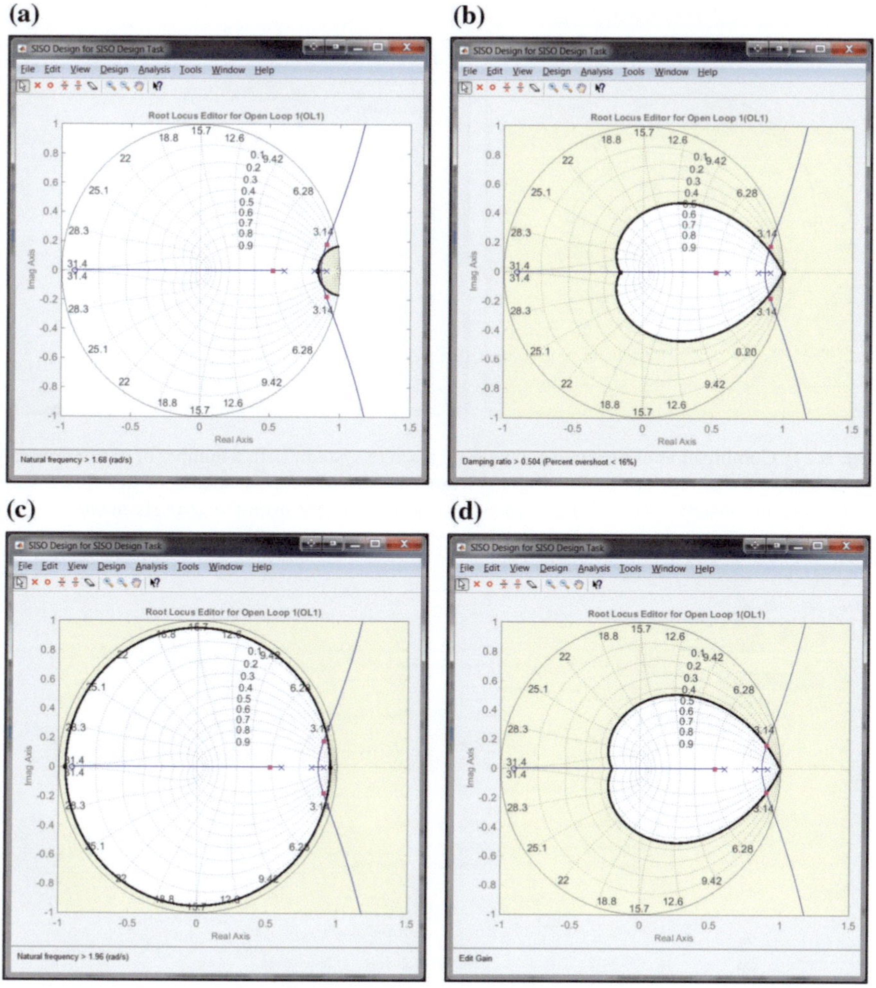

Fig. 5.20 Specification of closed-loop response constraints for a discrete system. **a** and **b** illustrate, respectively, constraints related to the natural frequency and damping ratio. **c** Specify a restriction about the settling time and **d** related to the maximum overshoot. Note the differences in the resultant areas, which come from the *s*–*z*-plane mapping

imposed for the continuous case are defined in the discrete case. The resultant areas are shown in Fig. 5.20c and d, respectively. Notice that the area for the settling time becomes a circle in the discrete case, which is the equivalent to the vertical frontier in the *s*-plane. Similarly, the percent overshoot constraint produces a region that resembles that for the damping ratio, given the direct relation between both.

The combination of these requirements, i.e. $t_{set} < 8$ s and $M_p < 20(\%)$, is depicted in Fig. 5.21 left over the root locus of the insulin pump system discretized with a sample time $T = 0.1$. In this case, the limit value for the compensator C that

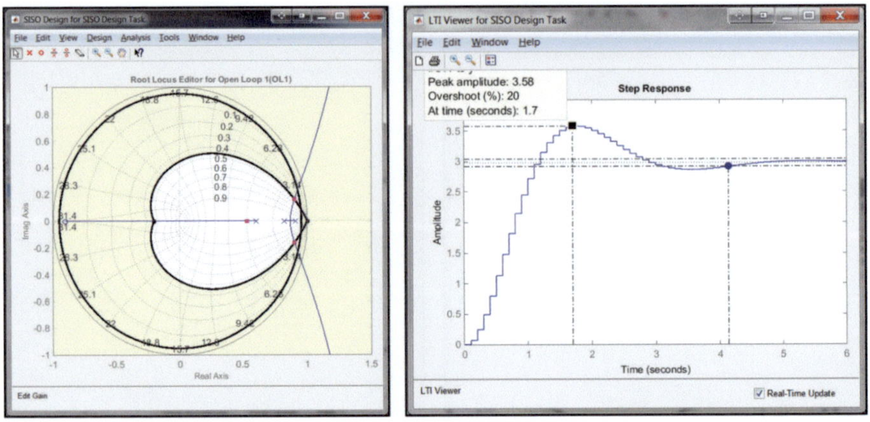

Fig. 5.21 Combination of time constraints for a discrete system. Left, a number of constraints can be defined, overlapping the resultant areas. Right, the discrete-time response of the closed-loop system can be observed through the response to step command from the analysis menu

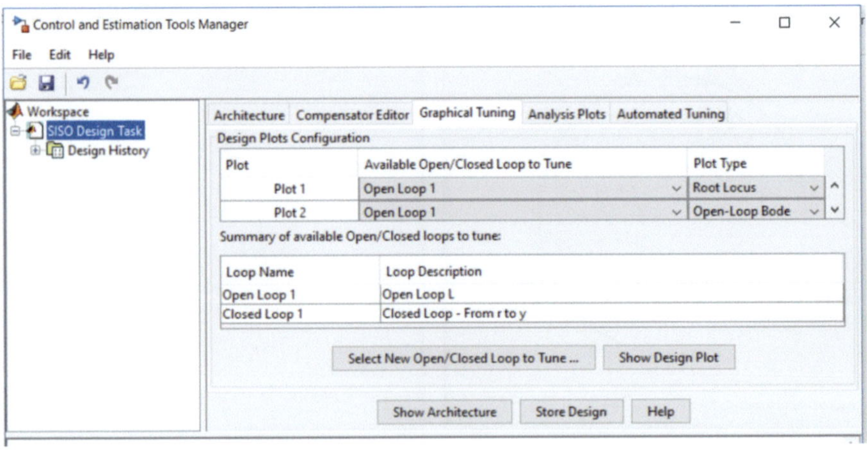

Fig. 5.22 Simultaneous visualization of root locus chart and open-loop Bode diagram. User should select both plots from the graphical tuning menu

fulfils both constraints is 14.7. Similarly, the time response is shown in Fig. 5.21 right.

The *rltool* also enables the designer to specify frequency response characteristics as alternative or complement to time response specifications. Constraint types as *Gain Margin, Phase Margin* for open loop and *Bandwidth* for closed loop can be visualized on Bode diagram together with the root locus diagram by selecting simultaneous visualization of root locus chart and open-loop Bode diagram (see Fig. 5.22).

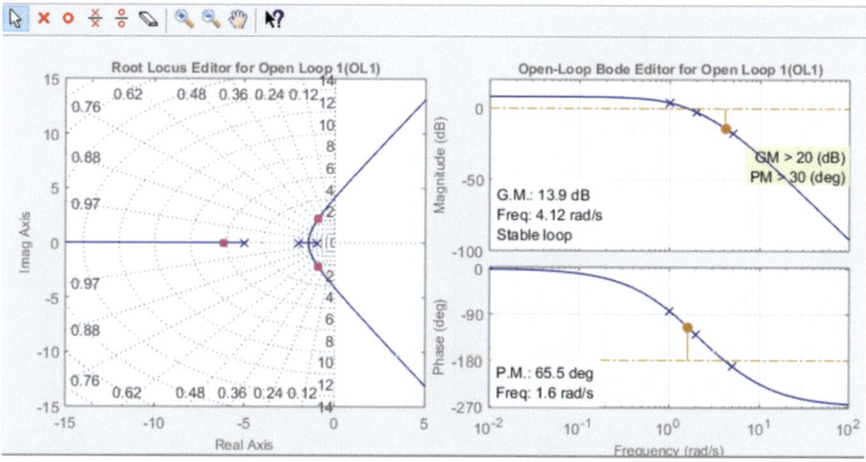

Fig. 5.23 Overall view of root locus diagram and open-loop Bode diagram. User can visualize time response characteristics and frequency response (PM, GM) altogether

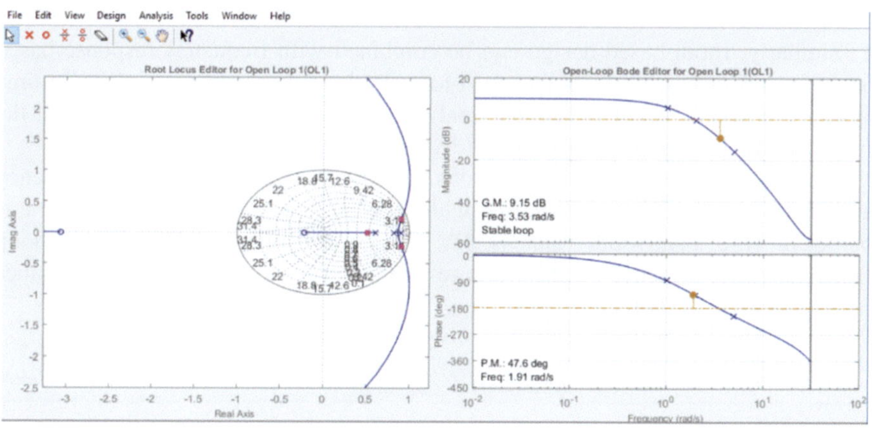

Fig. 5.24 Overall view of root locus diagram and open-loop Bode diagram for discrete-time systems. Frequency response specifications as GM and PM should be interpreted accordingly into z-plane

In this way, open-loop Bode specifications can be checked directly on gain and phase, while the controller is adjusted (Fig. 5.23). In this case, the figure shows the values corresponding to gain margin GM = 13.9 dB and phase margin PM = 65.5° where the poles and zeros of the open-loop system are superimposed on the Bode curves. It can be observed that the time response shown on the root locus (left-hand side plot) corresponds to that of Bode frequency response (right-hand side plot).

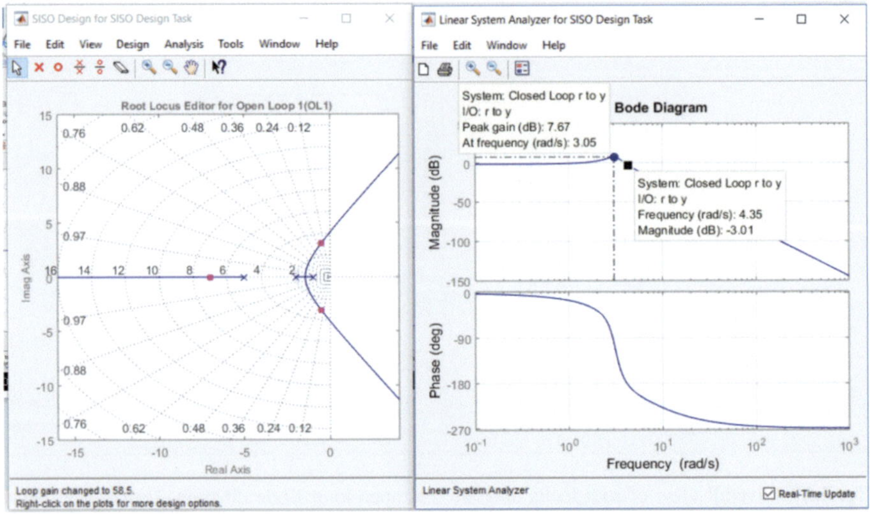

Fig. 5.25 Overall view of root locus diagram and closed-loop Bode diagram. User can visualize time response characteristics and frequency response (BW, M_p, ω_p) altogether

Similarly, rltool-based design can be combined with frequency response-based analysis and design through the open-loop Bode characteristics of discrete-time systems. It should be observed that analysis and design in z-plane must fulfil the frequency range limitation established by the sampling theorem (Fig. 5.24).

Also, closed-loop Bode specifications as peak resonance, resonance frequency and bandwidth can be easily checked with rltool as it is shown in Fig. 5.25. In this particular case, bandwidth value is BW = 4.35 rad/s obtained when closed-loop Bode gain curve reaches −3 dB while resonance peak $M_p = 7.87$ dB at resonance frequency $\omega_p = 3.05$ rad/s.

Solved Problems

Problem 5.1

The intravenous administration of anaesthesia continuously involves the incorporation of a liquid into the blood system at a certain rate. The rate of drug entry into the vein can be represented as the flowrate. Not all the drug that enters is kept in the blood system. An amount is expelled from the human body. The rate of drug clearance is proportional to the concentration of drug in the blood system. The relationship between the drug entry and the drug concentration can be expressed by the following equation:

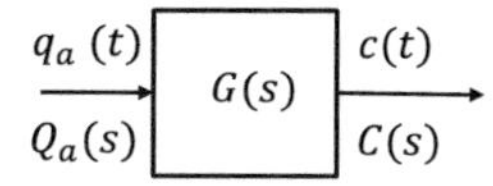

Fig. 5.26 Intravenous infusion model used in a general anaesthesia procedure

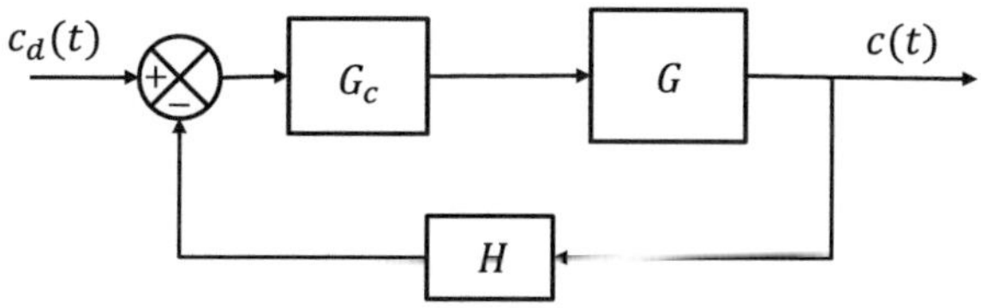

Fig. 5.27 Control system for a general anaesthesia procedure

$$q_a(t) = V_d \frac{\mathrm{d}c(t)}{\mathrm{d}t} + K_s c(t)$$

where K_s is the constant of proportionality between the amount of drug eliminated and the concentration of it in the blood system. V_d represents the apparent volume of distribution. This phenomenon can be considered as an LTI system ($G(s)$), Fig. 5.26.

In order to obtain the adequate concentration of anaesthesia in blood, the use of an intravenous infusion pump is considered. This pump allows a proportional control of the inflow rate of the drug. This pump is modelled as $G_c(s) = K$, being K, the constant of proportionality. The system is fed back to reach a desired concentration, $c_d(t)$. The monitoring of the actual drug concentration is inferred through the measurement of the bispectral index, BIS, a parameter that is developed from the bispectral analysis of the electroencephalogram. This sensor can be modelled as $H(s) = \mathrm{e}^{-Ts}$, with T, the time delay of the measurement. The architecture of the control system is shown in Fig. 5.27.

Data: $T = 0.5$ s, $K_s = 2$ and $V_d = 1$.

Note: It may consider a Pade approximation of order 1 for H.

Resolve the following questions for the feedback system:

(a) Indicate order and type of the system.
(b) Considering $K = 1$:

  i. Plot the map of poles and zeros. Justify if the system is stable.
  ii. Calculate, analytically, the overshoot and settling time for an 2% error.
  iii. Represent graphically the system response to a unit step input. Using this graph, calculate the overshoot and settling time for a 2% error. Compare the results with those obtained in the previous section. If there is any difference, justify it.

(c) Draw the root loci of the system with K as a variable. Identify the number of branches, the break points and the value of the critical stability gain as well as the oscillation frequency.

(d) Draw the specification places where the overshoot is less than 6% and settling time is below 4 s for a unit step input. Is there any value of K that satisfies the requirements?

(e) What is the steady-state error obtained with a unit step input, $c_d(t)$? Find the value of the gain, K, that makes the error less than 0.1.

(f) Which is the minimum modification of the controller, $G_c(s)$, that can be proposed to obtain a null steady-state error for a unit step input? Justify your answer.

(g) Find the root loci of the Discrete-Time Equivalent System (with sampling time, T_s, of 0.1 s and a zero-order hold) assuming the use of a digital gain controller, $G_c(z) = K$. Determine which value of the gain fulfils the detailed requirements in paragraph (d).

Solution 5.1

(a) The study of the transfer function of the system $G(s)$ can be obtained from the application of the Laplace transform to the following equation

$$q_a(t) = V_d \frac{\mathrm{d}c(t)}{\mathrm{d}t} + K_s c(t)$$

Applying the derivative property of the Laplace transform

$$\mathcal{L}\left[\frac{\mathrm{d}f(t)}{\mathrm{d}t}\right] = sF(s) - f(0^-)$$

And considering null initial conditions

$$Q_a(s) = V_d s C(s) + K_s C(s) = (V_d s + K_s) C(s)$$
$$G(s) = \frac{C(s)}{Q_a(s)} = \frac{1}{V_d s + K_s}$$
$$G(s) = \frac{1}{s+2}$$

For the sensor H (time delay function), it is possible to use the Padé approximation

$$H = e^{-Ts} = \frac{e^{-\frac{Ts}{2}}}{e^{\frac{Ts}{2}}} \approx \frac{1 - \frac{Ts}{2} + \frac{1}{2!}\left(\frac{Ts}{2}\right)^2 - \cdots}{1 + \frac{Ts}{2} + \frac{1}{2!}\left(\frac{Ts}{2}\right)^2 + \cdots}$$

Thus for an approximation of order 1

$$H \approx \frac{1-\frac{Ts}{2}}{1+\frac{Ts}{2}} = \frac{-s+4}{s+4} \triangleq H_{\text{pade}}$$

The feedback system can be represented by the following transfer function

$$G_{cl}(s) = \frac{G_c G}{1 + G_c G H_{\text{pade}}} = \frac{K\frac{1}{s+2}}{1+K\frac{1}{s+2}\frac{-s+4}{s+4}}$$

$$G_{cl}(s) = \frac{Ks + 4K}{s^2 + (6-K)s + 8 + 4K}$$

It can be seen that the feedback system is a second-order system. Similarly, it can also be seen that the open-loop system, $G_{ol} = G_c(s)G(s)H_{\text{pade}}(s)$,

$$G_{ol} = K\frac{1}{s+2}\frac{-s+4}{s+4}$$

is a 0 type system.

(b) The pole-zero map, considering $K = 1$, can be obtained using the `pzmap` command of MATLAB. The following code details the procedure to obtain the pole-zero map

```
s=tf('s');
G=1/(s+2);
T=0.5;
H=exp(-T*s);
Hpade=pade(H,1);
Gc=1;
Gcl=feedback(Gc*G,Hpade);
pzmap(Gcl)
```

which results in the graph shown in Fig. 5.28.

Since all the poles are in the left half-plane, the system is stable.

Given it is a second-order system, the following formulas to obtain the overshoot, M_p, and the settling time, t_s (for a 5% margin of error):

$$M_p = e^{\frac{-\zeta\pi}{\sqrt{1-\zeta^2}}}$$

$$t_s \approx \frac{4}{\zeta\omega_n}$$

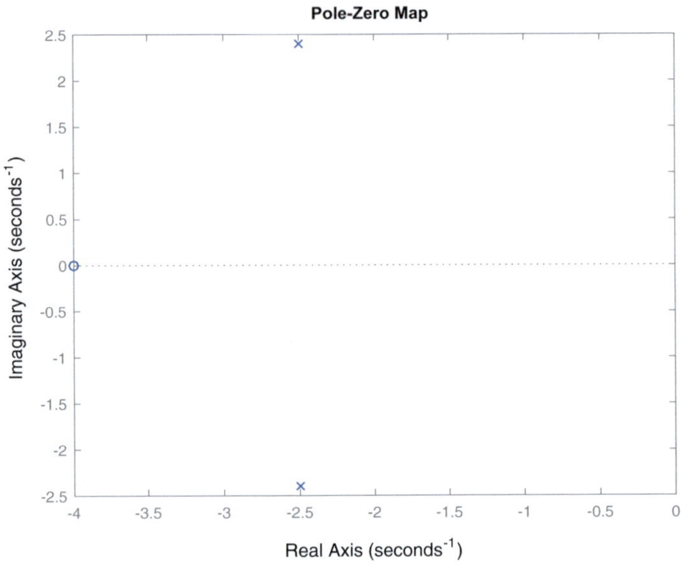

Fig. 5.28 Pole-zero diagram of the closed-loop transfer function $G_{cl}(s)$

Using the command `[wn,zeta]=damp(Gcl),` the natural frequency, $\omega_n = 3.4641$ rad/s, and the damping coefficient, $\zeta = 0.7217$, can be obtained. Thus, $M_p = 3,78\%$ and $t_s = 1,60$ s.

However, the system under study, although being a second-order system, does not correspond to the standard expression. The difference lies in the existence of a zero.

By applying the command `step(Gcl),` it can obtain the response shown in Fig. 5.29

where the following configuration has been made (2% convergence margin for the settling time) (see Fig. 5.30).

Thus, the overshoot and the settling time obtained are: $M_p = 7.68\%$ and $t_s = 1.49$ s. As can be seen, the settling time is similar to the calculated time in the previous section. However, the overshoot is significantly higher (7.68% vs. 3.78%). This is due to the presence of the zero in the actual response, compared to the approximated response by the analytical calculation based on a standard second-order system.

If we eliminate the effect of the zero, it can be seen that both responses (analytic and real without zero) are similar. Using the following code, the real answer can be obtained without the zero, where the gain value has been equalized. Two methods are used: elimination of the zero from the real function (`Gclaprox1`) and creation of the transfer function from the poles of the real function (`Gclaprox2`).

```
figure
step(Gcl)
```

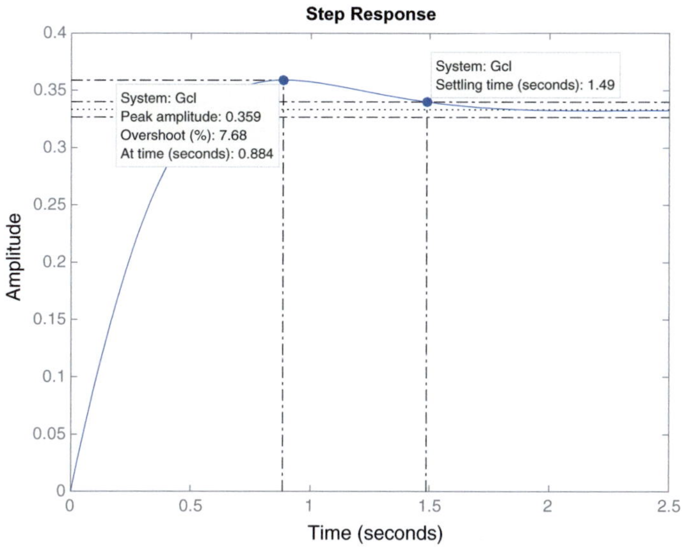

Fig. 5.29 Response to unit step input of the anaesthesia control system

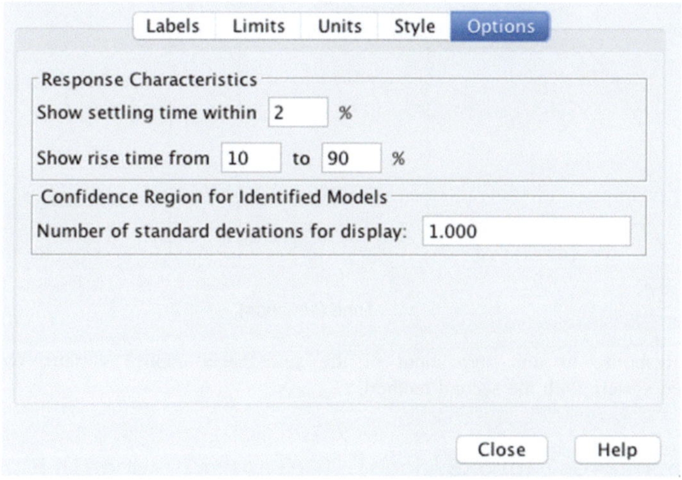

Fig. 5.30 Configuration of `step` tool metrics

```
p=pole(Gcl);
Gclaprox1=1/((s-p(1))*(s-p(2)))
Gclaprox1=Gclaprox1/dcgain(Gclaprox1)*dcgain(Gcl)
hold on
step(Gclaprox1)
Gclaprox2=minreal(Gcl/(s-zero(Gcl)))
```

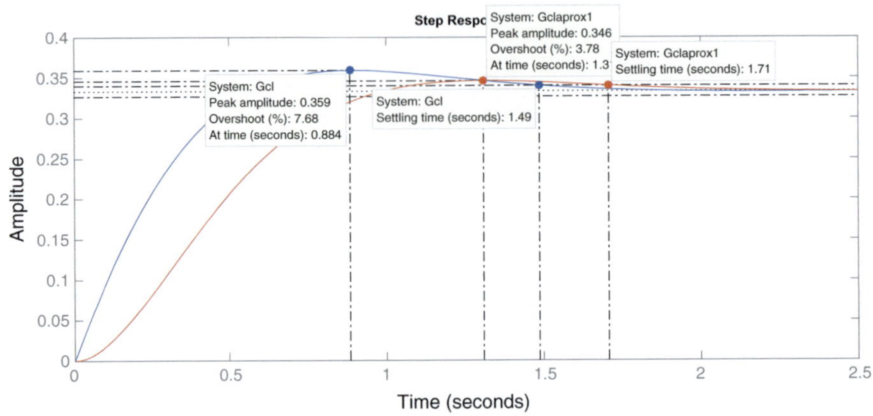

Fig. 5.31 Response to unit step input of the anaesthesia control system, $\boldsymbol{G_{cl}}$, and the approximated system with the first method, $\boldsymbol{G}_{\mathrm{claprox1}}$

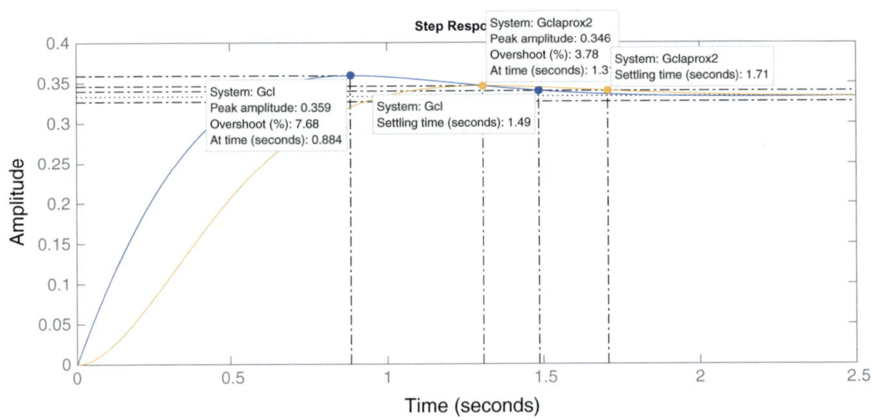

Fig. 5.32 Response to unit step input of the anaesthesia control system, G_{cl}, and the approximated system with the second method, $G_{claprox2}$

```
Gclaprox2=Gclaprox2/dcgain(Gclaprox2)*dcgain(Gcl)
step(Gclaprox2)
```

The graphs obtained are shown in Figs. 5.31 and 5.32.

As can be seen, `Gclaprox1` and `Gclaprox2` give identical results. The obtained overshoot, with this approximation, coincides with the analytical calculus, 3.78%. The settling time is similar to analytical result, 1.71 s. The difference lies in the method of analytical calculation of t_s, assuming an approximate expression.

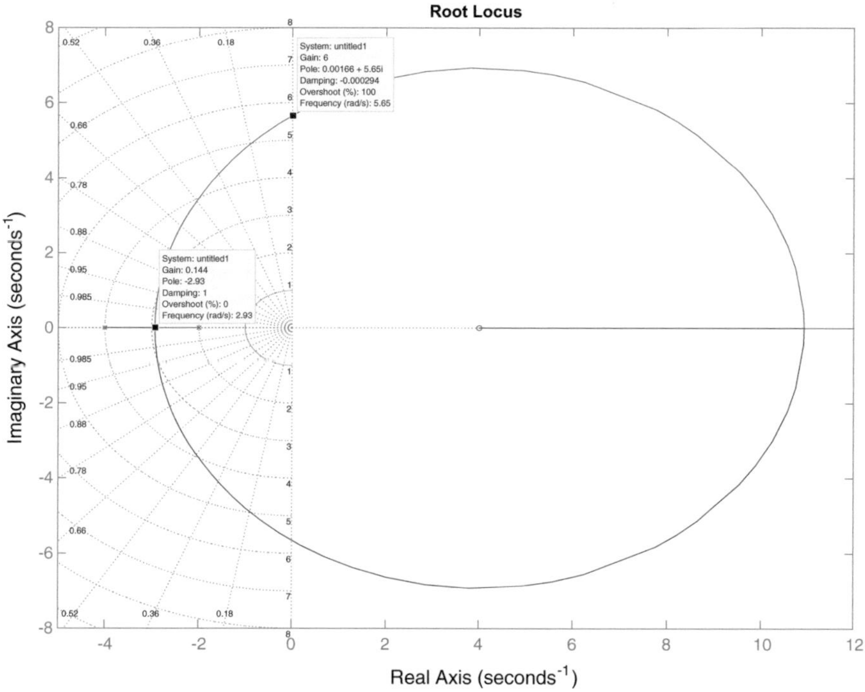

Fig. 5.33 Root loci of the anaesthesia control system

(c) In order to draw the root loci, it is possible to use the MATLAB `rlocus` tool, which presents a graphical interface with which the user can analyse the location of the poles of the closed-loop system as a function of its associated value of gain, K. The graphical tools `rltool` and `sisotool` integrate this graphical interface and allow, in addition, to study effects of the location of the poles and zeros in the unit step response.

To obtain the root loci of the proposed system, we use `rlocus` (considering $G_{ol} = G_c G H$ as open-loop transfer function, with $G_c = K$ and $H \approx H_{\text{pade}}$). The result is shown in Fig. 5.33, where the critical stability gain value ($K = 6$) and the break point (corresponding to $K = 0.144$ and oscillation frequency $\omega_n = 2.93$ rad/s) have been identified. Since the feedback system is a second-order system, there are two branches.

(d) The process to obtain the root loci that fulfil with the requirements of lower overshoot than 6% and lower settling time than 4 s can be realized with the aid of the tool rltool. Launching this tool with the command rltool (G_c*G*H) displays a warning message, suggesting the use of Pade approximation for time-delayed functions. In this case, it has selected a Pade approximation of first order (see Fig. 5.34).

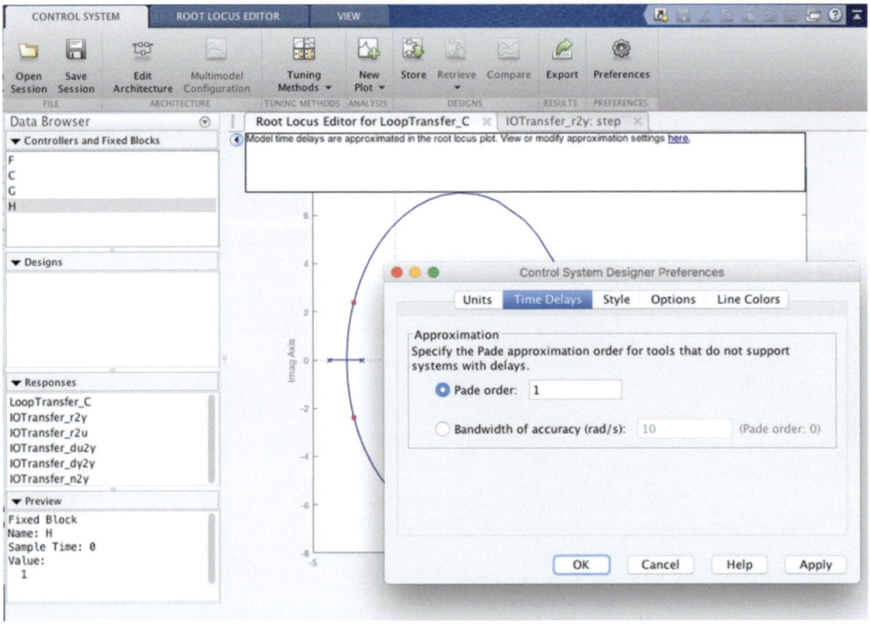

Fig. 5.34 Setting up Pade approximation in `rltool` tool

`rltool` allows to make a design based on requirements. Introducing the desired requirements, it is possible to obtain the root locus that satisfies the specified criterion of design. As an example, Fig. 5.35 shows the selection of a value of the gain of the controller ($K = 0.69$) that locates the poles in the permitted region according to the requirements.

(e) The steady-state error observed as a unit step input response can be obtained using two methods:

The first one is based on the use of the graphic representation obtained in the previous section. In the particular case of $K = 1$, see Fig. 5.36.

The final value is $c_\infty = 0.333$, so the steady-state error is $e_{ss} = 1 - c_\infty = 0.667$.

The second method consists of the analytical calculus of the steady-state error, $e_{ss} = \lim_{t\to\infty} e(t)$.

To obtain this limit, the final value theorem can be applied:

$$e_{ss} = \lim_{t\to\infty} e(t) = \lim_{s\to 0} sE(s)$$

Applying the expressions of the signals and the systems in the Laplace domain:

$$E(s) = C_d(s) - H(s)C(s) = C_d(s) - H(s)G_c(s)G(s)E(s)$$

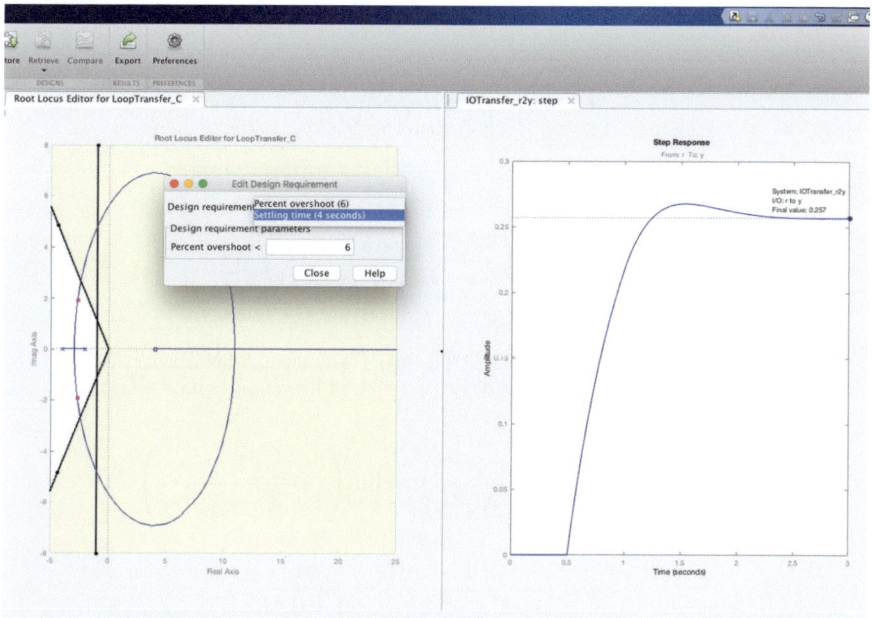

Fig. 5.35 Design based on `rltool`

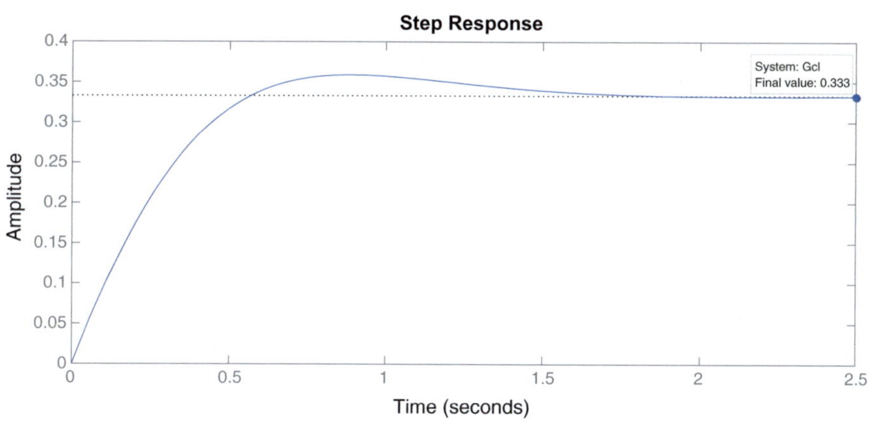

Fig. 5.36 Calculus of the steady-state error, using the graphical tool `step`

$$E(s)(1+H(s)G_c(s)G(s)) = C_d(s)$$

$$E(s) = \frac{C_d(s)}{1+H(s)G_c(s)G(s)}$$

yields:

$$e_{ss} = \lim_{s\to 0}\left(s\frac{C_d(s)}{1+H(s)G_c(s)G(s)}\right)$$

$$e_{ss} = \lim_{s\to 0}\left(s\frac{1/s}{(1+H(s)G_c(s)G(s))}\right) \approx \lim_{s\to 0}\left(s\frac{1/s}{(1+H_{\text{pade}}(s)G_c(s)G(s))}\right)$$

$$e_{\text{ss}} \approx \lim_{s\to 0}\left(s\frac{1/s}{1+\frac{-s+4}{s+4}K\frac{1}{s+2}}\right) = \lim_{s\to 0}\left(\frac{1}{1+K\frac{-s+4}{(s+4)(s+2)}}\right)$$

$$e_{ss} \approx \frac{1}{1+K\frac{-0+4}{(0+4)(0+2)}} = \frac{1}{1+0.5K}$$

In the case of $K = 1$,

$$e_{ss} \approx \frac{1}{1+0.5} = 0.667$$

that matches with the obtained value with the graphical tool.

The requirement that the steady-state error would be lower than 0.4 can be obtained by applying $e_{ss} \leq 0.4$; then

$$\frac{1}{1+0.5K} \leq 0.4$$

$$1 \leq 0.4 + 0.2K$$

$$K \geq \frac{1-0.4}{0.2} = 3$$

This result can be obtained using `rltool` (considering G_cGH as the open-loop transfer function, with Padé approximation of first order and $G_c = K$), by displacing the poles and analysing the effect on the unit step response; see Fig. 5.37.

(f) In order to obtain a null steady-state error when applying a unit step input, the type of the open-loop system must be greater or equal than 1. Since the open-loop system type is 0, the controller must be modified, including a pole at

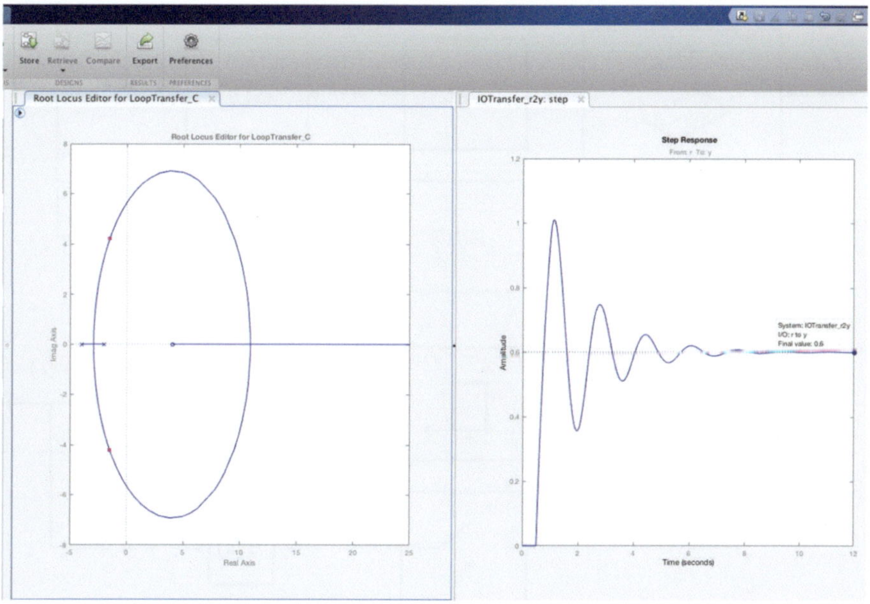

Fig. 5.37 Calculus of the steady-state error, using the graphical tool `step`

the origin (integrator). This way the system type is 1, and consequently, the position error constant, K_p, reaches an infinite value. Thus, the error is null

$$e_{ss} = \lim_{t\to\infty} \mathrm{e}(t) = \frac{1}{1+K_p}$$

Being

$$K_p = \lim_{s\to 0}(H(s)G_c(s)G(s)) \approx \lim_{s\to 0}\big(H_{\text{pade}}(s)G_c(s)G(s)\big)$$

$$K_p \approx \lim_{s\to 0}\left(\frac{-s+4}{s+4}\frac{K}{s}\frac{1}{s+2}\right) = \frac{-0+4}{0+4}\frac{K}{0}\frac{1}{0+2} = \infty$$

Therefore

$$e_{\text{ss}} = \lim_{t\to\infty} e(t) = \frac{1}{1+K_p} = \frac{1}{1+\infty} = 0$$

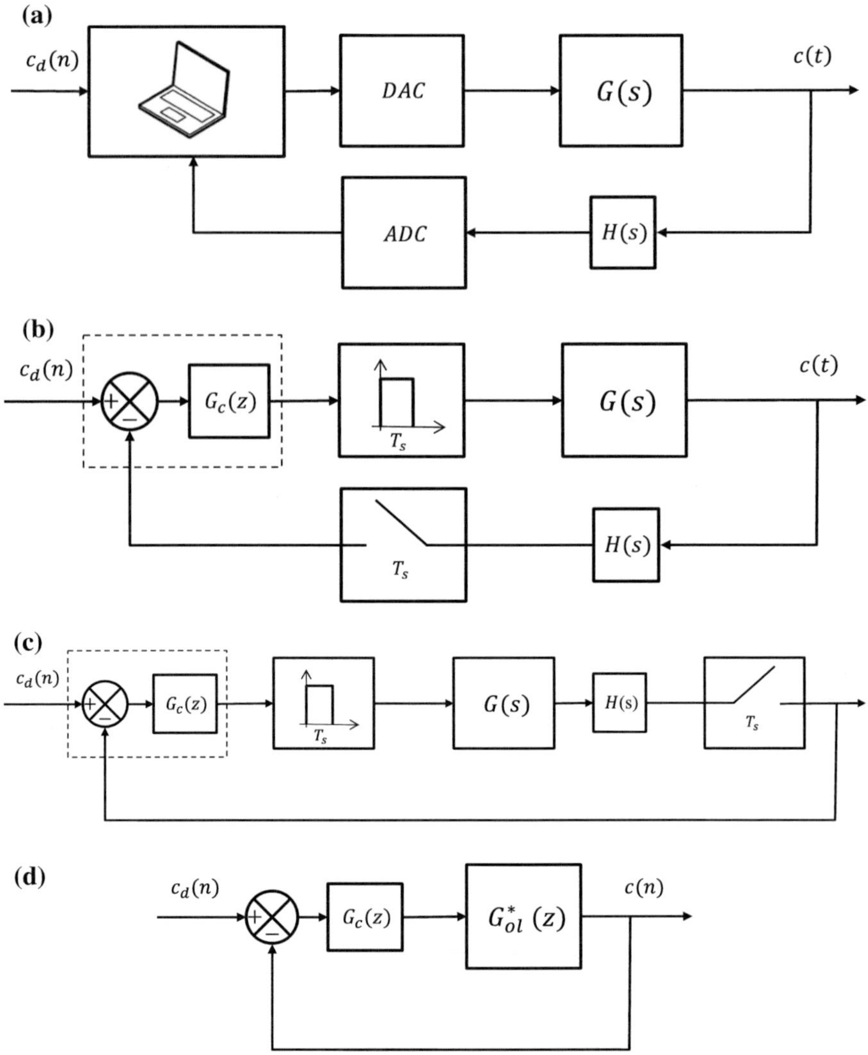

Fig. 5.38 Discretization process of an intravenous infusion model

as we wanted to demonstrate.

(g) The discretization of the system can be made assuming an analog-to-digital conversion process (ADC), implemented with a zero-order hold, and a digital-to-analog conversion (DAC) process, implemented with a sampler with a sampling time $T_s = 0.1$s, Fig. 5.38.

This discretization process can be described by the following MATLAB code.

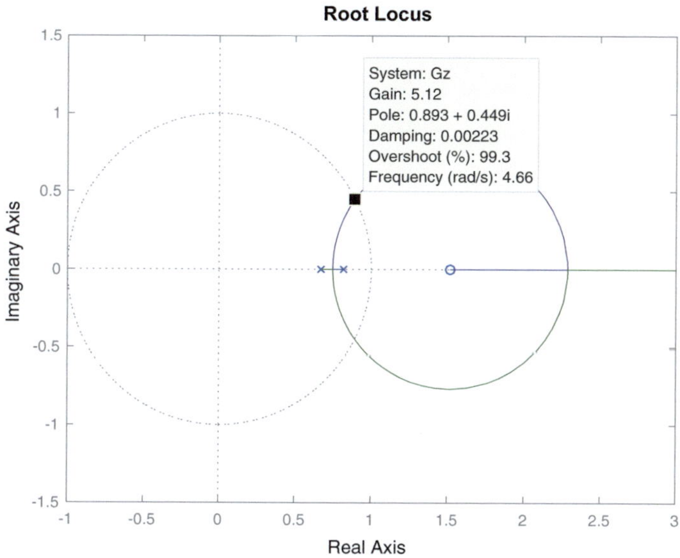

Fig. 5.39 Root locus of the discretized process

```
figure
Ts=0.1;
Golz=c2d(G*Hpade,Ts);
rlocus(Gz)
```

The result of the execution of the detailed code is depicted in Fig. 5.39, where it is shown that the controller must have a $K < 5.15$ so that the system is stable.

The design requirements of an overshoot below 6% and a settling time below 4 s can be done through the use of `rltool`. These requirements are fulfilled with $K < 0.95$ (see Fig. 5.40).

Problem 5.2

The pulmonary mechanics of the human respiratory system can be modelled according to the scheme shown in Fig. 5.41.

Air is considered to circulate along the trachea with an airflow, $q(t)$. This air input is associated with a pulmonary pressure, $p_{aw}(t)$, due to the occupied volume, $v(t)$.

The process of modelling has been done considering an approximation under the point of view of the existing pressure inside the pulmonary system. On the one hand, the trachea is considered as a cylindrical system that offers a resistance, R, at the entrance of air. On the other hand, the lungs are considered as an elastic

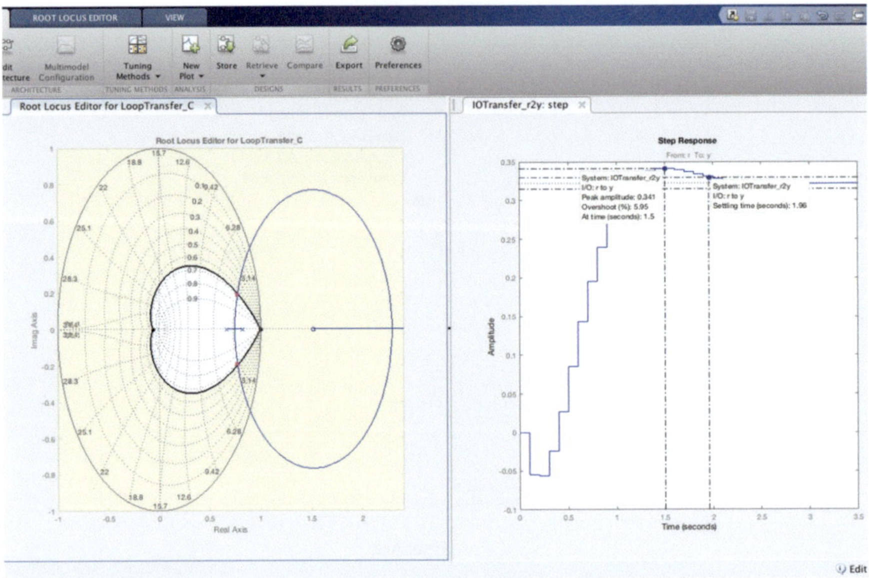

Fig. 5.40 Design of K based on the root locus of the discretized system

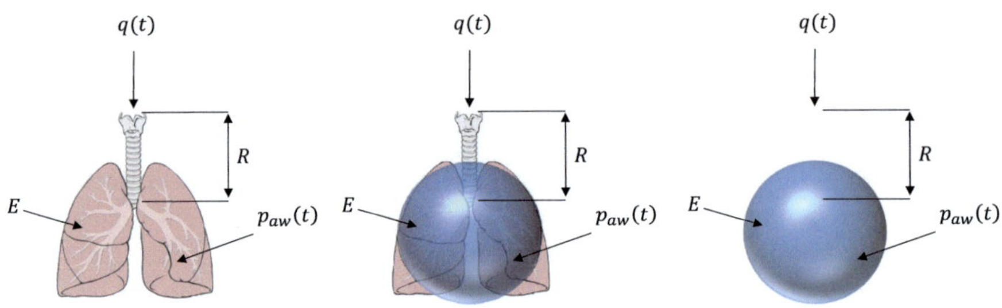

Fig. 5.41 Mechanical pulmonary ventilation system

compartment capable of a determined volume. The relationship between pressure and airflow is a property associated with the elasticity of the lungs, called elastance, E.

The behaviour of the system can be roughly described in terms of pressure with the following equation:

$$p_{\text{aw}}(t) = p_{\text{resistance}}(t) + p_{\text{elastance}}(t)$$

where the pressure effect, $\text{p}_{\text{resistance}}$, caused by the resistance of air while circulating through the system and the effect of elastic pressure, $p_{\text{elastance}}$, caused by the elastic characteristic of the lungs, have been considered.

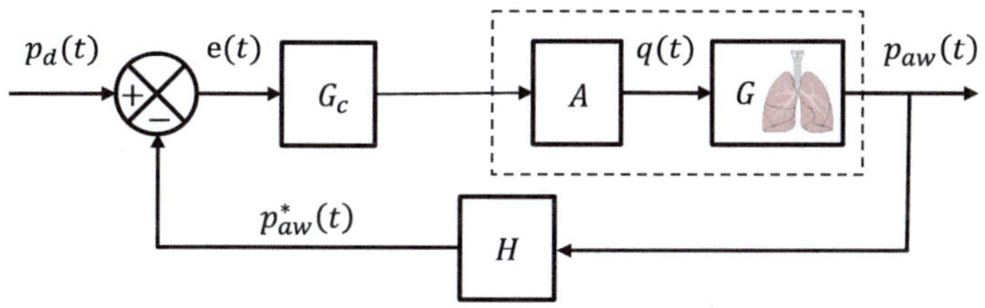

Fig. 5.42 Mechanical pulmonary ventilation control system

It is possible to relate the pressure, $p_{\text{aw}}(t)$, with the volume, $v(t)$, and the airflow, $q(t)$, by a lineal model:

$$p_{\text{resistance}} = Rv(t)$$

$$p_{\text{elastance}} = Eq(t)$$

$$p_{\text{aw}}(t) = Rv(t) + Eq(t)$$

Considering the relationship between the volume and the airflow $q(t) = \frac{\mathrm{d}v(t)}{\mathrm{d}t}$, and taking into account this model, a simple mechanical ventilation system consisting of a pressure sensor, H, a diaphragm piston pump, A, and a proportional controller, $G_c(s) = K$; see Fig. 5.42.

The pressure sensor and the pump can be modelled as first-order systems with gain 0.25 and 1, respectively, and time constants 0.25 and 1, respectively.

Consisting only of a proportional control effect on the pump, answer the following questions:

(a) Represent the evolution of air pressure, airflow and volume of the respiratory system during a 10 s inspiration process. Assume that the control is set up to have a desired unitary pressure, $p_d(t)$, with $K = 1$.
(b) Is it possible to determine a value of K that allows the following requirements: overshoot below 5% and settling time (with a tolerance error of 2%) below 5 s?
(c) Observe the effect of adding a real zero, z_a, around the point $(-1,\ 0)$, and plot the step response analysing whether it is possible to meet the requirements of the previous section.

In this study, the airway resistance (R) and the elastance (E) are set as constants, $R = 5$ cm H_2O_s/l, $E = 1$ cm H_2O/l.

Solution 5.2

(a) The linear model can be expressed considering the respiratory system as a transfer function that relates the airflow to the pressure, and using the Laplace transform, $G_{pq}(s)$:

$$\mathcal{L}[p_{aw}(t)] = \mathcal{L}[Rv(t) + Eq(t)]$$

$$P_{aw}(s) = RV(s) + EQ(s)$$

Considering $q(t) = \frac{dv(t)}{dt}$ and using the derivative property of the Laplace transform with null initial conditions:

$$Q(s) = sV(s) - v(0^-) = sV(s)$$

Therefore, substituting $V(s)$ in the equation of $P_{aw}(s)$ we obtain:

$$P_{aw}(s) = R\frac{Q(s)}{s} + EQ(s)$$

Consequently:

$$G_{pq}(s) = \frac{P_{aw}(s)}{Q(s)} = \frac{R + Es}{s}$$

The pressure sensor and the pump could be modelled as first-order systems with the following transfer functions:

$$H = \frac{0.25}{0.25s + 1} = \frac{1}{s + 4}$$

$$A = \frac{1}{s + 1}$$

The feedback system can be described analytically with the following expression:

$$G_{cl} = \frac{G_c A G_{pq}}{1 + G_c A G_{pq} H}$$

Considering $G_c(s) = K = 1$, it is possible to obtain the expression for the closed-loop transfer function of the system.

$$G_{cl} = \frac{K\frac{1}{s+1}\frac{R+Es}{s}}{1+K\frac{1}{s+1}\frac{R+Es}{s}\frac{1}{s+4}} = \frac{K(R+Es)(s+4)}{(s+1)s(s+4)+K(R+Es)}$$

$$G_{cl} = \frac{K\left[Es^2+(4E+R)s+2R\right]}{s^3+5s^2+(4+KE)s+KR}$$

The representation of the pressure observed on the respiratory system at the entrance of the trachea can be obtained executing the following MATLAB code:

```
s=tf('s');
E=1;
R=5;
Gpq=(R+E*s)/s;
A=1/(s+1);
H=1/(s+4);
Gc=1
Gcl=feedback(Gc*A*Gpq,H)

% Paragraph a
[paw,t]=step(Gcl,10);
q=lsim(1/Gpq,paw,t);
v=lsim(1/s,q,t);

figure
subplot(3,1,1)
plot(t,paw)
title('Pressure')
ylabel('p (cmH2O)')
xlabel('time (s)');
subplot(3,1,2)
plot(t,q)
ylabel('q (l/s)')
xlabel('time (s)');
title('Flow')
subplot(3,1,3)
plot(t,v)
title('Volume')
ylabel('v (l)')
xlabel('time (s)');
```

where the evolution of the airflow has been obtained from the feedback of the feedback system, $P_{aw}(s)$, and the transfer function of the pulmonary breathing system model, $G_{pq}(s)$:

$$P_{\mathrm{aw}}(s) = G_{pq}(s)Q(s)$$

Thus,

$$Q(s) = \frac{1}{G_{pq}(s)}P_{aw}(s)$$

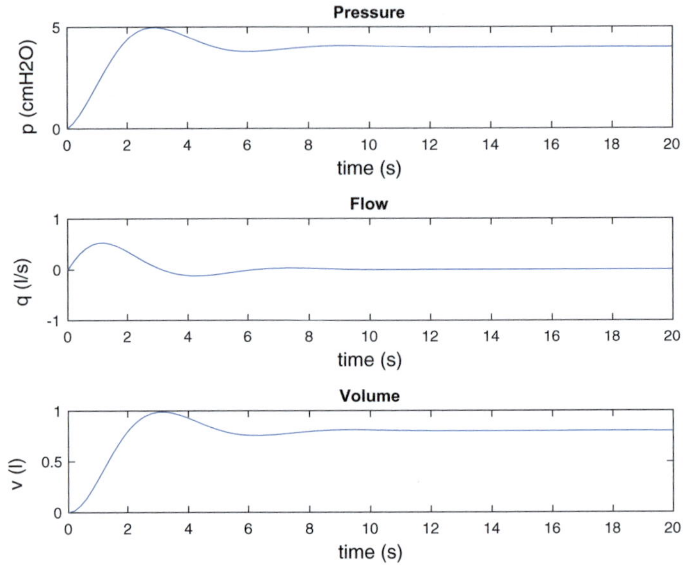

Fig. 5.43 Signals of the mechanical ventilation system

Using the MATLAB command `lsim` $q(t)$ could be obtained as a function of a given input, $p_{\mathrm{aw}}(t)$.

On the other hand, the volume, $v(t)$, can be obtained by integrating the flow, $q(t)$. In the Laplace domain:

$$V(s) = \frac{Q(s)}{s}$$

The result is depicted in Fig. 5.43.

It can be seen how initially the airflow starts from zero until it becomes maximum. As the lungs fill up, the volume and pressure of the air contained in the respiratory system increase. In the same way, the airflow decreases until it becomes zero at the moment in which the pressure and the volume are maximum.

(b) The design of the proportional controller, $G_c(s) = K$, is performed in order to fulfil with the desired requirements.

The regions correspond to the consideration of the behaviour of the feedback system as a second-order system (subdamped) where its dominant poles are located in

$$s_{1,\,2} = -\zeta\omega_n \pm j\omega_n\sqrt{1-\zeta^2}$$

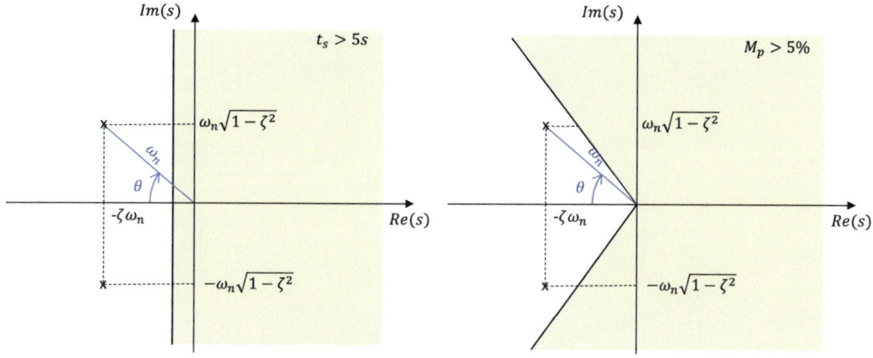

Fig. 5.44 Controller design based on the root locus

If we analyse, for this type of systems, the performance of the response to a single step input, the following expressions can be considered as valid:

$$M_p = e^{\frac{-\zeta\pi}{\sqrt{1-\zeta^2}}}$$

$$t_s(2\%) \approx \frac{4}{\zeta\omega_n}$$

It can be seen that, on one hand, the settling time depends inversely on $\zeta\omega_n$. Therefore, if it is required that it would be below a certain threshold, $\zeta\omega_n$ must be greater than that threshold multiplied by 4. This concludes that the valid region where the poles can be located is the one shown in Fig. 5.44.

On the other hand, the overshoot is a function only of the damping coefficient, ζ, which is related to the angle θ. Therefore, the requirement that the overshoot is below a certain threshold implies that the damping coefficient is below a certain value and consequently that θ is below a specific angle, as shown in Fig. 5.44.

For both specifications to be met, the poles must not be located in any of the prohibited zones (identified in yellow) in Fig. 5.44.

This design criterion is implemented in MATLAB with the use of the tool `sisotool` that allows the specification the plant, the controller, the sensor and the pre-filter as input of the function.

```
sisotool(A*Gpq,Gc,H,1)
```

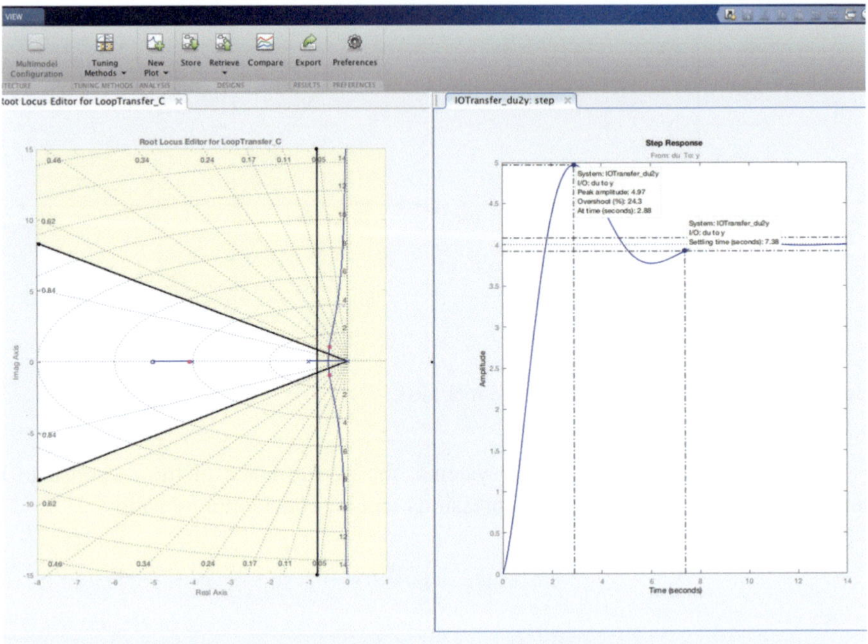

Fig. 5.45 Controller design based on the place of the root locus, using the tool `sisotool`

Figure 5.45 shows the graphical interface of `sisotool` where two prohibited regions are highlighted where the poles of $G_{cl}(s)$ cannot be located, in order to fulfil the two specified requirements: $M_p < 5\%$ and $t_s\ (2\%) < 5$ s.

As it can be seen, there is no geometric place where the poles of $G_{cl}(s)$ could be located so as to comply with the imposed specifications. Therefore, it can be concluded that there is no valid K value.

(c) Using the configuration utility of sisotool, it is possible to add a real zero with its associated gain. This way, the appropriate controller can be designed in order to fulfil the desired requirements; see Fig. 5.46.

The influence of the position of the zero can be analysed thanks to the help of MATLAB. The following code shows how to find the root loci with different values for the zero, z_a.

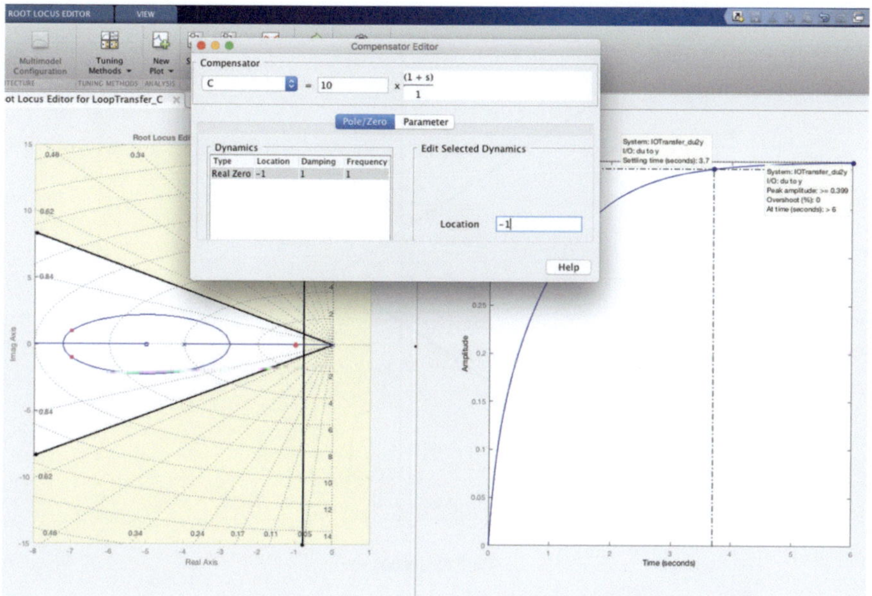

Fig. 5.46 Controller design based on the place of the root locus, using the tool `sisotool`. Analysis of the effect of adding a zero

```
% Construction of the root loci of the system
% for several values of a real zero

figure;
hold on;
G=A*Gpq*H;
h=rlocusplot(G,'b');
setoptions(h,'XLim',[-5,3]);
setoptions(h,'YLim',[-5,5]);

za=0.5;
Gc=(s+za); % zero
Ga=Gc*G;
h=rlocusplot(Ga,'m');

za=1;
Gc=(s+za); % zero
Ga=Gc*G;
h=rlocusplot(Ga,'g');

za=1.5;
Gc=(s+za); % zero
Ga=Gc*G;
h=rlocusplot(Ga,'r');
```

The result is shown in Fig. 5.47.

As can be seen, for the value of the real zero $(-1.5,\ 0)$, the root locus determines regions that allow to fulfil the requirements (specifically, for $K > 0.466$).

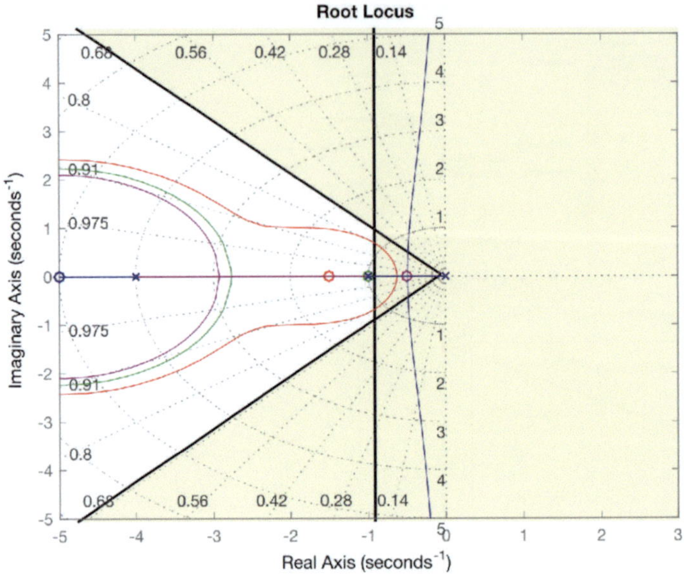

Fig. 5.47 Effect of adding a zero to a control system. Analysis through the root locus tool

Likewise, for the real zero value $(-1, 0)$ the regions that meet the requirements correspond with $K > 0.6$. Finally, for the selected value of the real zero $(-0.5, 0)$, there is no value of K that fulfils the requirements.

Control System Design

6

Control systems are aimed to modify the behaviour of an existing system in order to perform in a desired way. Several examples can be found in the real life in which certain control actions are needed to achieve the desired results. For example, the temperature of a room may change due to external and unexpected perturbations, e.g. somebody opens/closes a window, sunshine heats the walls. An air-conditioned system can be installed to help in keeping a given temperature in spite of any external change in the system. The air-conditioned system includes elements to modify the current temperature (also called actuators, since they act on the system), sensors to measure the current temperature and some control logic to determine the actuation to be carried out in order to achieve the target temperature. In this chapter, we will introduce the basis of the control systems, their influence on real-life systems and their design to modify the behaviour of a given system to fulfil particular specifications in the field of the biomedicine.

Control systems have been present in the society from the ancient past in an attempt to design and construct machinery with a particular purpose. For instance, the first records of the oldest control system date from the third century BC, when *Ktesibios* designed a water clock, knows as *Clepsydra*, based on a mechanism in charge of maintaining constant the water inflow of a tank. By means of a floating valve, the water volume of an auxiliary tank is maintained invariable and thus the inflow of water of the controlled tank constant. The water level of the latter indicates how much time is passing.

However, the real expansion and formal study of the control systems came into scene at the end of the twenty-eighth century with the arrival of the Industrial Revolution. Representative examples of this period are the *self-regulating wind machine* devised by E. Lee in 1745 which was aimed to automatically control the orientation and inclination of the windmill's sails according to the wind direction, improving thus the mill performance. Another example is the *steam engine centrifugal governor* envisaged by Watt in 1788 to regulate the speed of steam locomotives.

J. Fernández de Cañete et al., *Automatic Control Systems in Biomedical Engineering*,
https://doi.org/10.1007/978-3-319-75717-9_6

Nowadays, we can find control systems almost everywhere, from highly sophisticated and complex systems to maintain the speed and course of airplanes, to simple and cheap ones in charge of maintaining the temperature of an electric water heater. In biomedicine, control systems are also becoming crucial, for example in insulin pumps to keep constant the sugar level of a diabetic person or in left ventricular assist devices (LVADs).

With independence of their complexity, we can identify the following elements in any control system:

1. The desired output, also known as *reference*. For instance, the desired temperature in a typical temperature control system.
2. The *plant*, that is, the system whose output is controlled.
3. A *sensorial system* in charge of measuring the plant output.
4. An *acting system* that directly acts on the plant according to the inputs given by the controller.
5. *Perturbations*, that is, external and unforeseeable variations of the control system that can affect both the output of the plant, e.g. opening a door in a temperature control system, and the sensor measurements, e.g. a misleading sensor reading due to, for instance, electric interferences.
6. The *controller*, also known as *compensator*, responsible for calculating the proper inputs for the acting system (*control actions*) according to the difference between the reference and the plant output (*error*), in spite of perturbations.

The configuration of the components of a control system gives designers two different options: *open loop* and *closed loop*. The closed-loop configuration (see Fig. 6.1) is the most extended and useful configuration, probably also the most obvious way to control the output of a given plant since the continuous feedback provided by the sensorial system permits the controller to update its actions.[1] However, *open-loop* configuration (see Fig. 6.2) can be also found in certain simple and cheap control systems. In an open-loop configuration, no feedback is supplied (note that the sensorial system disappears) and the controller is wired to produce a fixed response to the given reference. Obviously, this type of control systems cannot manage perturbations and is only suitable for simple, cheap and deterministic systems.

When designing, the complexity and accuracy of the considered control system should be coherent with respect to the characteristics of the application at hand. Thus, in the temperature control example, the control system may be as simple as considering a closed-loop configuration in which the controller switches on/off the air-conditioned machine at a fixed power when the room temperature is higher/lower than a given threshold. This solution would be enough and cheaply implemented for the temperature control application in an apartment, for instance,

[1]The time required for completing a loop is critical and largely depends on the application at hand. For example, for controlling the sugar level of a diabetic patient, a couple of seconds (even minutes) can be enough, but for controlling the pump of a LVAD, it should be around tenths of seconds.

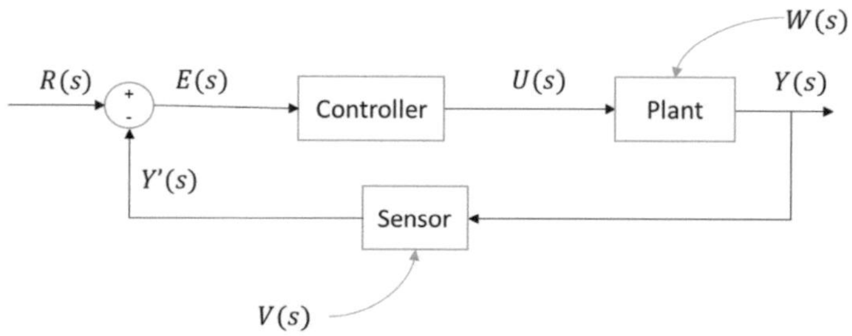

Fig. 6.1 Closed-loop control configuration. Perturbations in the plant and in the sensor are continuously managed by the controller, enabling the system to properly react

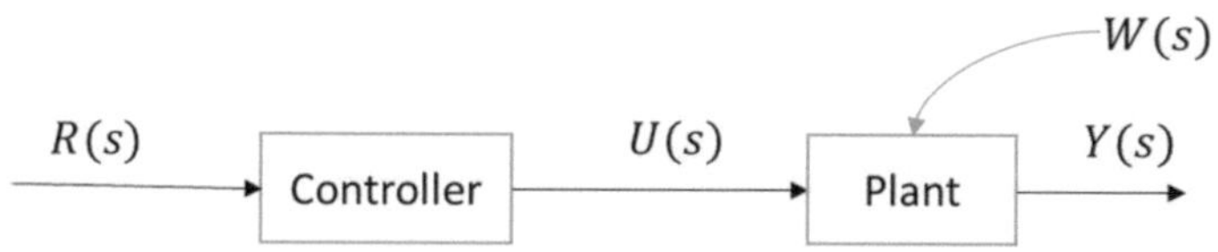

Fig. 6.2 Open-loop control configuration. The controller provides a fixed and pre-programmed control action based on the reference. Any perturbation on the plant will not be managed which would lead to catastrophic situations

but would be insufficient, even catastrophic for solving the same problem applied to a different context, e.g. controlling the temperature of the computing room of a bank, or a nuclear plant. That is, according to the special necessities and requirements, the control system to be designed should fulfil given specifications that normally are related to (a) the *swiftness* to achieve the desired output (or at least a bounded approximation to it) and (b) the *accuracy* of the obtained output with respect to the commanded reference.

Focussing on closed-loop systems, the most resorted type of compensators are as follows:

Bang-*bang compensator*. This strategy consists of providing two different and prefixed control actions according to the measured error. It is typically implemented by switching on/off the actuator when the error is higher/lower than a given threshold, thus the bang-bang control is also known as on/off control.

Proportional compensator. The proportional compensator produces a control action proportional to the error between the reference signal and the plant output.

$$G_c(s) = K_c$$

The higher the proportional constant, the lower the stationary error, but at the expense of higher oscillations during the transient.

Phase-lead compensator. This compensator considers a zero, s_z, and a pole, s_p, having that $|s_p| > |s_z|$

$$G_c(s) = \frac{K_c(s + s_z)}{s + s_p}$$

With this configuration if the pole is neglected with respect to the zero, i.e. $|p| \gg |z|$, then we obtain a proportional-derivative (PD) compensator.

Phase-lag compensator. Alternatively, the phase-lag compensator also includes a zero, s_z, and a pole, s_p, considering $|s_z| > |s_p|$. In the particular case that $|s_p| = 0$, we obtain a proportional-integrative (PI) compensator

Phase-lead-lag compensator. It combines previous cases, yielding a compensator with two zeros and two poles

$$G_c(s) = \frac{K_c(s + s_{z1})(s + s_{z2})}{(s + s_{p1})(s + s_{p2})}$$

fulfilling that $|s_{p1}| > |s_{z1}|$ and $|s_{z2}| > |s_{p2}|$.

These compensators can be physically implemented through on/off systems based on **relays** in the case of the bang-bang compensator, and using **operational amplifiers** for compensators based on the addition of zeros and poles.

A simple example of a bang-bang control based on a thermal relay can be the heating system for newborn incubators. A thermal relay is a bimetallic thermostatic switch that disconnects the power when a limit temperature is achieved and its internal metal components split up. When temperature goes down, the metal components of the thermostatic switch contract and they contacted, re-establishing the power again. That is, it works by switching on/off the heating system according to a pre-defined threshold.

Although this on/off control can be effective, the response of the controlled system exhibits severe inconveniences. The plant output, in this case the temperature of the incubator may slowly approach the reference, and what is worse, it may approach it in an oscillatory way. The following example illustrates these effects.

The SIMULINK diagram shown in Fig. 6.3 represents a closed-loop control system aimed to achieve the reference value, e.g. 36°, for the incubator temperature.

The control logic has been designed with a relay block configured to produce the control action *heat* when the error (the difference between the incubator temperature and the reference) is higher than a threshold, e.g. 0.5°, and *stop heating* when the error is less than −0.5°. The controlled response tends to the reference as shown in Fig. 6.4 left, although it exhibits a significant overshoot. This is due to the slow reactivity of the on/off controllers. Also, every time the relay switches the response oscillates which might be inadmissible for some critical systems, like the newborn incubator.

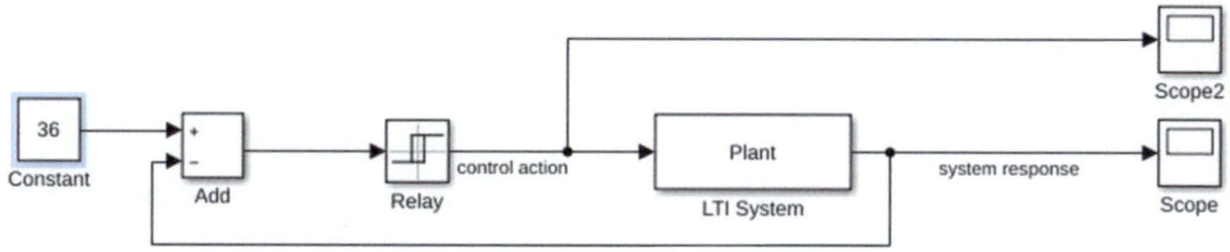

Fig. 6.3 Closed-loop control system based on a relay. The controller provides two fixed control actions depending on the error and on pre-established thresholds

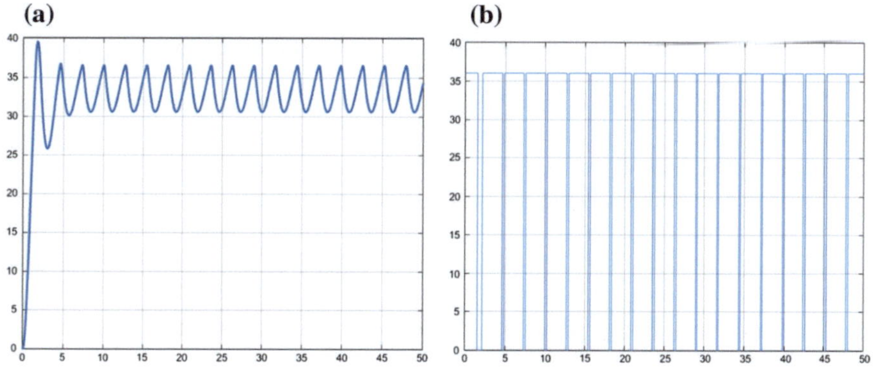

Fig. 6.4 Outputs from an on/off controlled system. **a** Controlled plant response. **b** Control action

The performance of on/off control systems is largely tight to the amplitude of the *hysteresis* area defined. In this example, the hysteresis occurs in the interval (−0.5, 0.5) producing a reduced number of relay's commutations at the expense of having appreciable oscillations of the controlled response at the steady state (see Fig. 6.4 right). From a theoretical point of view, a reduced interval for changing the control action is possible in order to improve the system response, even ideal relays can be considered where the commutation is produced at a crisp threshold, i.e. the switch on and off points are equal. However, from a practical point of view, the implementation of a control system through relays with a narrow hysteresis area produces the wear of the employed components due to the continuous activation/deactivation, reducing their operation life. On the other hand, considering a higher hysteresis area would bear in a higher steady time and oscillations in the response. In summary, on/off compensators provide cheap solutions to non-critical systems where non-precise responses are assumable, which actually is not the case of controlling the temperature of newborn incubators.

The physical implementation of compensators based on the addition to zeros and poles is normally carried out through operational amplifiers (see Fig. 6.5). Operational amplifiers are electronic devices with interesting properties that permit the design and implementation of a variety of operations.

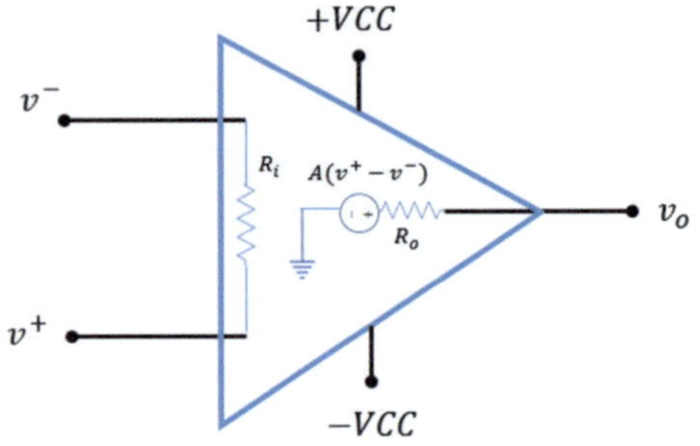

Fig. 6.5 Operational amplifier

The ideal operational amplifier shown in Fig. 6.5 exhibits the following features:

1. The input resistor R_i is infinite, and thus the voltage difference and the current between inputs v^+ and v^- are zero.
2. The output resistor R_0 is zero, i.e. ideal voltage source at the output.
3. The gain A is infinite, being the output voltage $A(v^+ - v^-)$.

Different configurations can be adopted, obtaining a particular behaviour. A simple one is shown in Fig. 6.6.

Applying Kirchhoff and assuming the properties aforementioned, we have

$$I_1 + I_2 = 0 \Rightarrow \frac{V_i}{R_1} + \frac{V_0}{R_2} = 0 \Rightarrow$$

$$V_0 = -\frac{R_2}{R_1} V_i$$

And thus, the transfer function of the ideal operational amplifier in the configuration shown in Fig. 6.6 is

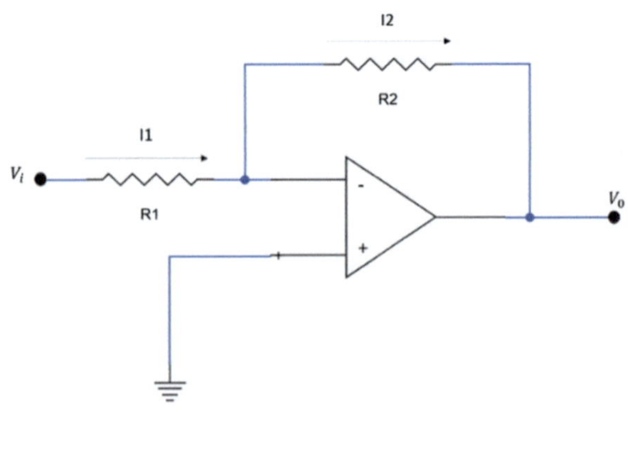

Fig. 6.6 Operational amplifier, implementing an inverter

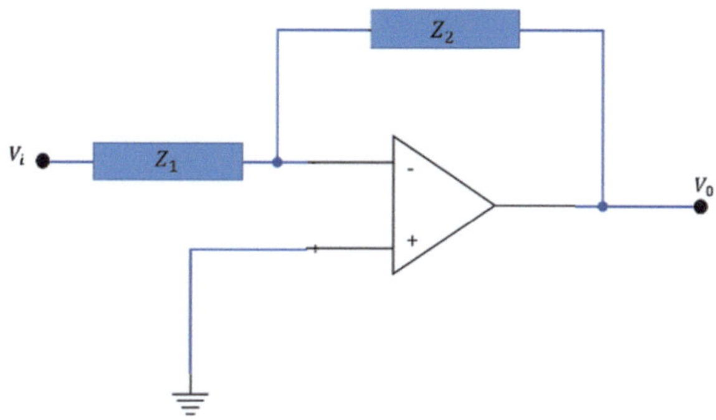

Fig. 6.7 Operational amplifier. A general configuration where Z_1 and Z_2 represent the complex impedances of an arbitrary circuit

$$G(s) = \frac{V_0(s)}{V_i(s)} = -\frac{R_2}{R_1}$$

That is, it generates an output voltage V_0, which is proportional to the input one, i.e. V_i.

In practice, the use of the complex impedance is preferred in order to obtain the transfer function of operational amplifiers. The complex impedance of an element or circuit, Z, is defined as the transfer function $Z(s) = V(s)/I(s)$. Having the complex impedances of each branch, Z_1 and Z_2 as shown in Fig. 6.7, it can be derived that the transfer function of the system is

$$G(s) = \frac{V_0(s)}{V_i(s)} = -\frac{Z_2(s)}{Z_1(s)}$$

According to the electrical equations of the ideal operational amplifier, different configurations lead to different transfer functions implementing useful operations as shown in Fig. 6.8. Notice it produces a negative output voltage, and thus, and additional amplifier in line is needed to obtain a positive value.

Control System Design by Root Locus Method in MATLAB

The design of a controller aims at modifying the behaviour of a closed-loop system to obtain a particular response, e.g. reducing the overshoot, gaining in stability or eliminating the steady error. A simple manner to do this is to add poles–zeros from the initial configuration up to achieve our goal, i.e. designing phase-lead-lag compensators. In this section, we analyse the effect of adding poles–zeros in the closed-loop by means of the root locus technique.

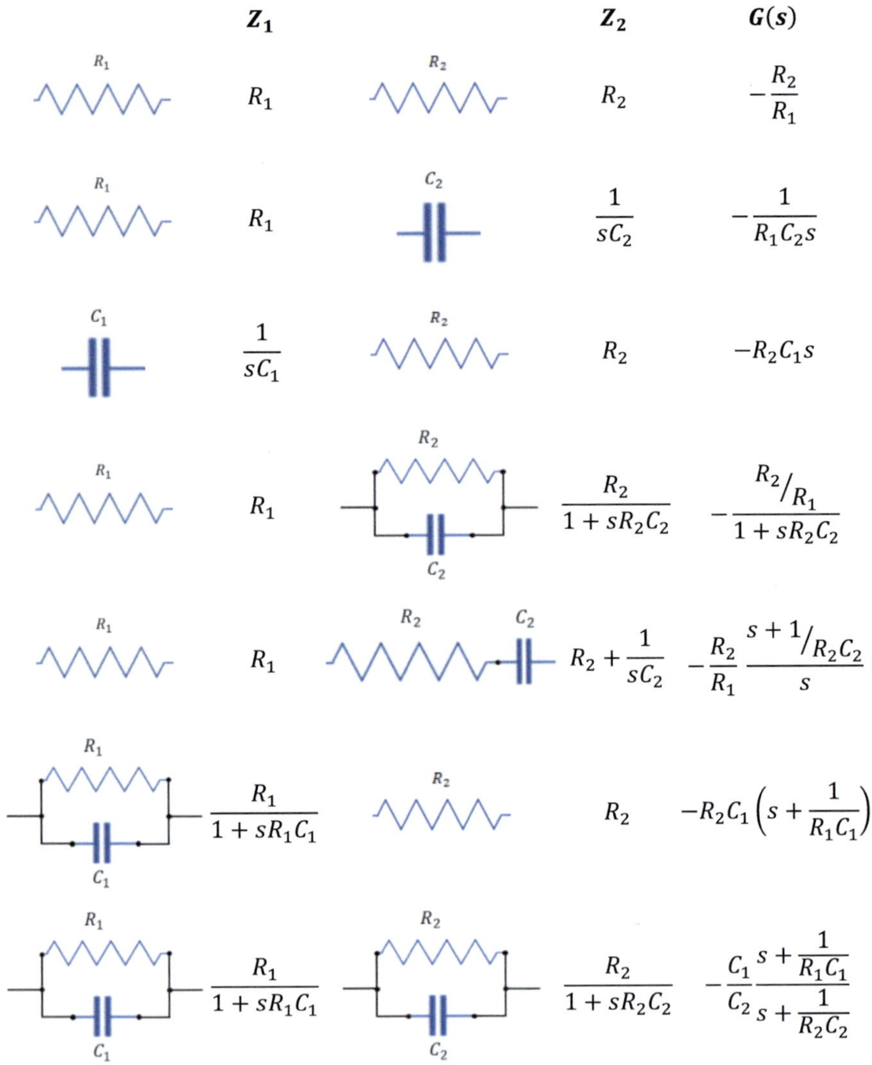

Fig. 6.8 Common configurations and the resultant transfer function

Let us begin with the simplest root locus, the root locus of a first-order system. The root locus for a system with a pole at $s = -5$ and the closed-loop response for different values of K is shown in Fig. 6.9. Notice that as long as K increases the pole moves over the negative part of the real axis, at the same time that the output reduces the permanent error. Note that this is a proportional compensator.

If we add now a pole, for instance at $s = -2$, the configuration of the root locus will change, appearing a rupture point from which the poles can move upward and downward. Note that the rupture point appears in the mid-point between the plant

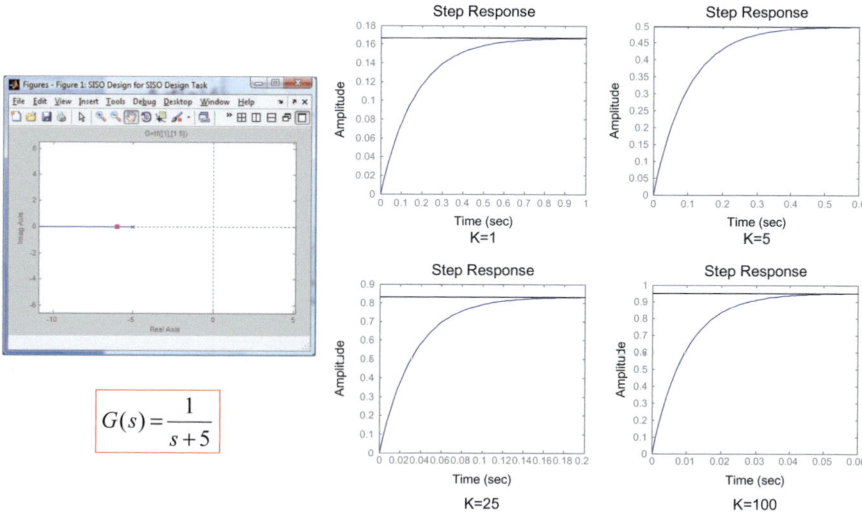

Fig. 6.9 Root locus for a first-order system. This is the simplest case where the root locus covers the real axis in the range $[-\infty, p_1]$

pole and the compensator pole, i.e. the newly added pole. Thus, with this new configuration, according to the value of K the system will be overdamped, critically damped or underdamped, exhibiting their correspondent behaviour (see Fig. 6.10).

In these examples, with only two poles the system is stable for any value of K. If we add another pole, for instance at $s = -7$, the root locus bends to the right, existing a value of K that makes the system unstable (see Fig. 6.11). Concretely, when $K = 850$, poles are located at $s = 0.177 \pm 8j$.

Summing up we can realize that increasing the number of poles may lead to instability and undesired effects like overshooting. Contrarily, by adding zeros to the closed-loop system we can obtain better results in the system response, even we can stabilize a system. Let us consider now the addition of a zero to the previous example.

When adding a zero, e.g. at $s = -10$, the root locus configuration drastically changes, as shown in Fig. 6.12. Now one of the segments ends at the new zero, while the other two exhibit behaviour similar to the case where only two poles are considered. Notice that the system is stable regardless the value of K, and thus we have added stability to the system.

Figure 6.13 shows the effect of adding one zero at different real locations, proving that in all cases the system keeps stable. The addition of poles and zeros to a given system is the basis for designing compensators which can be experimentally set up by inspecting the root locus of the closed-loop system and the contribution of the added poles–zeros. Next, a particular compensator, the PID controller that adds two poles and a zero is described.

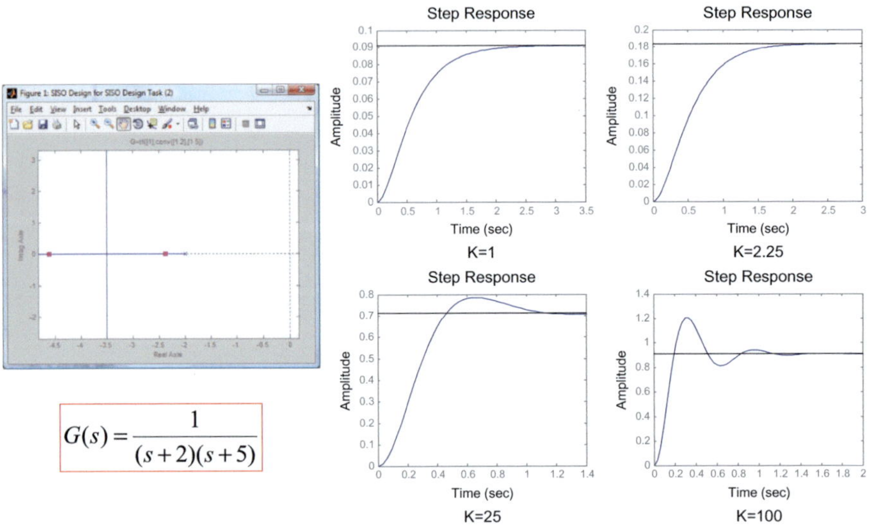

Fig. 6.10 Effect of adding a pole at $s = -2$. Note that the root locus turns into a cross making the poles to get a complex component. As long as K increases the damping ratio decreases, and thus, the system response oscillations increases

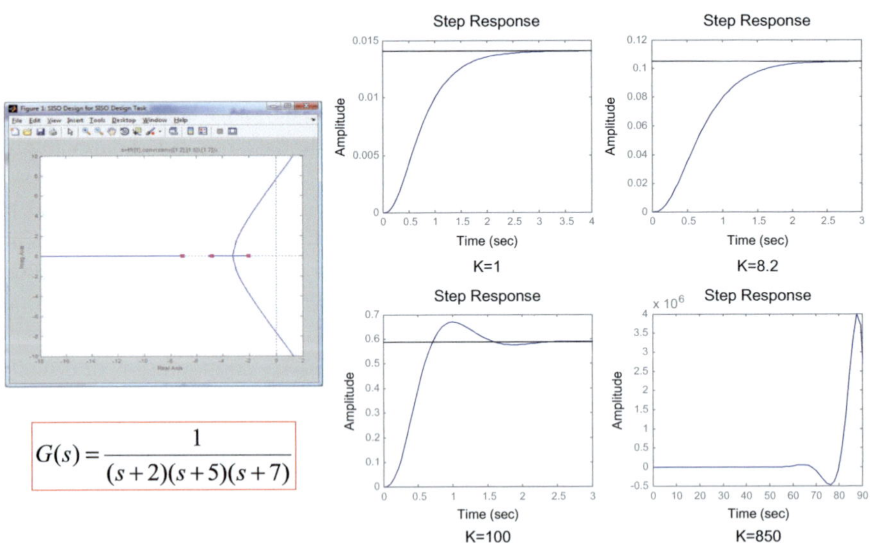

Fig. 6.11 A three-pole root locus example. Adding a new pole bends the root locus to the right. There is a critical value of K that moves the poles to the right part of the s-plane, destabilizing the system

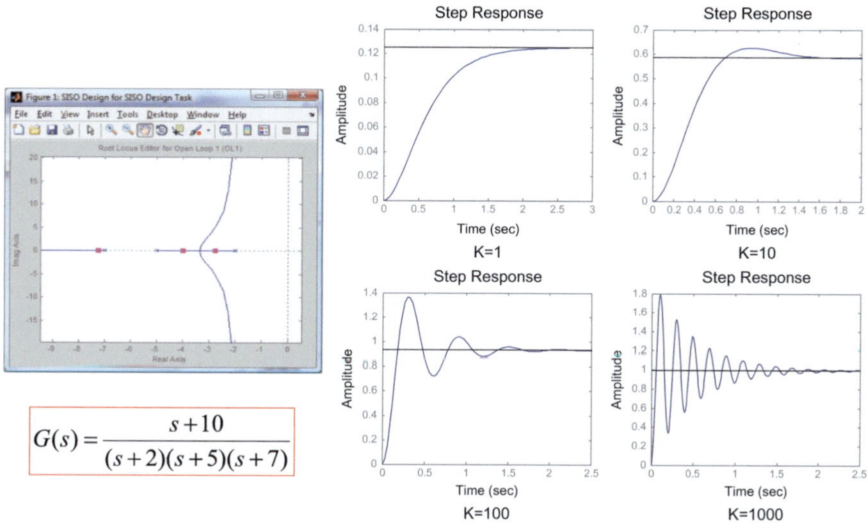

Fig. 6.12 Effect of adding a zero. The addition of a zero at $s = -10$ corrects the deviation of the root locus, avoiding the destabilization of the system

PID Continuous Controller

The PID controller is one of the most widespread control techniques in engineering and has being used since the 1930s. PID stands for Proportional, Integral and Derivative since this type of controller acts proportionally to the error, to the integral of the error and to the derivative of the error. The idea behind the PID control is that acting proportionally to the error will produce a promptly response of the controlled output; acting proportionally to the integral of the error will eliminate the error at the steady state, and finally, being proportional to the derivative of the error will reduce oscillations. Mathematically, the PID control action can be defined in the time domain as:

$$u(t) = K_p\left(e(t) + \frac{1}{T_i}\int_0^t e(t)\mathrm{d}t + T_d\frac{\mathrm{d}e(t)}{\mathrm{d}t}\right)$$

which is normally rewritten as:

$$u(t) = K_p e(t) + K_i\int_0^t e(t)\mathrm{d}t + K_d\frac{\mathrm{d}e(t)}{\mathrm{d}t}$$

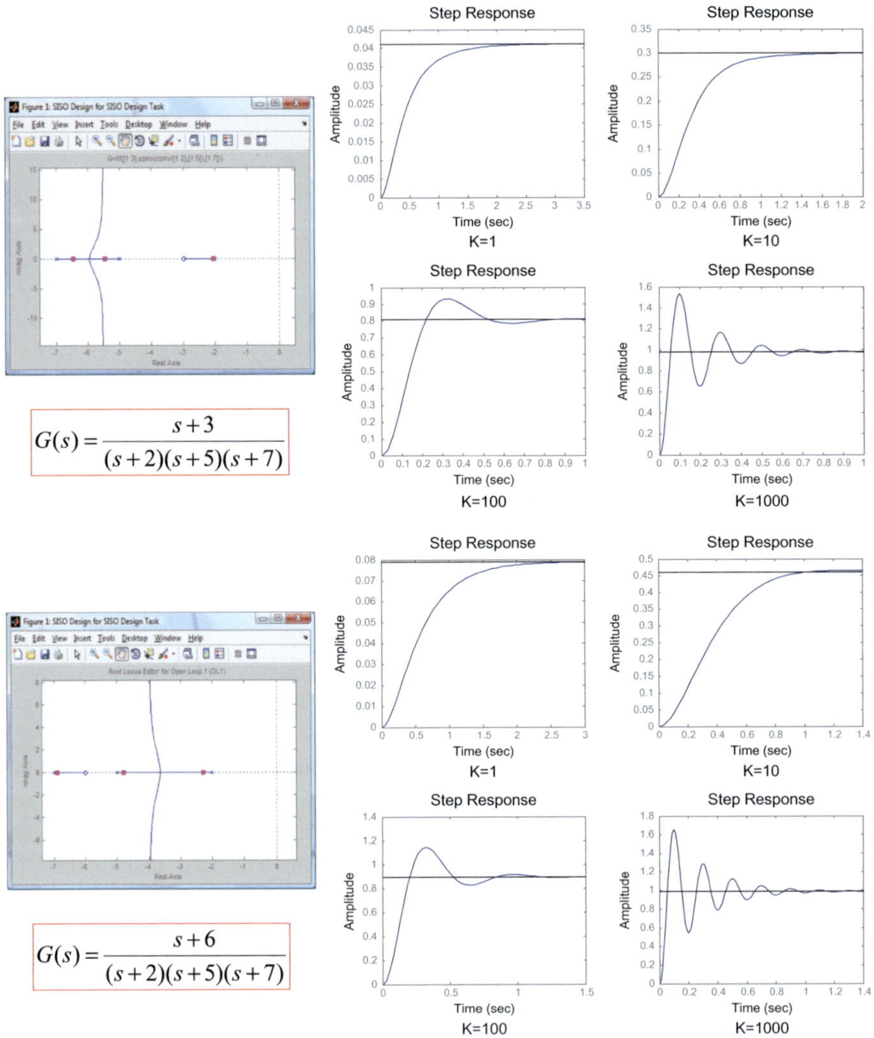

Fig. 6.13 Effect of adding a zero to the three-pole system at different locations. Notice that in any case, the resulting system gains in stability

$$K_i = \frac{K_p}{T_i}; \quad K_d = K_p T_d$$

With this, the transfer function of a PID is as follows (see Fig. 6.14):

$$G_{\text{PID}}(s) = \frac{U(s)}{E(s)} = K_p + \frac{K_i}{s} + K_d s = \frac{K_d s^2 + K_p s + K_i}{s} \tag{6.1}$$

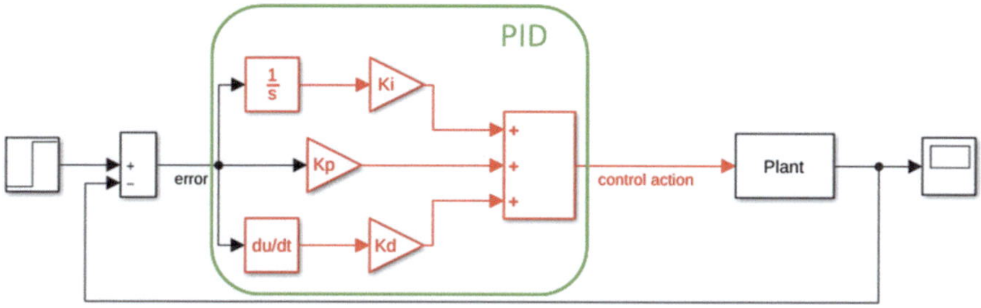

Fig. 6.14 The PID control. The PID control, shown in red, acts over the plant proportionally to the error, its integral and its derivative

Note from Eq. (6.1) that the ideal PID controller adds two zeros and one pole to the plant, and therefore it is not implementable, given it represents a non-causal system, i.e. it exhibits more zeros than poles. The implementable version of the PID has an additional pole according to Eq. (6.2)

$$G_{\mathrm{PID}}(s) = \frac{U(s)}{E(s)} = K_p + \frac{K_i}{s} + K_d \frac{N * s}{s + N} \tag{6.2}$$

That is, a first-order filter is added to the derivative term, resulting in a two-pole two-zero system. The additional pole, determined by the filter coefficient, N, is normally placed to be non-dominant, having thus an insignificant effect in the system response. Next, the components of the PID control and their effects are individually described in more detail.

Proportional Control

The proportional control relies on acting proportionally to the measured error. That is, the higher the magnitude of the error is, the stronger the control action is applied to the plant. This is an elemental and intuitive control action that improves the plant output but also may lead to undesirable effects.

Actually human reactions usually follow this type of control. Let us think, for example, you are driving your car, trying to be at the centre of the lane. This is actually a control system where your brain is the controller, your hands directly act on the vehicle (the plant) through the steering wheel, your eyes are the sensors, and the error, computed by your brain, is the distance of the car to the centre of the lane. Small deviations are corrected via small actuations, however an extremely high error caused, for example, by a sudden gust could make you react excessively, swerving to keep the track in an impulsive way that may lead the car to swerve from side to side, requiring additional actuations on the steering wheel. The control action in this case is directly proportional to the instant error. Small errors cause slight corrections, but a sudden high error causes an instinctive and totally out of proportion reaction. In the worst case, the system will turn into unstable, i.e. a traffic accident in this example.

The proportional controller is configured through the parameter, K_p, which indicates how strong is the control action according to the current error. The contribution of the parameter K_p to the controlled system response is, in fact, the same as the one studied in Chap. 5 for sketching root loci. Recalling the results presented in that chapter, varying the parameter K_p, implies relocating the closed-loop poles along the strokes of the root locus, which in general reduces the steady-state error and makes the response faster, but could also result in worsening the transient response, i.e. increasing oscillations, even in turning the system unstable.

Example 6.1

Let's consider the insulin pump system from Chap. 2, in which, in this case the flowrate of insulin exhibits a first-order dynamic. The set of equations is presented again for the convenience of the reader. Note the relation of the flowrate and the motor speed in Eq. (6.7).

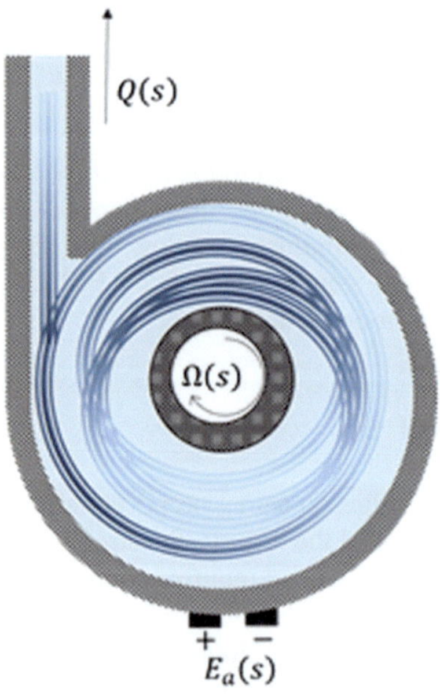

$$L_a s I_a(s) + R_a I_a(s) + E_m(s) = E_a(s) \tag{6.3}$$

$$E_m(s) = K_m \Omega(s) \tag{6.4}$$

$$T(s) = K_m I_a(s) \tag{6.5}$$

$$Js\Omega(s) + B\Omega(s) = T(s) \tag{6.6}$$

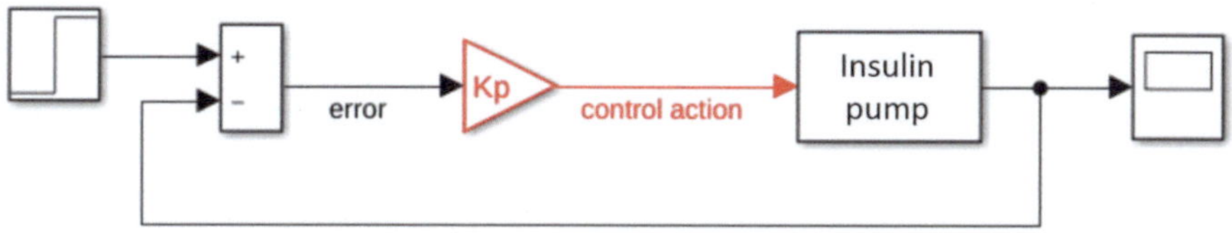

Fig. 6.15 The proportional control. The control action is proportional to the error, (***control*** $=$ ***error*** $*$ $\boldsymbol{K_p}$), enabling the system to quickly react against perturbations and/or changes in the reference, but at the expense of producing oscillations, even instability, when the error and/or $\boldsymbol{K_p}$ are sufficiently high

$$sQ(s) + K_f = K_f\Omega(s) \tag{6.7}$$

From Eqs. (2.101)–(6.7), the transfer function of the system can be derived as[2]

$$G(s) = \frac{5}{s^3 + 9s^2 + 20s + 10}$$

We will study the effect of adding a proportional controller using different K_p values.

The gain of the system is 5/10 = 1/2, and thus, the step response in open loop will achieve that value, while we pretend to produce the insulin flowrate commanded through the input to the controlled system, i.e. the reference, in this case, for simplicity, a unitary step (Fig. 6.15).

The closed-loop control, considering a proportional controller, is represented by the transfer function:

$$G_{\mathrm{CL}}(s) = \frac{5K_p}{s^3 + 9s^2 + 20s + 10 + 5K_p}$$

Note that when $K_p = 1$ the final value of the response is $5/15 = 0.333$. As long as K_p becomes higher, the error at the steady state decreases. In theory when $K_p \to \infty$, the error at the steady state tends to 0. However, increasing the value of K_p produces oscillations in the output response when the magnitude of the error is considerably high. At an extreme situation, the system may become unstable given a certain value of K_p. Figure 6.16 shows the plant output for different values of K_p where considering a value of $K_p = 25$ yields a final value of $125/135 = 0.926$, but at the expense of extremely high oscillations.

[2]For clarity and pedagogical sake, the parameters have been conveniently chosen to obtain rounded figures.

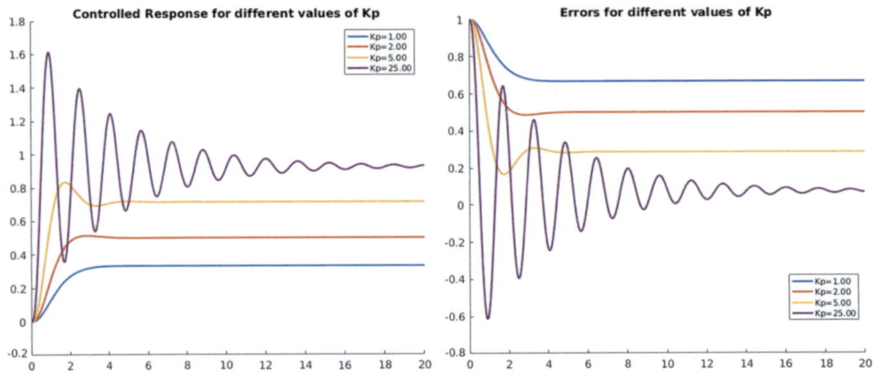

Fig. 6.16 Influence of the K_p parameter in the proportional control. As long as K_p increases, the system reacts more impulsively against the error, reducing the error at the steady state but causing more oscillations

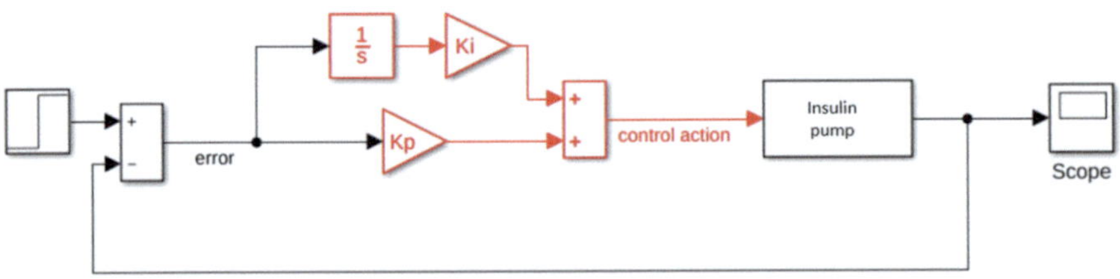

Fig. 6.17 The PI control. The control action considers now both the error and its integral. Taking into account the integral of the error permits this controller to eliminate the error at the steady state

Integral Control

Considering only the instant error to control the plant is not a good option to achieve a highly accurate and smooth response in plant output. The proportional control has no records of past experiences, i.e. of the evolution of the response along the time, and thus its exhibits the aforementioned limitations.

An improvement on the proportional control is the addition of the integral control, obtaining the PI controller, as shown in Fig. 6.17. A control action proportional to the integral of the error is added to eliminate the error at the steady state.

Example 6.2
For the system proposed in Example 6.1, consider also an integral action over the error and check its effect for different values of K_i. The closed-loop transfer function is now:

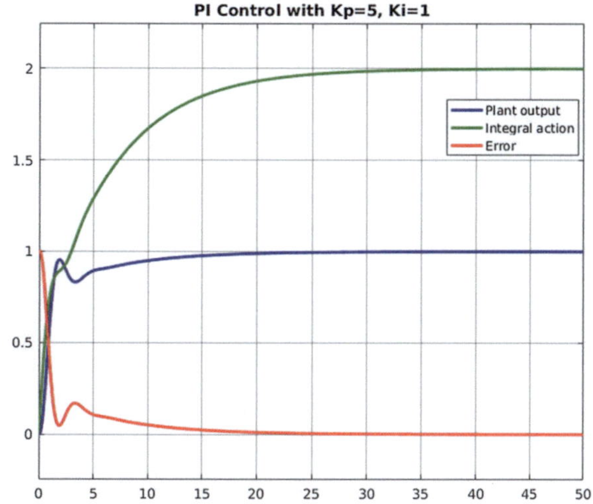

Fig. 6.18 The PI control. After the first peak the response tends to the reference. Note how the error (red line) tends to zero, while at the steady state, the input to the plant is only produced by the integral component

$$G_{\text{CL}}(s) = \text{feedback}\left(\text{series}\left(K_p + \frac{K_i}{s}, \frac{5}{s^3 + 9s^2 + 20s + 10}\right), 1\right)$$

$$G_{\text{CL}}(s) = \frac{5K_p s + 5K_i}{s^4 + 9s^3 + 20s^2 + 10s + 5K_p s + 5K_i}$$

That is, the system gain is 1, i.e. $\frac{5K_i}{5K_i}$.

The integral component of the PI adds an integrator, and thus the type of the system increases, eliminating the error at the steady state when the input reference is a step.

Let's observe this effect considering the previous example with $K_p = 5$ and $K_i = 1$ (see Fig. 6.18). Note that the system response (blue line) has slightly increased the overshoot but, most importantly, the error at the steady state has been completely eliminated.

The effect of varying the parameter K_i reduces the steady-state time and the settling time at the expense of largely increasing the overshoot of the system as shown in Fig. 6.19. Note how the red response reaches its settling time at time $t = \sim 10$ s (and the blue one around 20), but the former exhibits a 40% of overshoot.

Derivative Control

Finally, the third component of the PID controller comes into scene to reduce response oscillations. The two other components have demonstrated their ability to produce a fast and precise response, but at the expense of causing oscillations and

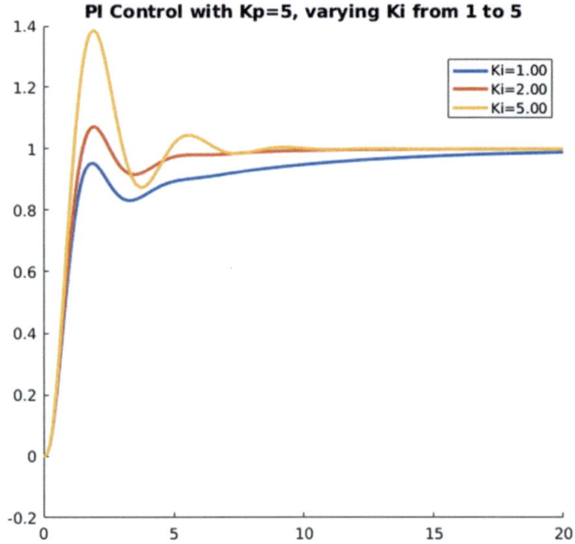

Fig. 6.19 The PI control. Notice that increasing the integral parameter the output achieves the steady state faster but at the expense of a high overshoot

high overshoots. This is because in both cases the controller has no information about the change of the error, but only its instant value and its integral along the time. The derivative of the error provides information about its changes along the time. The derivative control will firmly act against sharp changes of the error values, in an attempt to moderate it.

Example 6.3
Repeat the previous example including a derivative action with parameters $K_p = 5, K_i = 5$ and $K_d = 1$. Try different values for K_d and observe the effects on the system output. In this case, we do not consider the inclusion of the filter coefficient, studying the ideal PID. Figure 6.20 depicts the system output for the example considered in the previous section with $K_p = 5$ and $K_i = 5$ (the situation illustrated by the red line of Fig. 6.19) adding a derivative action with $K_d = 1$. Notice that including a derivative control, an additional control action opposes the impulsive action of the proportional and integrative components. This occurs when the derivative of the error is negative, which coincides with the overshoot.

Varying the value of K_d will cause a slower and smoother response, where the response oscillations are flatten. See Fig. 6.21 for $K_d = 2$. However, note that higher values for this parameter will have the opposed effect; it causes an increment in the overshoot. This is because as long as the parameter increases, the control action is stronger and stronger, producing an excessive reaction to the controlled variable, i.e. the derivative of the error.

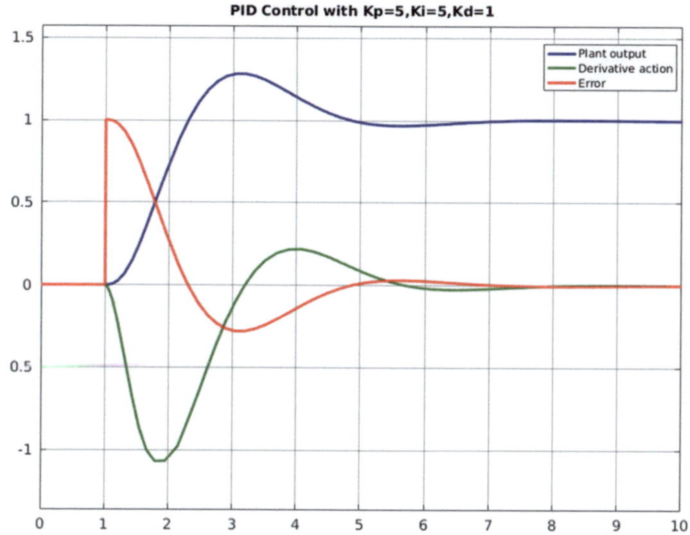

Fig. 6.20 The PID control. Adding a derivative control the response is smoothed, reducing the overshoot and oscillations. (To produce this plot, the derivative action has been limited to be properly shown since when the step is raised, the derivative of the error tends to $-\infty$)

Fig. 6.21 The PID control. The derivative control strives for reducing the overshoot; however, high values of the parameter $\mathbf{K_d}$ may cause the opposed effect, even make the system unstable

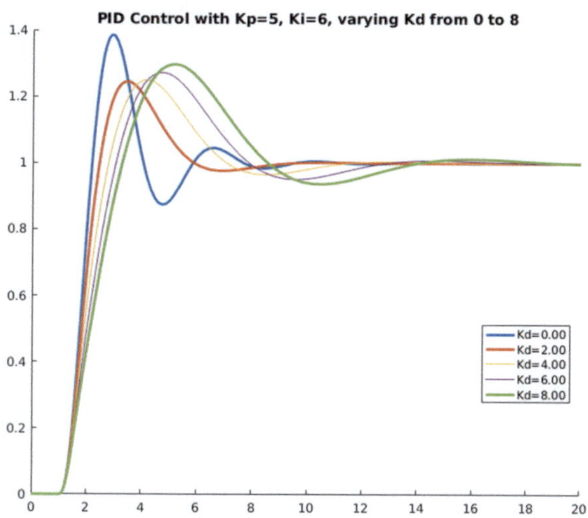

PID Summary

As stated in the previous sections, each component of the PID control exhibits valuable advantages but also undesirable disadvantages. The contribution of the different parameters of a PID is summarized in Table 6.1.

Table 6.1 Contribution of each component of the PID to the system response

	Rise time	Overshoot	Setting time	Error
K_p	Diminishes	Increases	Slight change	Diminishes
K_i	Diminishes	Increases	Diminishes	Disappears
K_d	Slight change	Diminishes	Diminishes	Slight change

Note that the effects are sometimes conflictive since, for instance, trying to reduce the rise time by increasing the value of K_p, will lead to an increment in the overshoot, which could be solved by modifying the value of K_d, that in turn, could modify the rise time. That is, the selection of the parameters of a PID controller is a trade-off between the response characteristics and it is responsibility of the designer, based on his experience, to tune it accordingly.

PID Design Using RLTOOL

As commented, a PID controller adds an integrator and two zeros to the controlled system, so a straightforward way to design a PID is to conveniently decide the positions of such two zeros in order to obtain the desired behaviour. In previous sections, we have analysed the contribution of adding zeros and poles to the root locus, which serves as a base for designing controllers.

Example 6.4
Let's illustrate the design of a PID controller a biomedical system that follows the differential Eq. (6.8)

$$\frac{d^2y}{dt} + \frac{dy}{dt} - 2y(t) = 10u(t) \tag{6.8}$$

which results in the transfer function of (6.9) as

$$G(s) = \frac{10}{(s+2)(s-1)} \tag{6.9}$$

For this system, we impose the following restrictions: (i) settling time under 2 s, (ii) maximum overshoot under 5% and (iii) null steady-state error for step inputs. Note that the type of the open-loop system is 0, and thus it exhibits a steady error for step inputs. Moreover, the system is even unstable, since it has a pole at the right part of the s-plane. Figure 6.22 shows the original system root locus considering only a proportional controller. The system becomes stable for $K_p > 0.2$, and the maximum overshoot constraint (ii) is met for $K_p < 0.25$. However, the constraint (i) about the settling time (not shown in the figure) cannot be met for any value of K_p.

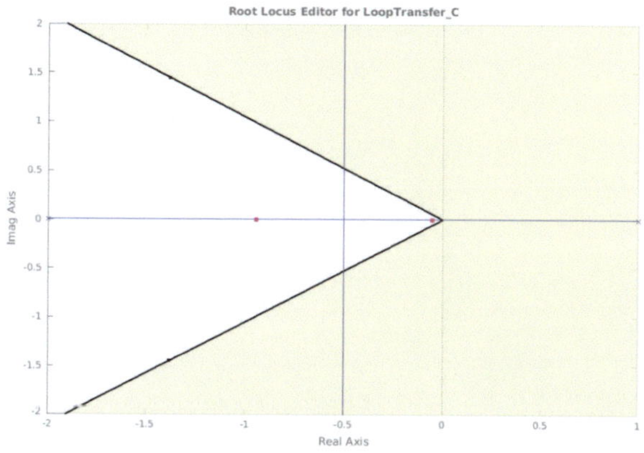

Fig. 6.22 Root locus of the thermal system using a P controller. The restriction of overshoot under 5% has been imposed. The closed-loop poles must be in the white area in order to fulfil such constraint

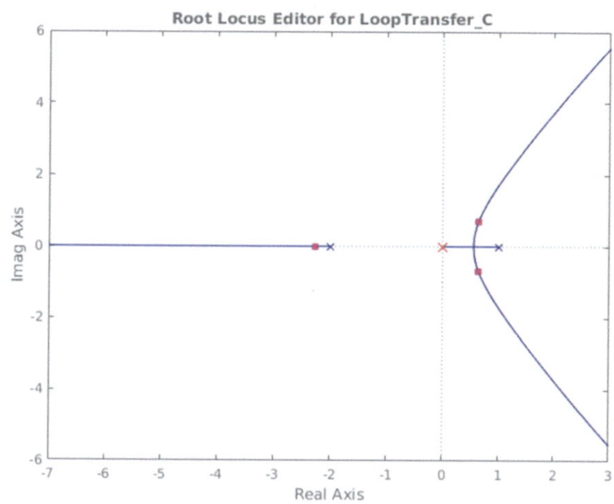

Fig. 6.23 Root locus of the thermal system using a PI controller. Note that the system is unstable

By considering a P controller, we can stabilize the system and meet the second constraint. In order to fulfil the restriction (iii), we can add an integrator, i.e. PI controller, and adjust properly the proportional gain.

Figure 6.23 shows the root locus of the closed-loop system when considering a PI controller. In this case, the type of the system has been increased, but it is unstable for any proportional and/or integral gains. Recalling Fig. 6.11, the addition of poles to the closed-loop system tends to destabilize the system. Thus, it is clear in this case that the derivative component must be included. Adding two zeros (at the left part of the s-plane) will stabilize the system as shown in the discussion illustrated in Fig. 6.13. Let's add two real zeros, one in the interval $[0, -2]$ and another one placed on the left of the plant pole on $s = -2$.

There is no unique solution, and the selection of the zeros' position depends on external issues to the control itself, like the cost of physically implementing the

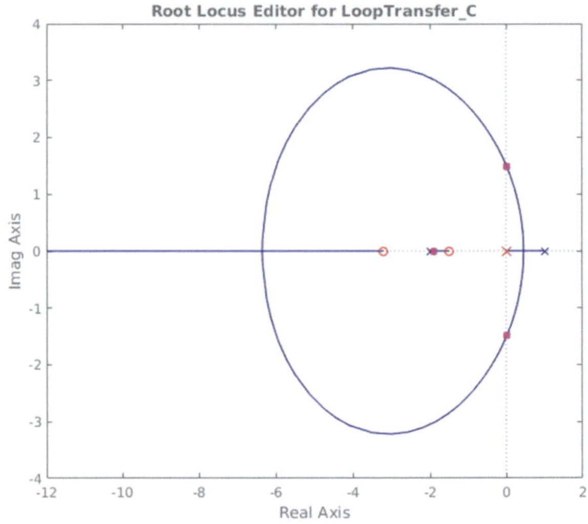

Fig. 6.24 Root locus of the thermal system using a PID controller. Note that the system is now unstable for a small range of values of C. Actually the system is stable for $C > 0.45$

resultant PID. In this case, we have placed one zero at $s = -1.5$ and the other at $s = -3.2$, which drastically modified the root locus as shown in Fig. 6.24.

The result of imposing the two restrictions to the root locus can be shown in Fig. 6.25 left. Note that for a certain value of C, i.e. $C = 3.36$, two closed-loop poles move to the white area; however, it does not suffice to fulfil the requirements, since requirements' areas are only valid for pure second-order systems. Certainly,

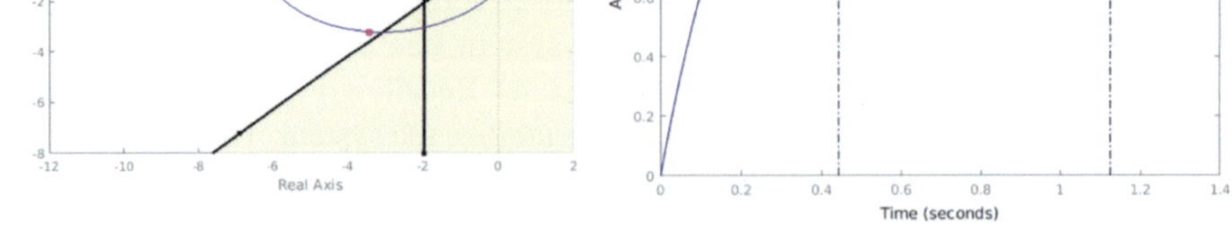

Fig. 6.25 Requirements' areas depicted by rltool (left). Areas shown are only valid for pure second-order systems, so the system response must be analysed to assure the restrictions are met. In this case, the step response (right) meets the restriction related to the settling time but not the one related to the maximum overshoot

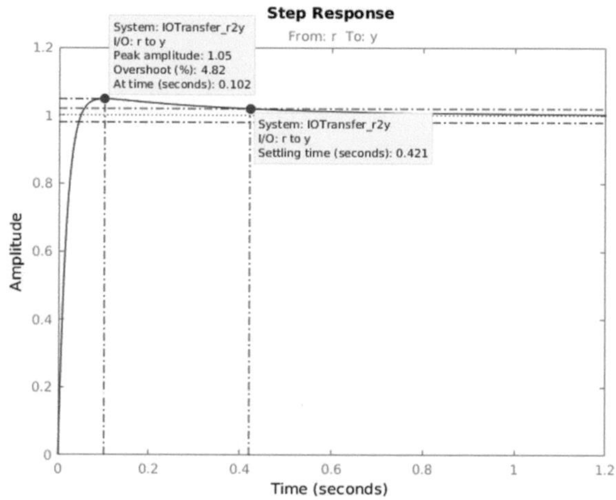

Fig. 6.26 Time response for $C = 30.3$. Notice that both parameters, i.e. overshoot and settling time, are under the imposed constraints

as shown in Fig. 6.25 right, the system response, for $C = 3.36$, meets the settling time requirement but not the maximum overshoot. In order to fulfil both constraints, the designer has to increase the value of C, moving the closed-loop poles towards the zeros. In this example, selecting $C = 30.3$ produces a control system that fully fulfil the imposed constraints as shown in its step response (see Fig. 6.26).

The resultant PID controller can be exported as shown in Fig. 6.27. The transfer function is stored in a variable of the workspace, C in this case, which can be accessed to, for example, compute the parameters K_p, K_i and K_d of the PID.

As previously commented, the tuned PID controller exhibits two zeros and only one pole, making the system non-causal, and therefore, not physically implementable. Thus, an additional term is required, known as a derivative filter, that is a first-order filter exhibiting a quick dynamic. In this case, we chose the filter coefficient $N = 450$, minimizing thus its effect in the controlled system response. Thus, the transfer function of the PID controller is shown in Eq. (2.2), and the output of the controlled system for a 36.5 amplitude step is depicted in Fig. 6.28.

$$\begin{aligned} &\gg \mathtt{C= C*(450/(s+450))} \\ &\mathtt{C=}\frac{6.312\mathtt{s}^2+29.67\mathtt{s}+30.3}{\mathtt{s}^2+50\mathtt{s}} \end{aligned} \tag{6.10}$$

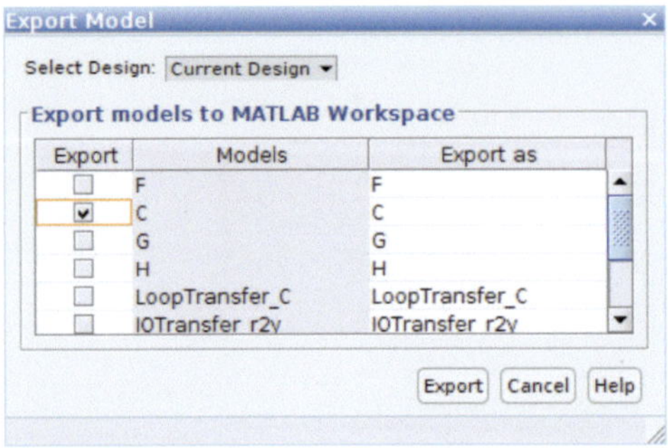


```
>> C

C =

  6.3125 (s+1.5) (s+3.2)
  ---------------------
           s

>> pid(C)

             1
  Kp + Ki * --- + Kd * s
             s

with Kp = 29.7, Ki = 30.3,
Kd = 6.31

>> step(36.5*feedback(C*G,1))
```

Fig. 6.27 Exporting the designed controller. Left, the export option permits us to export any of the models involved in rltool. Once exported the controller, C, a variable is added to the workspace with its value. (Right) Showing the controller in the workspace of MATLAB. The command `pid` extracts the parameters of K_p, K_i and K_d. Finally, plotting the response of the controlled system confirms the suitability of the step response

Fig. 6.28 Step response of the controlled system through the `step` command

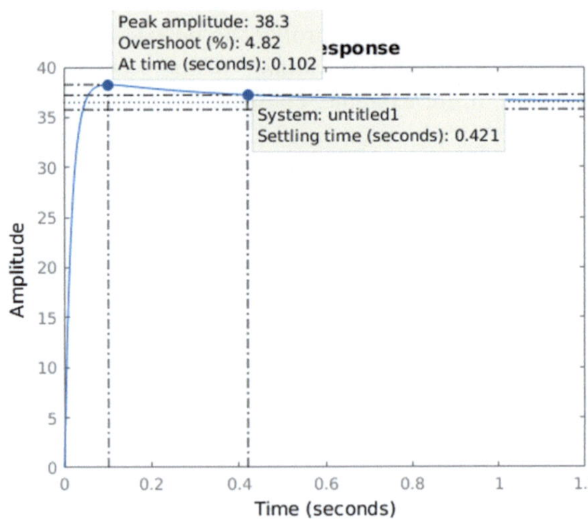

Discretization of Continuous Controllers

Differently to the PID continuous controller, the discrete version can be modelled differently according to the discretization method used (recall chapter 2). Thus, following the discrete transfer functions of Table 6.2, for the most common discretization methods, the process is similar to the one presented in the previous section but within the Z-plane, i.e. properly adding poles and zeros, obtaining the values for the parameters of the PID.

Example 6.5
Design a discrete PID controller for the biomedical system from Eq. (6.9), considering the backward Euler method and a sample time $T = 0.01$. In this case, the unique restriction imposed is that the settling time must be under 1 s.

The first step is to discretize the plant, that is, obtaining the discrete version of the transfer function

$$G(s) = \frac{10}{(s+2)(s-1)}$$

This can be done through the c2d command. The normal usage of the c2d command requires three arguments: the continuous system, the sample time and the discretization method used. By default, the zero-order hold method is used, although other methods are also implemented which can be used adding a third argument: "foh" for the first-order hold, "tustin" for the trapezoidal method, "matched" for the matched pole–zero method, among others.

$$\begin{aligned}
&\gg \mathtt{Gz} = \mathtt{c2d(G, 0.01)}\\
&\mathtt{Gz} =\\
&\frac{\mathtt{0.0004983z + 0.0004967}}{\mathtt{z^2 - 1.99z + 0.99}}\\
&\mathtt{Sampletime : 0.01seconds}
\end{aligned} \tag{6.11}$$

Table 6.2 Discrete PID transfer functions for different discretization methods

Forward Euler	$K_p + K_i T_s \frac{1}{z-1} + \frac{K_d}{T_s}\frac{z-1}{z}$
Backward Euler	$K_p + K_i T_s \frac{z}{z-1} + \frac{K_d}{T_s}\frac{z-1}{z}$
Trapezoidal	$K_p + \frac{K_i T_s}{2}\frac{z+1}{z-1} + \frac{K_d}{T_s}\frac{z-1}{z}$

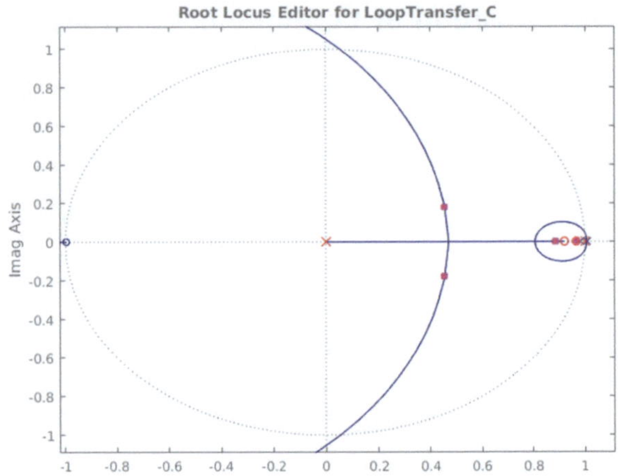

Fig. 6.29 Discrete root locus for the controlled system. Zeros have been added at $z = 0.919$ and $z = 0.965$

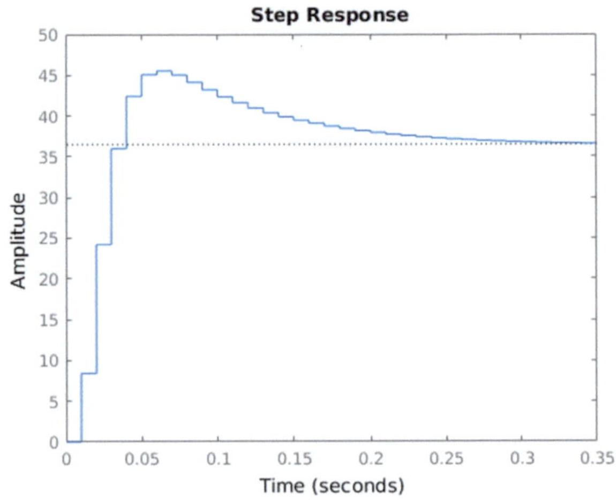

Fig. 6.30 Discrete response of the controlled discrete system

The backward Euler method implies the PID discrete controller which exhibits two poles ($z = 0$ and $z = 1$) and two variable zeros. Thus, we can proceed similarly to the continuous case, invoking the rltool command and adding the needed poles and zeros, as shown in Fig. 6.29. The discrete transfer function of the designed controller is shown in Eq. (6.12), while the system response is depicted in Fig. 6.30.

$$\frac{463.04(z - 0.9194)(z - 0.9653)}{z(z + 1)} \tag{6.12}$$

A different and interesting option to design a digital controller is to firstly obtain a tuned continuous controller and then proceed with its discretization. This option may not be possible in some cases in which the responses of the continuous and discrete systems are slightly different, as well as it largely depends on the selected discretization time. In the next paragraphs, we will proceed to discretize the PID controller tuned in the previous section and test its performance for different discretization time and methods.

Starting from the controller from Eq. (6.13) and invoking the c2d command (with the Tustin method in this case), we have:

$$\gg \mathtt{Cz} = \mathtt{c2d(C, 0.01,'tustin')}$$

$$\mathtt{Cz} =$$

$$\frac{894.7\mathtt{z}^2 - 1748\mathtt{z} + 853.6}{\mathtt{z}^2 - 0.6154\mathtt{z} - 0.3846} \tag{6.13}$$

$$\begin{array}{l}\mathtt{Sampletime : 0.01seconds} \\ \mathtt{Discrete - timetransferfunction.}\end{array}$$

The discretized version of the system to be controlled is shown in (6.11); we can simply obtain the response as (considering a reference $r(t) = 36.5$):

$$\gg \mathtt{step(36.5 * feedback(Cz * Gz, 1))}$$

The obtained response (see Fig. 6.31) is fairly similar to the continuous case, being a convenient procedure to produce a discrete PID controller.

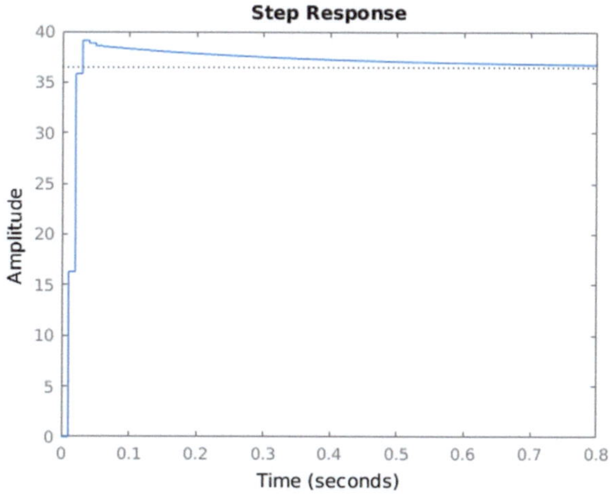

Fig. 6.31 Discrete step response of the controlled system. Response for the tuned controller discretized using the Tustin (trapezoidal) method and a sampling time of $T = 0.01$

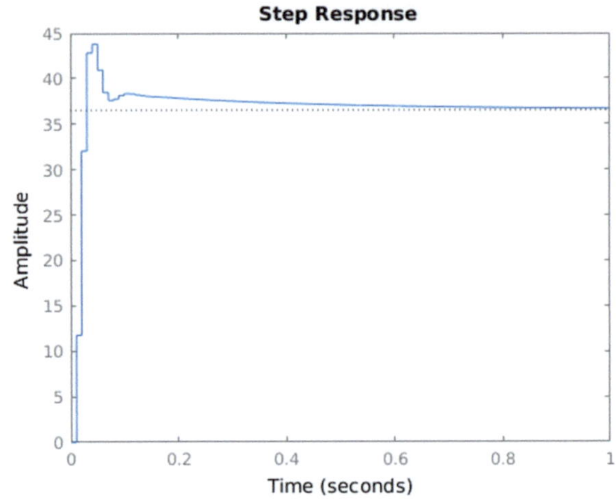

Fig. 6.32 Discrete step response of the controlled system. Response for the tuned controller discretized using the first-order hold (foh) method and a sampling time of $T = 0.01$

The use of different discretization methods for the controller may largely vary the output, even making the system unstable. For instance, using the "foh", i.e. first-order hold, method:

$$\gg \mathtt{Cz} = \mathtt{c2d}(\mathtt{C}, 0.01, '\mathtt{foh}')$$

$$\mathtt{Cz} =$$

$$\frac{647.5\mathtt{z}^2 - 1265\mathtt{z} + 618.1}{\mathtt{z}^2 - 1.011\mathtt{z} + 0.}$$

The controlled response (see Fig. 6.32) is similar to the one produced by the Tustin method, albeit it exhibits a larger overshoot.

Similarly, we can produce the discretization of the controller using the zero-order hold method

$$\gg \mathtt{Cz} = \mathtt{c2d}(\mathtt{C}, 0.01, '\mathtt{zoh}')$$
$$\mathtt{Cz} =$$

$$\frac{2841\mathtt{z}^2 - 5652\mathtt{z} + 2811}{\mathtt{z}^2 - 1.011\mathtt{z} + 0.01111}$$

But in this case the resultant system becomes unstable as revealed the response shown in Fig. 6.33. Besides the effect produced by the different discretization methods, the selection of the sampling time is also critical. In the examples shown in this section, a common and reduced sample time is considered, i.e. 0.01 s, since larger values quickly make the system unstable, independently to the discretization method. Obviously, in real applications, the smaller the better, approximating the

Fig. 6.33 Discrete step response of the controlled system. It considers the tuned controller discretized using the zero-order hold (zoh) method and a sampling time of T = 0.01

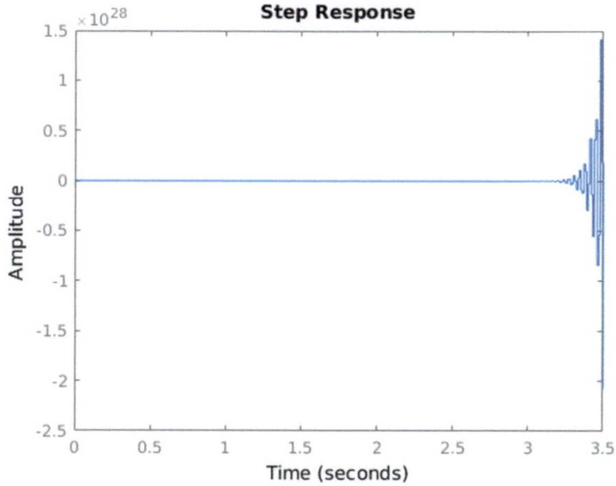

result to the continuous case. However, when implementing a digital controller, the sample time may not be freely chosen, but it will depend on the computer or microcontroller computation capacity, the latency of sensors and actuators, etc.

Implementation of a Continuous PID Controller

The physical implementation of a continuous PID controller can rely on a variety of elements and able to implement the proportional, integrative and derivative effects. However, the most resorted solution is to use operational amplifier for this purpose. Certainly, as shown in Fig. 6.8 there are a variety of configurations combining electrical elements, i.e. resistors, capacitors and operational amplifiers to obtain the desired result. Thus, a continuous controller can be built with the circuit shown in Fig. 6.34.

Using the relations shown in Fig. 6.8, the transfer function is:

$$G(s) = \frac{R_{\text{inv2}}}{R_{\text{inv1}}}\left[\frac{R_{p1}R_iC_iR_dC_ds^2 + R_iC_iR_{p2}s + R_i}{R_{p1}R_iC_is}\right]$$

And thus the relation between the values of the different resistors and capacitors and the PID parameters, i.e. K_p, K_i and K_d, is:

$$K_p = \frac{R_{p2}}{R_{p1}} \quad K_i = \frac{1}{R_iC_i} \quad K_d = \frac{R_{\text{inv2}}}{R_{\text{inv1}}}$$

The implementation shown in Fig. 6.34 conveniently separates the three actions of the PID, however other more compact implementations are also possible, lie the one shown in Fig. 6.35.

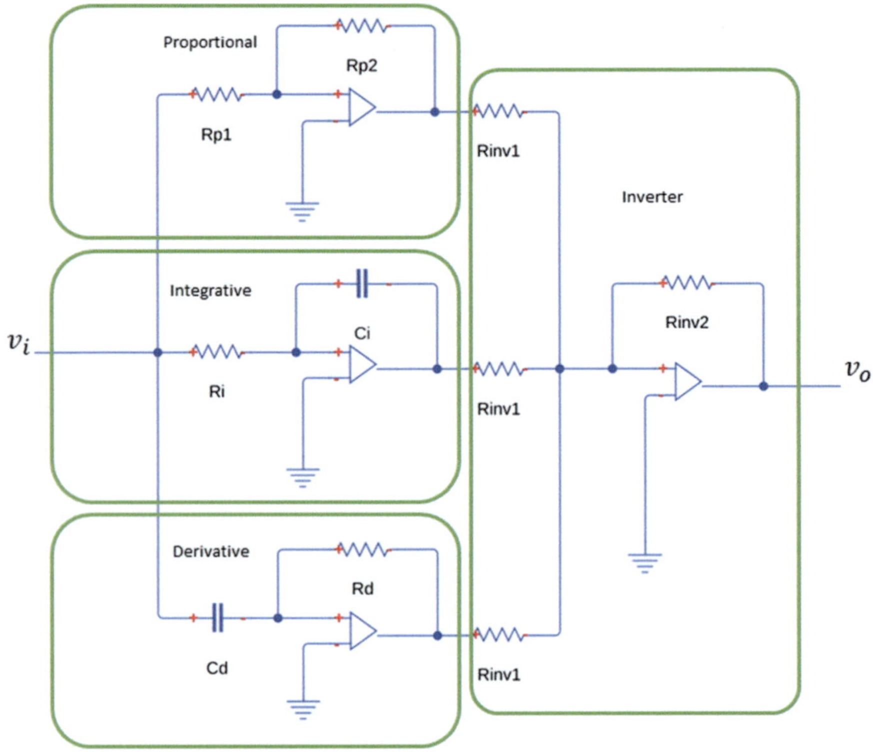

Fig. 6.34 Continuous PID controller implemented with operational amplifiers

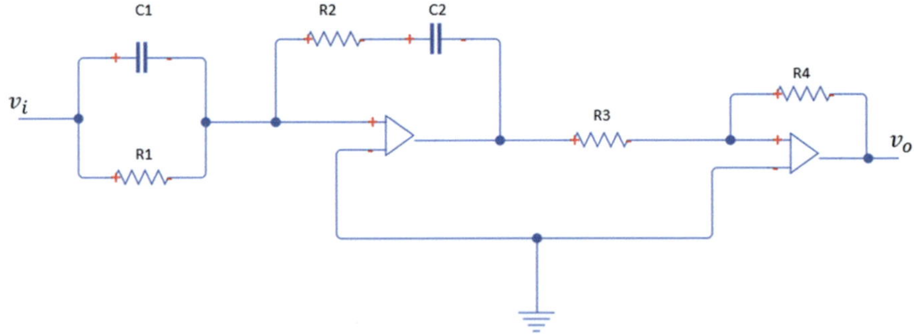

Fig. 6.35 Alternative implementation of a continuous PID based on operational amplifiers

In the second configuration, the transfer function is

$$G(s) = \frac{R_1 R_2 C_1 C_2 s^2 + (R_1 C_1 + R_2 C_2)s + 1}{R_1 C_2 s}$$

And the relation between the electrical elements and the parameters of the PID is

$$K_p = \frac{R_1 C_1 + R_2 C_2}{R_1 C_2} \quad K_i = \frac{1}{R_1 C_2} \quad K_d = R_2 C_1$$

Implementation of a Digital PID Controller

Most of current systems are controlled through any computational system, ranging from high-performance computers to control surgical robots and to reduced microcontrollers that can be used to control small devices, like LVADs, perfusion pumps. In any case the chosen control method must be implemented in a computational system which is inherently discrete.

Two main possibilities can be adopted for implementing a digital PID controller. In the one hand, the physical meaning of the PID equation can be considered which implies to reckon the integral and derivative of the error to provide the control signal. On the other hand, the difference equation of the PID, based on past, discrete information of the error and control signals, can be directly implemented. The next two examples show the implementation of a digital PID controller using MATLAB and SIMULINK.

Example 6.6
Implement a digital version of the PID controller tuned for Example 6.4. In this case, the system is modelled by the open-loop unstable transfer function

$$G(s) = \frac{10}{(s+2)(s-1)}$$

and the tuned PID controller discretized by the Tustin method with Ts = 0.01 as shown in (6.13) is

$$\frac{894.7z^2 - 1748z + 853.6}{z^2 - 0.6154z - 0.3846}$$

```
function control = fcn(error)
 %persistent variables
      persistent integral;
      persistent derivative;
      persistent prev_error;
 % initalize the persistent variables
      if isempty(integral)
          integral = 0;
      end
      if isempty(derivative)
          derivative = 0;
      end
      if isempty(prev_error)
          prev_error=error;
      end

 % sample time
      T=0.01;
 % PID parameters
      Kp=29.7;Ki=21.7;Kd=6.25;
 % numerically computing the integral and derivative
      integral=integral + (prev_error*T);
      derivative= (error-prev_error)/T;
 % the control action:
      control = Kp*error + Ki*integral + Kd*derivative;
      prev_error=error;
end
```

Fig. 6.36 Digital PID implemented as a MATLAB function. The input of the function is the error, from which the action signal is computed in terms of the PID parameters

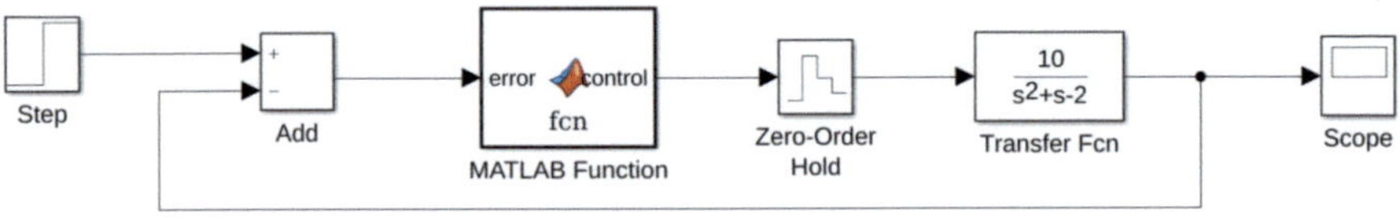

Fig. 6.37 SIMULINK diagram including a digital PID

whose parameters obtained through the PID command are $K_p = 29.7, K_i = 21.7$ and $K_d = 6.25$ (with a derivative filer with Tf = 0.00722 which is neglected in this example). The implementation of such controller can be done in MATLAB as shown in the code of Fig. 6.36.

The implemented function can be tested using the SIMULINK diagram shown in Fig. 6.37.

Running the system, it yields the system response depicted in Fig. 6.38.

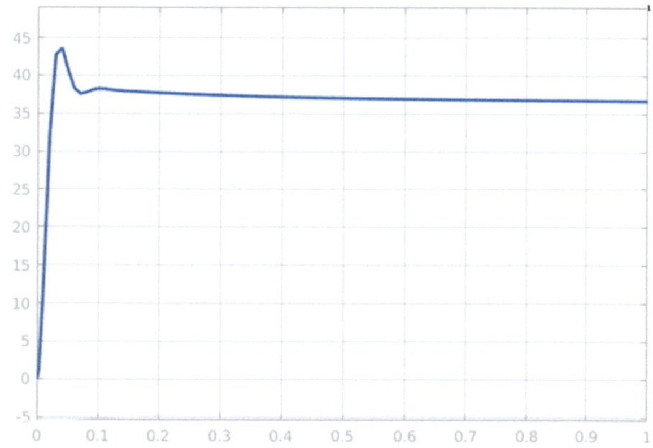

Fig. 6.38 Controlled system response using the digital PID implemented as a MATLAB function

Example 6.7
The second option is based on the meaning of the difference equation of the controller. That is, for the discrete controller (6.13)

$$G(z) = \frac{Y(z)}{U(z)} = \frac{894.7z^2 - 1748z + 853.6}{z^2 - 0.6154z - 0.3846}$$

The difference equation is given by

$$\begin{aligned} &y(k+2) - 0.6154y(k+1) - 0.3846y(k) \\ &\quad = 894.7u(k+2) - 1748u(k+1) + 853.6u(k) \end{aligned}$$

That is,

$$\begin{aligned} y(k) = &\ 0.6154y(k-1) + 0.3846y(k-2) + 894.7u(k) \\ &- 1748u(k-1) + 853.6u(k-2) \end{aligned} \tag{6.14}$$

The expression (6.14) models the PID output, $y(k)$, in terms of the current input, i.e. the error $u(k)$, and the past inputs and outputs. This expression can be directly implementable as (Fig. 6.39):

Independently to the method chosen for implementing a PID, it is important to highlight the situation known as integral wind-up, in which the integral component produces a large control action on the system. In real implementations, this must be taken into account in order to limit such action, typically a voltage, to the system to be controlled. Different anti-wind-up methods can be applied, being limiting the

```
function control = fcn(error)
% Implementation of a digital controller based on its
% difference equation: y(k+n)+....y(k)=u(k+m).....+u(k)
% in this example n=m=3
% vector for outputs, store the last n values
  persistent y;
%vector for inputs, store the last m values
  persistent u;
% iteration
  persistent k;
% initalize the persistent variables
 if isempty(y)
    y=zeros(3,1); k=0;
 end
 if isempty(u)
    u=zeros(3,1);
 end
 u(1)=error;
 k=k+1;

 y(1)= 0.6154 * y(2) + 0.3846 * y(3)  +   894.7 * u(1)-1748 *
      u(2) + 853.6 * u(3);

 control=y(1);

 u=circshift(u,1)
 y=circshift(y,1)
end
```

Fig. 6.39 Digital PID implemented as a MATLAB function through its difference equation

controller output the most direct one. For example, the controlled system depicted in Fig. 6.37 can be modified with a saturation block[3] to limit the output to ±24 (see Figs. 6.40 and 6.41).

The system output and the saturated action of the PID are shown in Fig. 6.42.

Tuning PID Controllers

Tuning a PID controller is not a trivial issue given the interdependence of the three involved parameters, i.e. K_p, K_i, K_d, the wide nature of the systems to be controlled, and the specific response requirements according to the applications. Therefore, instead of a unique, infallible method to select the PID parameters, in the literature there are a number of different techniques and recipes to choose them based on

[3]The effect of this saturation block could be also easily implemented in the MATLAB function block.

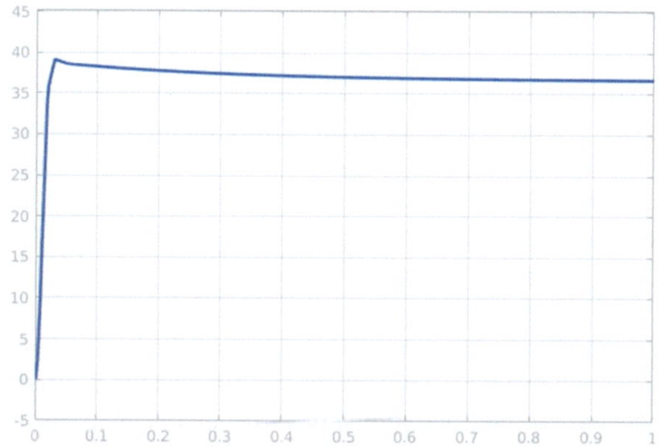

Fig. 6.40 Controlled system response using the digital PID implemented as a difference equation

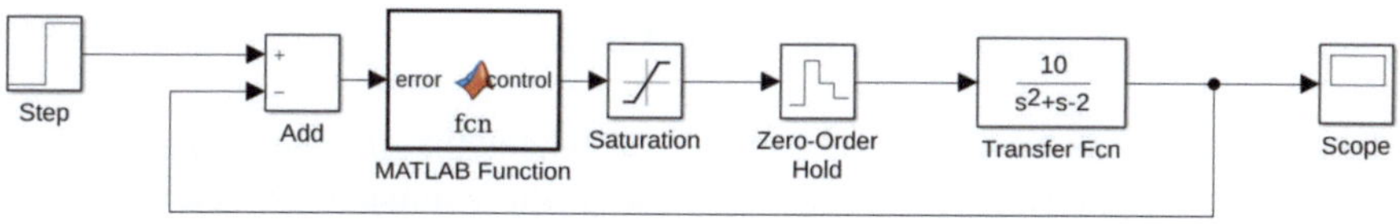

Fig. 6.41 Controlled system with a saturated PID

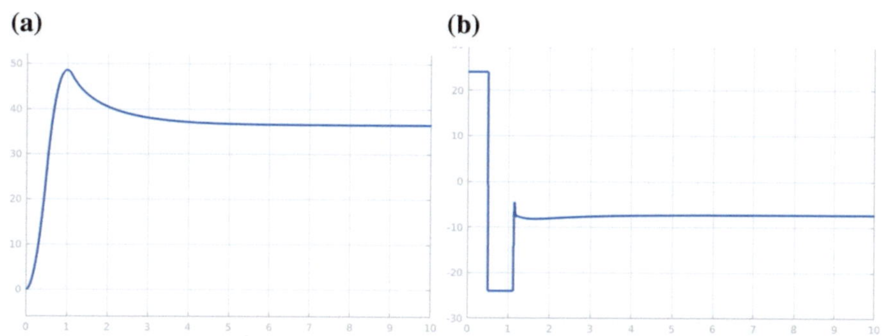

Fig. 6.42 Saturated controlled system. **a** System output, note how in this case the system is slower when compared to the non-saturated implementation. **b** Saturated PID output

experimental and specific techniques. In this book, we will focus on the Ziegler–Nichols techniques to mathematically obtain a starting set of parameters, some recent techniques like SIMC and the AMIGO techniques. This section ends describing the automatic tuning tool provided by MATLAB to ease and automatize the process.

Ziegler–Nichols Techniques

Ziegler and Nichols described in 1942 two simple but effective recipes for helping engineers to properly tune PID controllers. These methods are completely experimental and do not require to model the system, which in practice may be an arduous task. Certainly, the tuning method consists on considering the system as a black box that yields a certain output when it is excited with a given input. By analysing such outputs, the Ziegler–Nichols techniques advise the gain values for the proportional, integrative and derivative components of the PID. It is important to highlight that this method is intended to be general enough to be applied to a variety and diversity of systems and thus the results are not always as good as the desired ones. Thus, the solution provided by Ziegler–Nichols should be taken as a starting point in the tricky chore of PID tuning, which should be then properly adjusted to achieve the required output characteristics.

Ziegler–Nichols Technique in Open Loop

The Ziegler–Nichols technique in open loop, also known as *first method of Ziegler–Nichols*, can be applied when the output of a system exhibits a "*S*" shape for step inputs. That is, it is applicable when the system response resembles a first-order output with, probably, a small delay. Systems containing integrators or dominant complex poles will not exhibit this behaviour and thus, this recipe is not applicable for them.

The steps to tune a PID controller following this method are as follows:

1. Consider the system in open loop and excite it with a step input of amplitude A (typically $A = 1$)
2. Approximate the response to the typical first-order system output, identifying the time constant, τ, and the response delay, L.
3. Use these values to calculate the gains for each controller component according to Table 6.3.

Example 6.8
Design a PID controller for a perfusion pump whose step response in open-loop is as shown in Fig. 6.43.

Table 6.3 PID gains provided by the first method of Ziegler–Nichols

Controller	K_p	T_i	T_d
P	$\frac{\tau}{KL}$	–	–
PI	$\frac{0.9\tau}{KL}$	$\frac{L}{0.3}$	–
PID	$\frac{1.2\tau}{KL}$	$2L$	$0.5L$

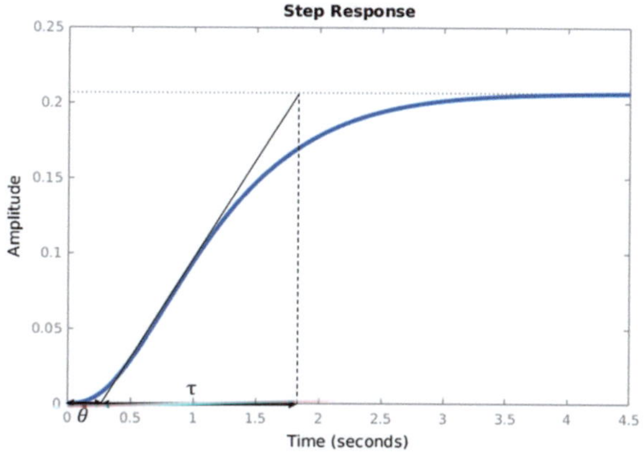

Fig. 6.43 Obtaining the parameters for the open-loop Ziegler–Nichols method

Note that the system response resembles the output of a first-order system with a delay, which is the assumption to be met in order to apply this method. Thus, we can model it by computing the time delay, θ, and the time constant, τ. These parameters can be calculated by tracing the tangent to the output at the inflection point. Crossings with the x-axis and the final value determine the required parameters. Note that this computation of τ is only valid for pure first-order system. Since we are assuming an arbitrary system as a first-order one, it would be more precise to rely on the formal definition of the time constant, as the moment in which the output achieves the 63.3% of the final value.

Also note that the resultant PID is not implementable, in the sense it exhibits two zeros and one pole. This is the theoretical PID transfer function which requires the addition of the derivative filter to be implementable.

In this example, $\theta = 0.2754, \pi = 1.7605 - \theta = 1.4851$. With these values, a PID controller tuned through this method will be:

$$\boldsymbol{K_p} = \tfrac{1.2\tau}{\theta} = 1.2\frac{1.4851}{0.2754} = 6.471$$

$$T_i = 2\theta = 0.558, \left\{K_i = \tfrac{K_p}{T_i}\right\} \Rightarrow \boldsymbol{K_i} = \tfrac{6.471}{0.558} = 11.5968$$

$$T_d = 0.5\theta = 0.1377, \left\{K_d = K_p T_d\right\} \Rightarrow \boldsymbol{K_d} = 6.471 * 0.1377 = 0.8911$$

Figure 6.44 illustrates the benefits of using the tuned PID. While the non-controlled system presents a very slow response with a significant error at the steady state, when the obtained PID controller is considered, the output eliminates such error (what is normal because of the integral component), at the expense of some oscillations and a maximum overshoot close to 30%. Although the use of the tuned PID controller enhances the system response, it may be desirable to improve it. For that, and starting with the values obtaining through the Ziegler–Nichols method, a manual tuning should be necessary.

A final consideration should be taking into account when using this method. The prerequisite of approximating the system response to a first-order response with a delay makes us to calculate the time constant τ through its geometrical definition,

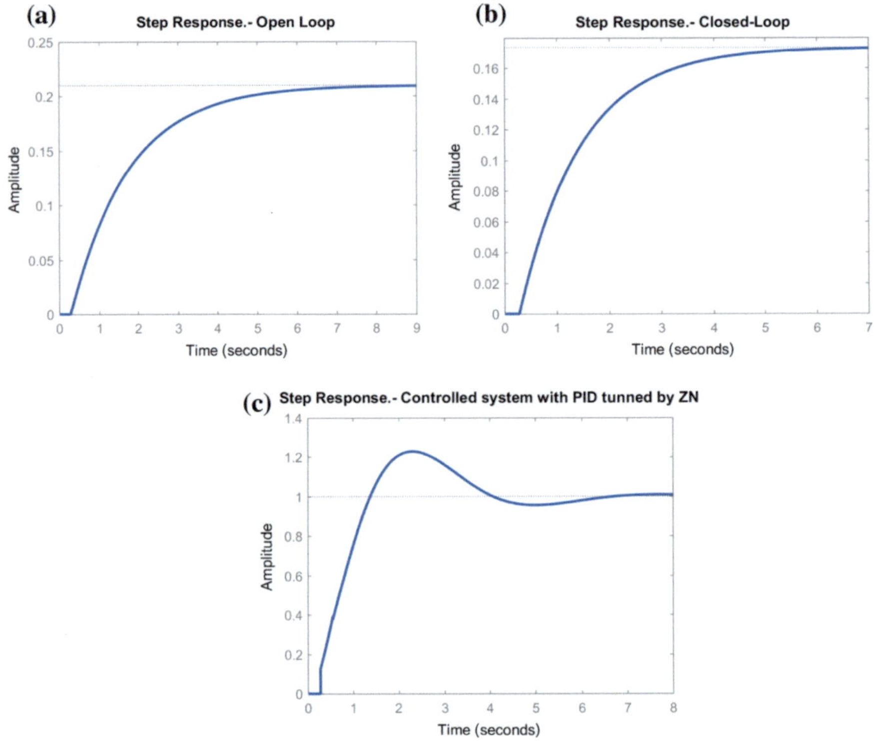

Fig. 6.44 System responses. **a** Open-loop output, used for obtaining the parameters θ and τ. **b** Closed loop without controller, i.e. a proportional controller with $K_p = 1$. **c** PID controlled response. Note that the output achieves the reference, although it exhibits a remarkable overshoot

i.e. tracing the tangent line at the inflexion point. However, it is only valid for (real) first-order systems. In order to eliminate this problem, it is recommendable to consider the formal definition of τ as the moment when the output achieves the 0.63% of its final value by simple inspection of the response.

Ziegler–Nichols Technique in Closed Loop

The second recipe for tuning PID controllers, known as the *closed-loop technique*, can be only applied to systems which can be turned into unstable under proportional control. This method was designed to obtaining a diminishing of the overshoot between the first and the second oscillations in 25%. Obviously, such result may not be assumable in most of cases, especially when the maximum oscillation is excessive, and thus this method is normally used as a preliminary step to a manual tuning.

Table 6.4 PID gains provided by the second method of Ziegler–Nichols

Controller	K_p	T_i	T_d
P	$0.5K_{\mathrm{cr}}$	–	–
PI	$0.45K_{\mathrm{cr}}$	$\frac{P_{\mathrm{cr}}}{1.2}$	–
PID	$0.6K_{\mathrm{cr}}$	$\frac{P_{\mathrm{cr}}}{2}$	$\frac{P_{\mathrm{cr}}}{8}$

The steps to be followed in this second method of Ziegler–Nichols are as follows:

1. Consider only proportional control increasing the value of K_p up to a certain value, the so-called critical value K_{cr}, for which the controlled output oscillates at the limit of the stability.
2. Analyse the system output to assess the period of the oscillation, P_{cr}, when considering the critical proportional gain K_{cr}.
3. Use the values of P_{cr} and K_{cr}, to calculate the gains for each PID parameter according to Table 6.4.

Example 6.9

Design a PID controller for the pump of a LVAD using the second method of Ziegler–Nichols. Note that Ziegler–Nichols methods do not require to know the transfer function, which makes them useful in a variety of cases. Next for the clarity sake, we assume we have such information.

Assuming that the transfer function of the pump is:

$$G(s) = \frac{10}{(s+2)(s+3)(s+7)}$$

The first step is to compute the critical proportional gain that leads the system to the limit of the stability. This value can be obtained through the root locus tool by moving the closed-loop poles to the imaginary axis (see Fig. 6.45). In this example, the critical proportional gain $K_{\mathrm{cr}} = 45$.

The next step is to calculate the period of the output oscillations. We can select two consecutive maximums directly on the image and check the time between them. As shown in Fig. 6.45, in this case the period is one second, i.e. $P_{\mathrm{cr}} = 1$.

Having K_{cr} and P_{cr}, the parameters needed for this recipe, the step 4 can be performed to obtain the K_p, K_i and K_d gains for a PID controller:

$$\boldsymbol{K_p} = 0.6 * K_{cr} = 0.6 * 45 = 27$$
$$T_i = \tfrac{P_{cr}}{2} = \tfrac{1}{2}; \left\{K_i = \tfrac{K_p}{T_i}\right\} \Rightarrow \boldsymbol{K_i} = \tfrac{27}{0.5} = 54$$
$$T_d = \tfrac{P_{cr}}{8} = \tfrac{1}{8}; \{K_d = K_p * T_d\} \Rightarrow \boldsymbol{K_d} = 27 * 0.125 = 3.3750$$

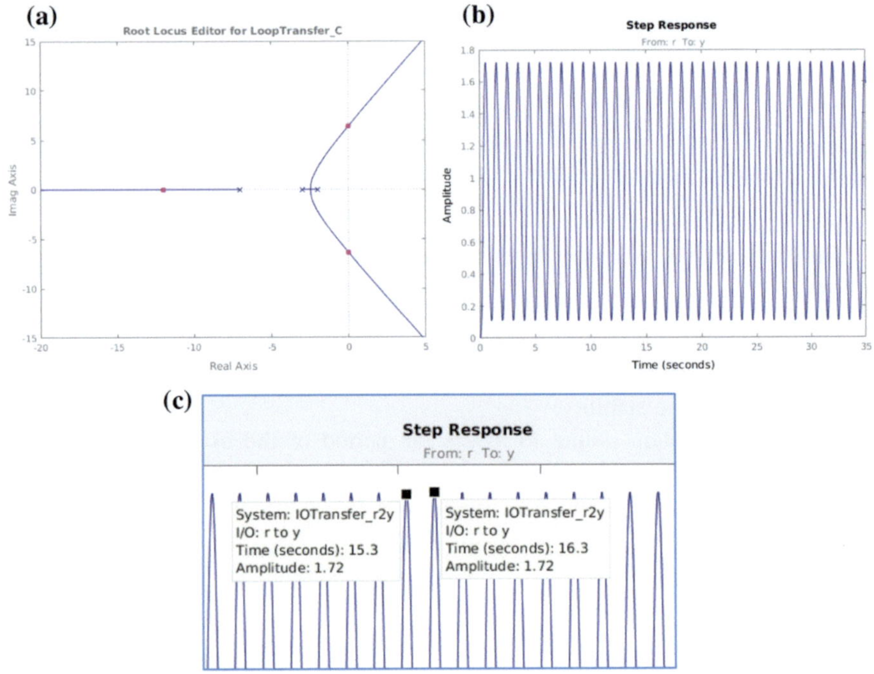

Fig. 6.45 Maintained oscillations caused by setting a proportional control that leads the system to the stability limit. **a** Root locus at the limit of the stability. **b** Closed-loop system response. **c** Detail of the step response reckoning the period of the oscillation

which results in the transfer function

$$\mathrm{PID}(s) = \frac{3.375s^2 + 27s + 54}{s}$$

Note that this is the ideal, theoretical PID transfer function, where an additional pole, i.e. the derivate filter, must be included. The controlled system response, i.e. using the tuned PID, is shown in Fig. 6.46. Note that the output might not be as good as desired, but it is what the Z-N method promised: it obtains a diminishing of the overshoot between the first and the second oscillations in a 25%.

In order to produce a more effective control system, we can adjust manually the parameters. For that we can rely on SIMULINK to easily represent the controlled system as shown in Fig. 6.47, and bearing the contribution of each component of the PID controller to the response (recall the summary of Table 6.1).

For instance, in order to reduce the maximum overshoot and oscillations exhibited in Fig. 6.46, we can reduce the K_p and K_i gains and slightly increase the K_d. Thus setting $K_p = K_i = 25$, $K_d = 6$ the resultant controlled response is shown in Fig. 6.48.

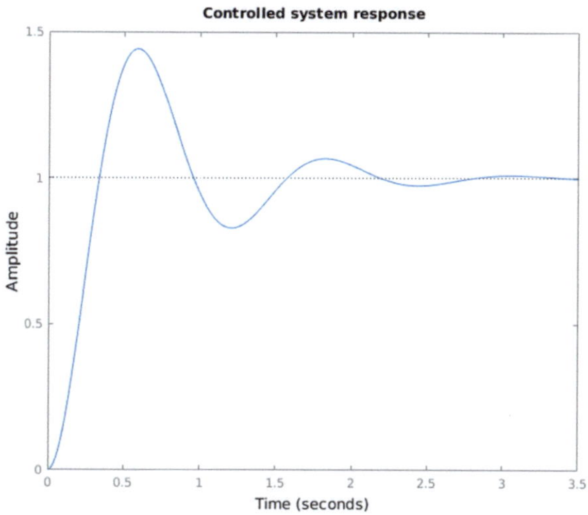

Fig. 6.46 Controlled system response using the parameters yielded by the Z-N second method

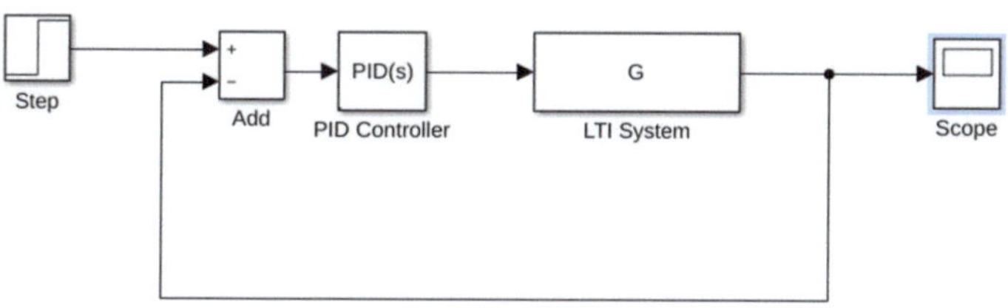

Fig. 6.47 SIMULINK model for the continuous controlled system

Note how the maximum overshoot has been significantly reduced while keeping a reduced response and settling time. A modification of the parameters to better adjust the response can be $K_p = 18, K_i = 17, K_d = 5.5$ in order to reduce even more the overshoot and improve the transient response. The result is shown in Fig. 6.49. Notice that the system output will not overpass the final value, at the expense of a small increment in the settling time.

The SIMC Tuning Rules

The SIMC tuning technique was presented by Skogestad in 2003. It requires that the system to be controlled can be modelled as a delayed first- or second-order system. The advantage of this method is that only one parameter has to be tuned,

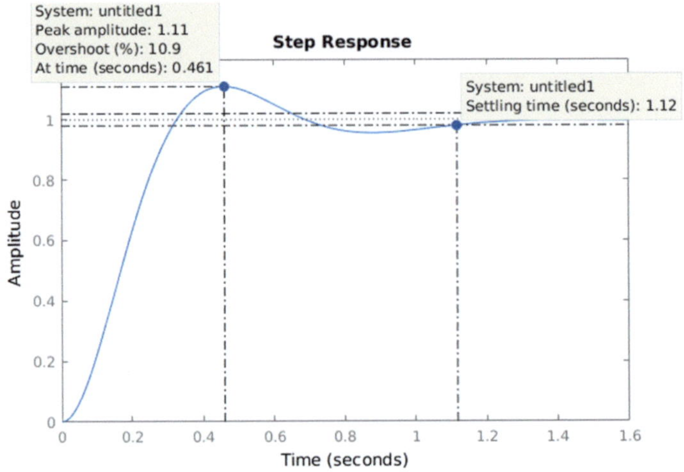

Fig. 6.48 Controlled response of the system with $K_p = K_i = 25$, $K_d = 6$

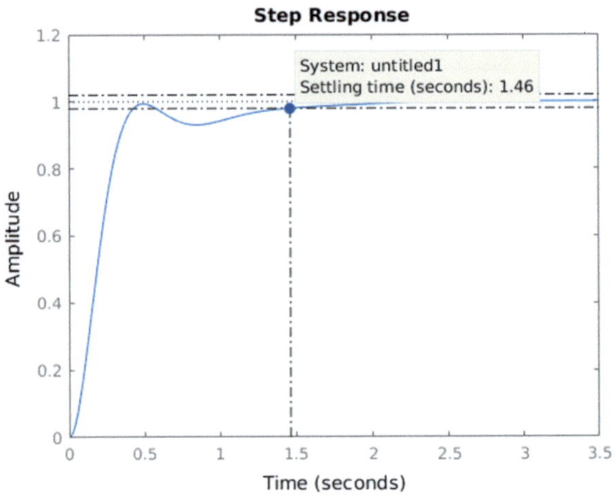

Fig. 6.49 Controlled response of the system with $K_p = 18, K_i = 17, K_d = 5.5$

i.e. τ_c. In the case the system is modelled as a delayed first-order system, a PI controller is derived from the SIMC rules, while for a second-order system, a PID is obtained.

The SIMC for First-Order Systems

In this case, the plant to be controlled must be approximated as a delayed first-order system, that is

$$G(s) = \frac{K}{\tau s + 1} e^{-\theta s}$$

These parameters can be easily obtained from the procedure illustrated in Fig. 6.43, i.e. the system response in open loop. That is, the system gain, K, the time constant τ, and the system delay, θ. Once these parameters are obtained, the SIMC rule for designing a PI controller is

$$\begin{aligned} K_c &= \tfrac{1}{K} \cdot \tfrac{\tau}{\tau_c + \theta} \\ T_i &= \min\{\tau, 4(\tau_c + \theta)\} \end{aligned} \tag{6.15}$$

where τ_c, the desired closed-loop time constant, is the unique parameter to be tuned.

Example 6.10
For the system tuned using Ziegler–Nichols in the Example 6.8, obtain a PI controller following the SIMC rules and compare the result.

Starting from the first-order parameters previously obtained, $\tau = 1.4851, \theta = 0.2754$, and $K = 0.21$, the adjusted parameters of the PI can be computed from Eq. (6.15), in terms of τ_c as

$$K_c = \frac{1}{0.21} \cdot \frac{1.4851}{\tau_c + 0.2754}$$

$$T_i = \min\{1.4851, 4(\tau_c + 0.2754)\}$$

The selection of the parameter τ_c depends on the control requirements, although there is a recommendation to choose $\tau_c = \theta$ in order to have a good output performance. Thus, setting $\tau_c = 0.2754$, we obtain

$$K_c = 12.8463$$

$$T_i = 1.4851$$

Therefore, the series PI transfer function is

$$\mathrm{PI}_{\mathrm{series}} = K_c\left(\frac{T_i s + 1}{T_i s}\right) = \frac{K_c T_i s + K_c}{T_i s} = \frac{19.08s + 12.85}{1.485s}$$

which produces the controlled response shown in Fig. 6.50. Note the difference with the results obtained with Ziegler–Nichols (see Fig. 6.44). In this case, the response exhibits a lower overshoot, reaching the steady state in 2 s.

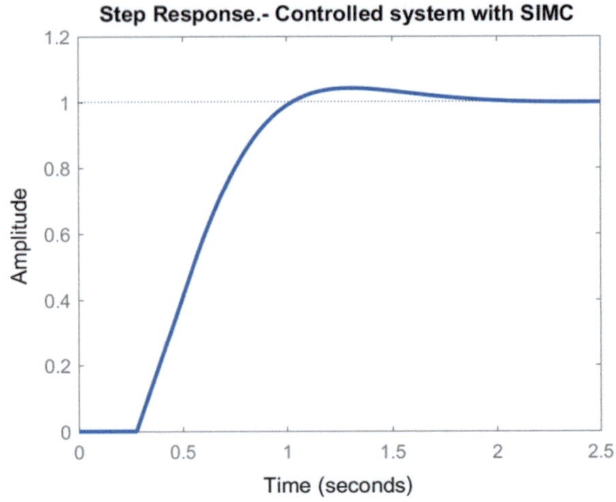

Fig. 6.50 Controlled response of the system using the SIMC rule

The SIMC for Second-Order Systems

In this case, the plant to be controlled must be approximated as a delayed second-order system, that is

$$G(s) = \frac{K}{(\tau_1 s + 1)(\tau_2 s + 1)} \mathrm{e}^{-\theta s} \tag{6.16}$$

In order to approximate an arbitrary plant to a second-order system, the *half rule* can be applied. Having a general plant given by

$$G(s) = \frac{K \prod_j (z_j s + 1)}{\prod_i (\tau_i s + 1)} \mathrm{e}^{-\theta s} \tag{6.17}$$

where all time constants follow a decreasing order, i.e. $\tau_i > \tau_j, i < j$, the half rule establishes that the general transfer function (6.16) can be approximated to a second-order system with the following parameters

$$\begin{aligned} \tau_1^* &= \tau_1 \\ \tau_2^* &= \tau_2 + \tfrac{\tau_3}{2} \\ \theta^* &= \theta + \tfrac{\tau_3}{2} + \sum_{i \geq 4} \tau_i + \sum_j -z_j \end{aligned} \tag{6.18}$$

For instance, the third-order transfer function from Example 6.9 rewritten as

$$G(s) = \frac{\frac{10}{2 \cdot 3 \cdot 7}}{\left(\frac{1}{2} s + 1\right)\left(\frac{1}{3} s + 1\right)\left(\frac{1}{7} s + 1\right)} \tag{6.19}$$

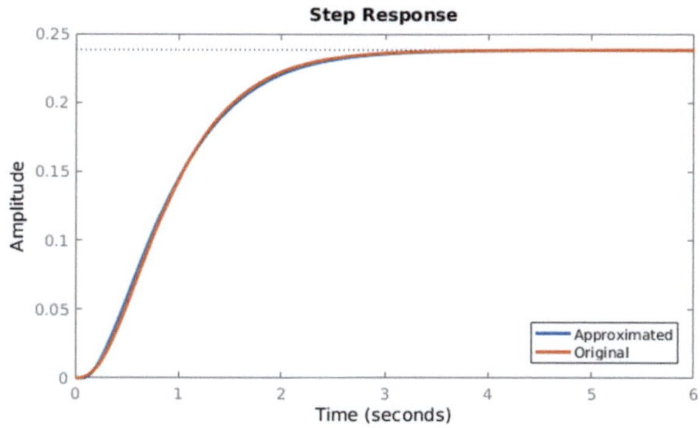

Fig. 6.51 Responses of a third-order system and its approximation using the half rule

can be approximated following Eq. (6.18) as

$$\tau_1^* = \frac{1}{2} = 0.5$$

$$\begin{aligned} \tau_2^* &= \frac{1}{3} + \frac{1/7}{2} = 0.4048 \\ \theta^* &= 0 + \frac{1/7}{2} = 0.0714 \end{aligned} \tag{6.20}$$

being thus the second-order approximated function

$$G(s) = \frac{0.2381}{(0.5s+1)(0.4048s+1)} \mathrm{e}^{-0.0714s} \tag{6.21}$$

Step response comparison between both expressions (6.19) and (6.21) is illustrated in Fig. 6.51.

Once a convenient method has been followed for approximating the plant to be controlled, the SIMC rule suggests the following PID parameters.

$$\begin{aligned} K_c &= \tfrac{1}{K} \cdot \tfrac{\tau_1^*}{\tau_c + \theta} \\ T_i &= \min\{\tau_1^*, 4(\tau_c + \theta)\} \\ T_d &= \tau_2^* \end{aligned} \tag{6.22}$$

which in our example become

$$K_c = 14.7056, \quad T_i = 0.5, \quad T_d = 0.4048$$

yielding the series PID transfer function

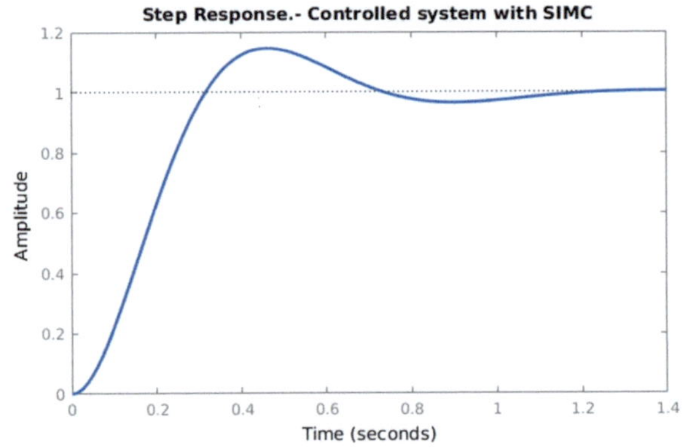

Fig. 6.52 Controlled response for a PID tuned through the SIMC rule

$$\mathrm{PID_{series}} = K_c\left(\frac{T_i s+1}{T_i s}\right)(T_d s+1) = \frac{K_c(T_i T_d s^2 + (T_i + T_d)s + 1)}{T_i s}$$
$$= G_{\mathrm{PID}} = \frac{2.976s^2 + 13.31s + 14.71}{0.5s}$$

The controlled system response using this PID is shown in Fig. 6.52. Note the smooth response obtained with respect to the one produced by the closed-loop Z-N method (shown in Fig. 6.46).

The AMIGO Tuning Rules

The tuning method called Approximated MS-constrained Integral Gain Optimization (AMIGO) relies on maximizing the integral gain while keeping the maximum overshoot under 20%. The AMIGO rules are based on the approximation of the system to be controlled to a first-order system, and thus, only the system gain, time constant and system delay parameters are needed. In some cases, the maximum slope of the response, i.e. K_v, is also needed, which can be easily obtained as $K_v = K_p/\tau$. Adjusting a PI controller depends on the relation between the time constant and the system delay as shown in Table 6.5, while for the case of a PID, the parameters are directly obtained by the expression.

Table 6.5 AMIGO rules for adjusting PI controllers

	$\theta < \frac{\tau}{6}$	$\frac{\tau}{6} < \theta < \tau$	$\tau < \theta$
K_c	$\frac{0.35}{K_v\theta} - \frac{0.6}{K_p}$	$\frac{0.25\tau}{K_p\theta}$	$\frac{0.1\tau}{K_p\theta} + \frac{0.15}{K_p}$
	$\theta < 0.11\tau$	$0.11\tau < \theta < \tau$	$\tau < \theta$
T_i	0.7θ	0.8τ	$0.3\theta + 0.5\tau$

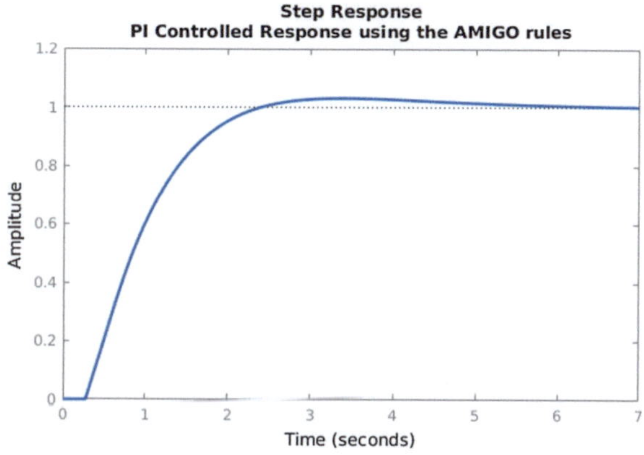

Fig. 6.53 Controlled response for a PI tuned through the AMIGO rule

$$K_c = \frac{1}{K_p}\left(0.2 + \frac{0.45\tau}{\theta}\right) \quad T_i = \frac{0.4\theta + 0.8\tau}{\theta + 0.1\tau}\theta \quad T_d = \frac{0.5\theta\tau}{0.3\theta + \tau} \tag{6.23}$$

Example 6.11
For the system tuned using Ziegler–Nichols in the Example 6.8, obtain a PI and a PID controller following the AMIGO rules.

Starting from the first-order parameters previously obtained, $\tau = 1.4851, \theta = 0.2754$, and $K_p = 0.21$, the adjusted parameters of the PI can be computed from Table 6.5 as $K_c = 6.4197, T_i = 1.1881$. The PI transfer function is therefore

$$\mathrm{PI_{series}} = K_c\left(\frac{T_i s + 1}{T_i s}\right) = \frac{K_c T_i s + K_c}{T_i s} = \frac{7.627s + 6.42}{1.188s}$$

which yields the controlled response illustrated in Fig. 6.53.

In order to adjust a PID controller, the parameters obtained through Eq. (6.23) are $K_c = 12.5078, T_i = 0.8434, T_d = 0.1304$, and the PID transfer function is given by

$$\mathrm{PID_{series}} = \frac{K_c(T_i T_d s^2 + (T_i + T_d)s + 1)}{T_i s} = \frac{1.376s^2 + 12.18s + 12.51}{0.8434s}$$

which produces the controlled response illustrated in Fig. 6.54.

Tuning a PID Using MATLAB

Although the aforementioned technique (and others) is valid to tune a PID controller, largely requires the expertise of the designer. A direct and quick alternative is to rely on the automatic tuning methods implemented in MATLAB that visually

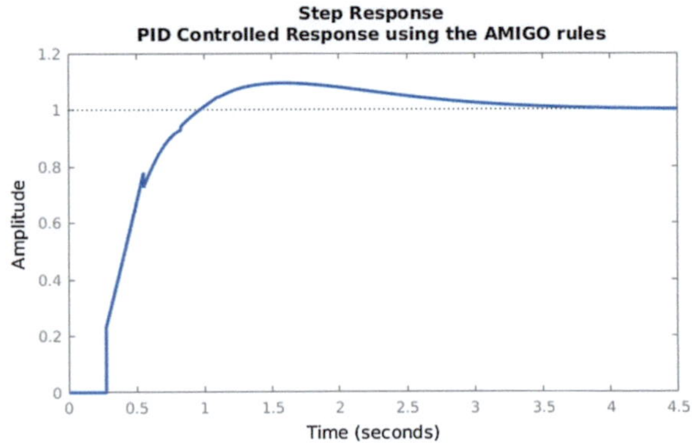

Fig. 6.54 Controlled response for a PID tuned through the AMIGO rule

provides the controlled response and the computed parameters of the PID. The dialog box of the PID block is shown in Fig. 6.55 (and accessed by double clicking on it within the SIMULINK diagram).

In this dialog, the designer can enter the chosen parameters, including the derivative filter that adds an additional pole to the controller making it implementable. Moreover, the tune button[4] enables the automatic tuning features of MATLAB, opening the PID Tuner window as shown in Fig. 6.56. Mainly in this utility, the response considering the original PID controller (in this case the one proposed by Z-N) and the adjusted one by MATLAB are visualized. In this case, the utility suggests PID parameters: $K_p = 9.72, K_i = 12.74, K_d = 1.77, N = 298.3$, as a trade-off between time response and maximum overshoot. The balance of these parameters can be adjusted in the upper panel, in order to produce a faster (or slower) response and/or a robust or aggressive transient behaviour.

The output saturation can be also selected in the PID Advanced Tab as shown in Fig. 6.57.

A similar procedure can be followed for discrete systems in which the SIMULINK diagram can be constructed as shown in Fig. 6.58.

Within the PID block dialog box, the user can select discrete time, the sample time and the discretization method to tune a discrete controller as shown in Fig. 6.59a, b, where in this case the suggested parameters are $K_p = 8.74, K_i = 9.9, K_d = 1.4, N = 7.92$.

[4]This graphical utility can be also accessed from the MATLAB workspace through the `pidTuner` command.

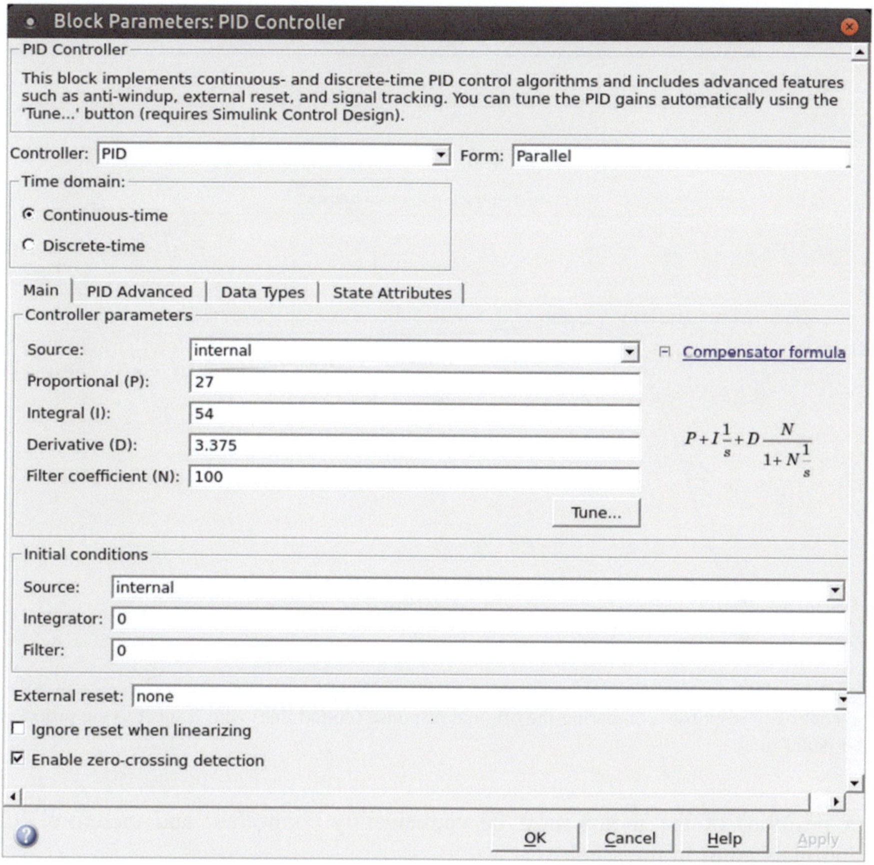

Fig. 6.55 PID controller dialog box. The PID block from SIMULINK enables the designer to select a number of parameters and different configurations

Applications in Biomedical Engineering

Many applications in biomedical engineering rely on PID controllers in order to accurately provide the desired output. Along this book, different examples have been considered for applying the theory behind automated control. This section highlights and summarizes some of the most relevant ones.

Insulin pumps (and in general any automated drug administered system). Broadly speaking, insulin pumps consist of a motor acting on the device that provides the insulin (see Fig. 6.60). The flow administered to the patient can be seen as proportional to the speed or, in some cases to the position, of the motor. Thus, in order to control the right flow to be administered to the patient, the speed

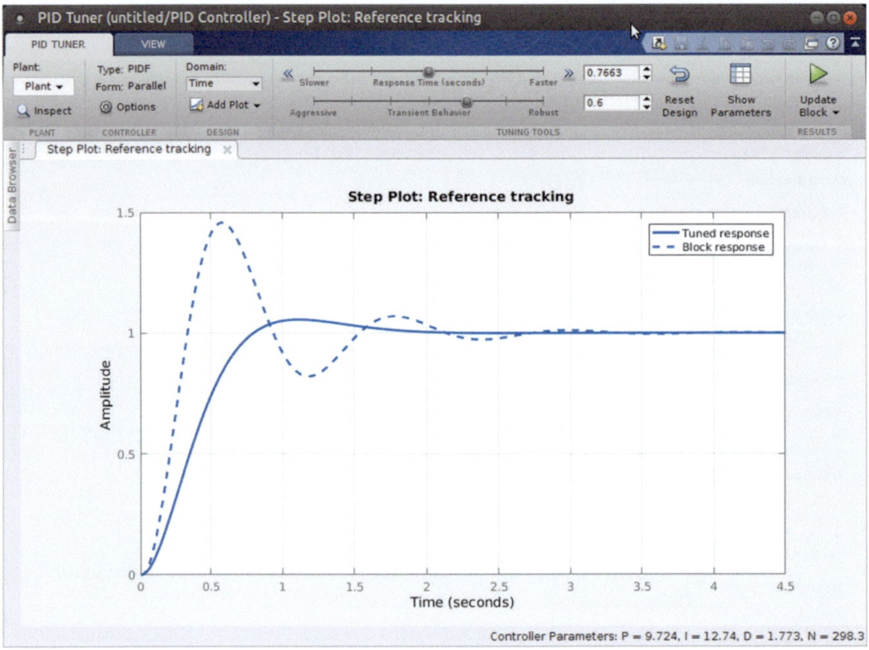

Fig. 6.56 PID automatic tuner application. SIMULINK provides different techniques to adjust the controlled response, comparing the original response (dotted line) with respect to the proposed one (solid line)

(or position) of the motor must be conveniently controlled, and therefore PID controllers enter in the scene.

Similarly, *dialysis machines* normally also act as a pump, extracting the blood from the patient, filtering it and then returning it back. In this case, the variable to be controlled is the input/output blood flow. Thus a controller, like a PID, is needed where in this case the reference signal is the flow to be maintained during the process, and the sensory system entails flowrate sensors (see Fig. 6.61).

Another example in which the control system plays a paramount role is in the case of LVAD (see Fig. 6.62). These devices largely help people with serious heart failures while waiting for a heart transplant. It requires an invasive surgery to place a pump in the left ventricle of the patient. This pump, when conveniently controlled, will assist the damage heart to pump the enough blood flow through the aorta.

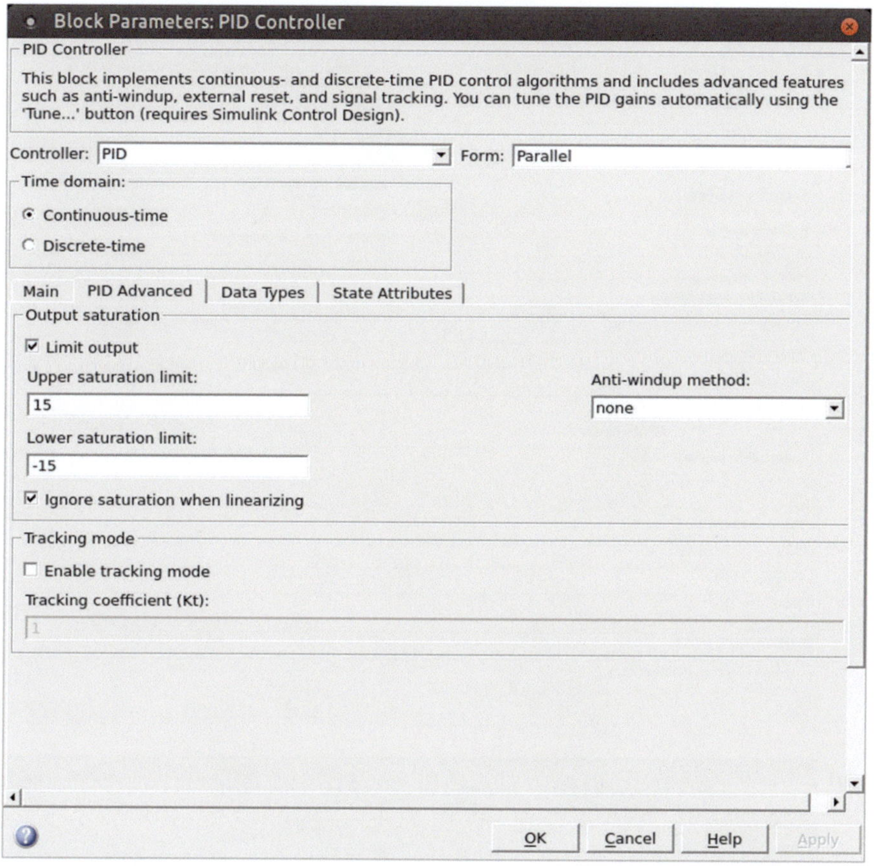

Fig. 6.57 PID output saturation

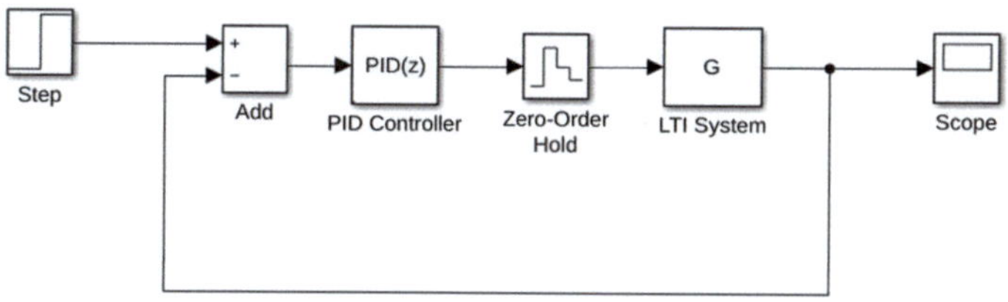

Fig. 6.58 SIMULINK model for the discrete controlled system

(a) (b)

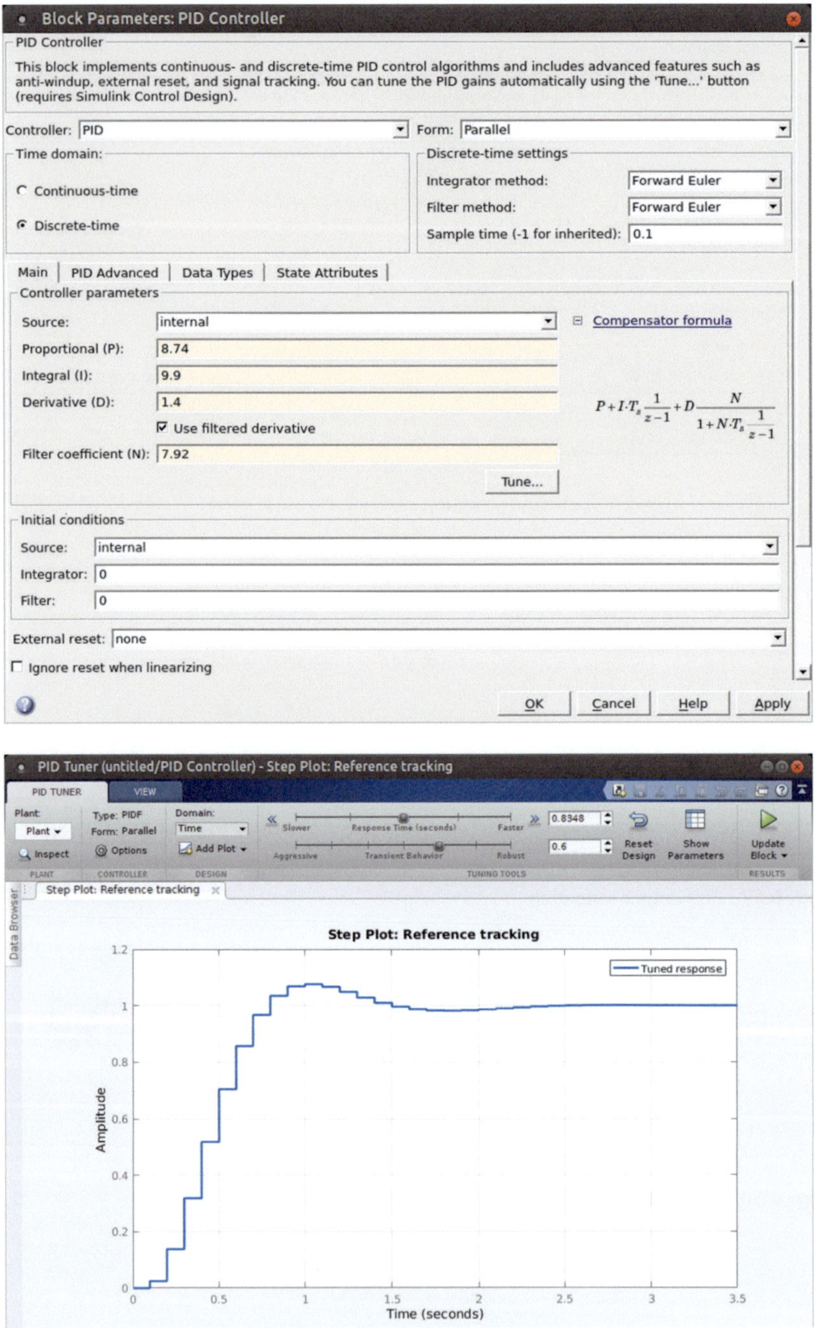

Fig. 6.59 Discrete controlled response. **a** The selected parameters within the PID block. **b** The resultant output

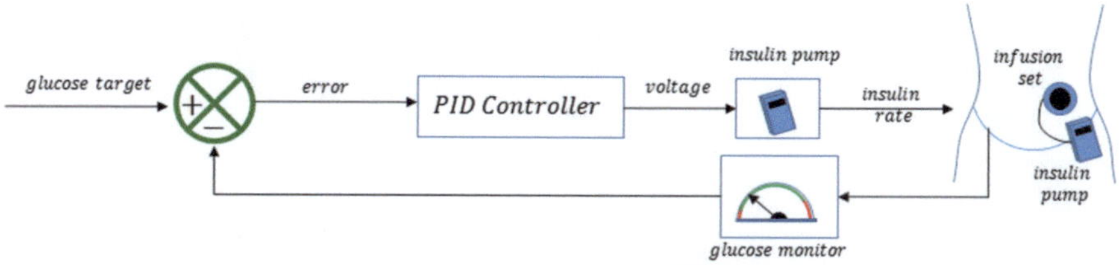

Fig. 6.60 PID controlled insulin pump. The PID controller computes the required input voltage to the motor according to the reference and the measurement. Notice that both must be conveniently converted to voltage signals

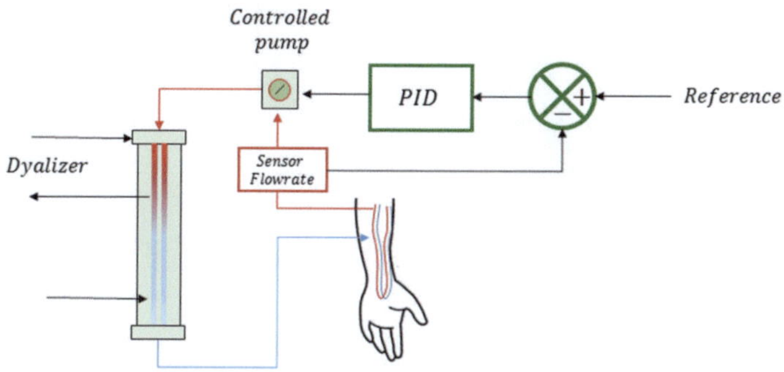

Fig. 6.61 A dialysis machine controlled by a PID

Fig. 6.62 Scheme of a left ventricular assist device (LVAD)

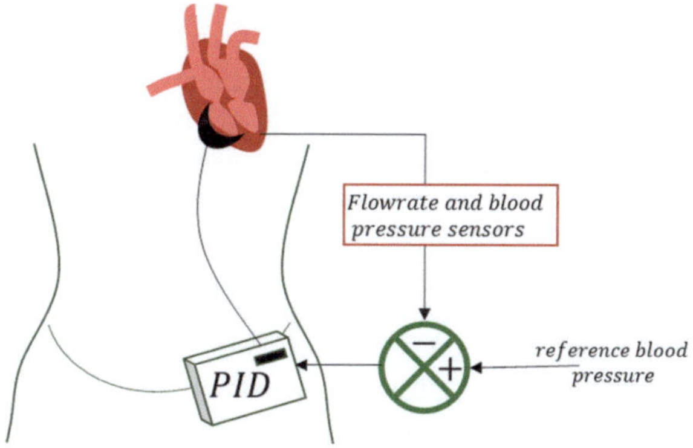

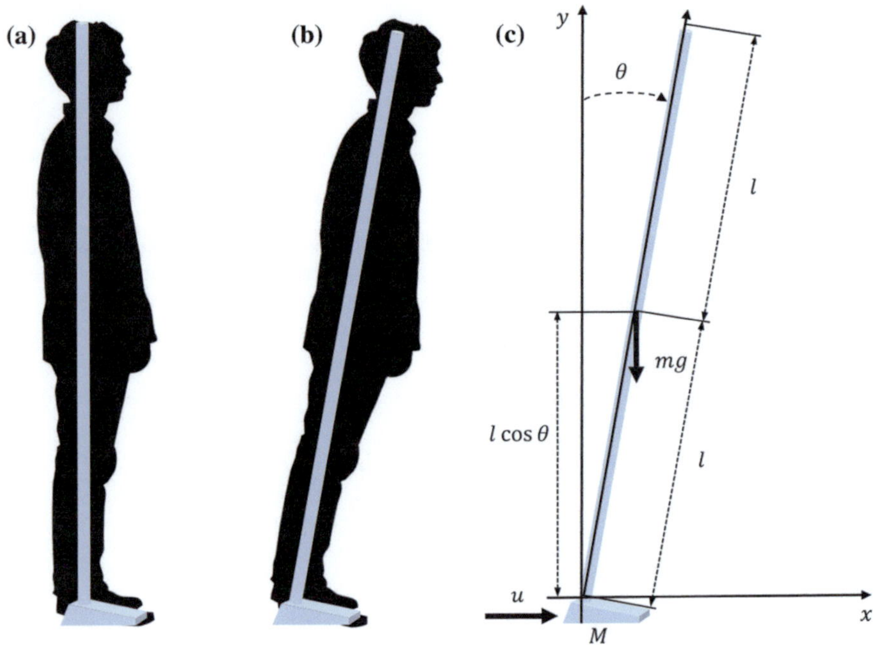

Fig. 6.63 Human bipedal posture. **a** No inclination, **b** with inclination and **c** mechanical model

Solved Problems

Problem 6.1

The human bipedal posture is inherently unstable. Standing up requires skeletal adaptations to changes in posture to maintain balance and prevent a fall.

Postural control or balance is the ability to maintain equilibrium in a gravitational field by maintenance of the centre of mass of a body over the base in which it is supported.

The balance must be adapted to different situations. Both the sensory information acquisition and the processing of a neural level involve the interpretation, conceptualization, planning, activation and execution of movement patterns that aim to react quickly and in a coordinated way, through muscular activation facing the disturbances of the environment.

From a local point of view, ankle control can be understood as a strategy used in standing up to maintain balance. In the presence of small disturbances, the control system acts by bringing the centre of mass to a stable position by moving the body through the ankle joint, behaving like an inverted pendulum.

Figure 6.63a represents the interpretation of bipedal balance of a person with a system formed by a cylinder attached by an axis to a base with mass M. If the person tilts slightly, the system can be represented according to Fig. 6.63b. The mechanical model of bipedal balance is based on the balance of forces. The representation of the person can be taken as valid by a cylinder of mass m and length $2l$, with its centre of gravity located at a distance l from the ends. Figure 6.63c represents this model, where an inclination θ in a plane xy with origin of coordinates located in the union between the cylinder and the base has been identified.

The balance can be understood as a position control system, θ, with the aim of maintaining the vertical position. This objective can be identified with the control of an inverted pendulum, the default state of which is unstable since it can rotate at any time and in any direction, unless a convenient control force is applied.

If the person tends to make a shift in the direction x, it can be understood that he has applied, in that direction, a force u. In this situation, the mathematical model for this system can be expressed as follows:

$$(M+m)\ddot{x}+ml\ddot{\theta}=u$$

$$\left(I+ml^2\right)\ddot{\theta}+ml\ddot{x}=mgl\theta$$

Being:

I, the moment of inertia of the primary system. It can be shown that

$$I=\frac{1}{3}mL^2$$

g, the acceleration due to terrestrial gravity.

By developing the proposed system of equations, we can easily obtain the following relationship between u and θ

$$\left[-(M+m)\frac{7}{3}l+ml\right]\ddot{\theta}(t)+g(M+m)\theta(t)=u(t)$$

The considered system is illustrated according to the functional diagram of Fig. 6.64.

Data: $m=70\,\text{kg}$, $M=1\,\text{kg}$, $g=9.81\,\text{m/s}^2$, $l=0.5\,\text{m}$.

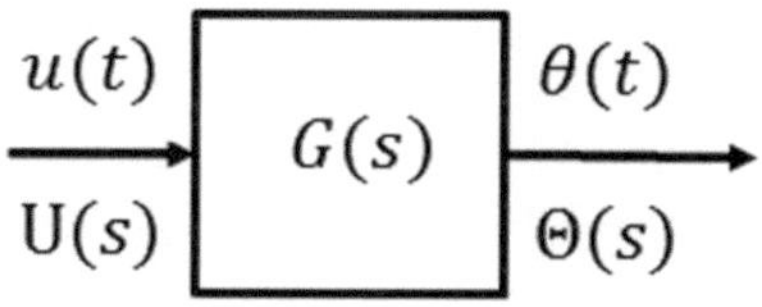

Fig. 6.64 Model of a biped system

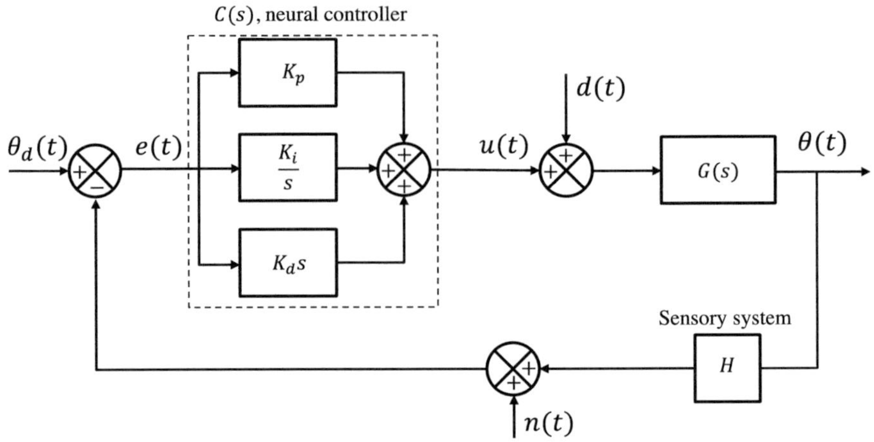

Fig. 6.65 Control system for a model of human balance

(a) Analysis of the system without controller, $G(s) = \Theta(s)/U(s)$.

i. Determine the system transfer function, $G(s)$.
ii. Plot the map of poles and zeros. Justify if the system is stable.

(b) Analysis of the controlled system.
The posture control system (Fig. 6.65) that the person performs can be modelled by identifying a pressure measurement system located at the feet (ideally $H = 1$) and a neural controller based on the inclination error ($\theta_d - \theta$). The controller not only considers the current error, but also considers the past and the one to be committed, according to the trend. This controller can be identified as a PID system. The complete control system includes the effect of disturbance ($d(t)$) and noise ($n(t)$) signals.

i. Would it be possible to perform a controller tuning using any of the Ziegler–Nichols methods?
ii. Using the `sisotool`, design which would be a suitable controller, which fulfils the following requirements: overshoot 10%, settling time ($\varepsilon = 2\%$) $< 2s$ and null steady-state error for unit step input. Consider that there are neither disturbance, $d(t)$, nor noise, $n(t)$, signals.
iii. If the person tilts a certain degree forward, this effect can be considered as a disturbance signal, d, of unit step type. In this case, there is neither reference signal, $\theta_d(t)$, nor noise, $n(t)$. Determine the angle of the system, $\theta(t)$, after the transient effect has elapsed, that is, only in the steady state.
iv. In dizziness, the measure of inclination, based on perceived pressure on the sole of the foot, may be altered. This alteration could be considered as a noise, $n(t)$, added to the actual reading of the inclination, $\theta(t)$. How would the controlled system react to the input of a unit-impulse-type noise.

Solution 6.1

(a) From the relation

$$\left[-(M+m)\frac{7}{3}l+ml\right]\ddot{\theta}(t)+g(M+m)\theta(t)=u(t)$$

The transfer function can be obtained by applying the Laplace transform, using the second derivative property with null initial conditions:

$$L\left[\frac{\mathrm{d}^2}{\mathrm{d}t}f(t)\right]=s^2F(s)-sf(0^-)-\dot{f}(0^-)$$

$$\left[-(M+m)\frac{7}{3}l+ml\right]s^2\Theta(s)+g(M+m)\Theta(s)=U(s)$$

Thus:

$$G(s)=\frac{1}{\left[-(M+m)\frac{7}{3}l+ml\right]s^2+g(M+m)}$$

Substituting the data:

$$G(s)=\frac{-0.0209}{s^2-14.56}$$

The calculus in MATLAB can be made using the following code:

```
s=tf('s');
M=1;
m=70;
g=9.81;
l=0.5;
G=1/((-(M+m)*7/3*l+m*l)*s^2+g*(M+m))
```

The system is of order 2 and type 0.

The transfer function has two real poles and no zero. One in the positive half-plane ($s_1=3.8159$) and one in the negative half-plane ($s_2=-3.8159$). In this way, the map of poles and zeros can be easily obtained (see Fig. 6.66).

The criterion for identifying whether the system is stable in the presence of input patterns, such as a unitary step input, is based on the determination of the poles of the transfer function. When there are poles in the positive half-plane, the system is unstable. In this case, since there is one, the system is unstable. To illustrate this feature, the following code can be used in MATLAB, obtaining the graph represented (Fig. 6.67), where it is observed how the value of the output diverges:

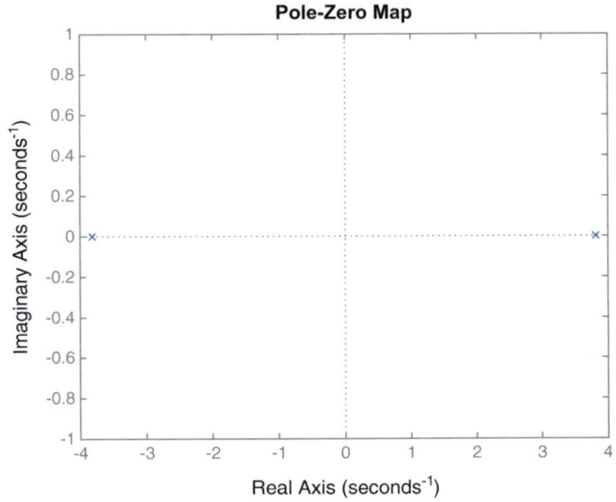

Fig. 6.66 Maps of poles and zeros of $G(s)$

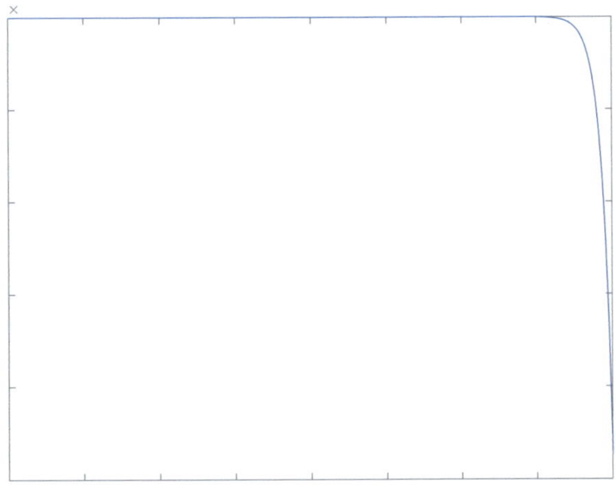

Fig. 6.67 Response to unit step input of $G(s)$

```
step(G)
```

(b) According to the map of poles and zeros (Fig. 6.66), the tuning by the first Ziegler–Nichols method is not possible to be applied since the response for a unit step input is not stable.

On the other hand, if it analysed the root locus (Fig. 6.68) of the feedback system, it can be seen that there is no value of the gain that causes the system to be critically unstable. Therefore, it is not possible to tune the controller using the second Ziegler–Nichols method.

By using the `sisotool`, you can choose the controller tuning.

Fig. 6.68 Root locus of the feedback system

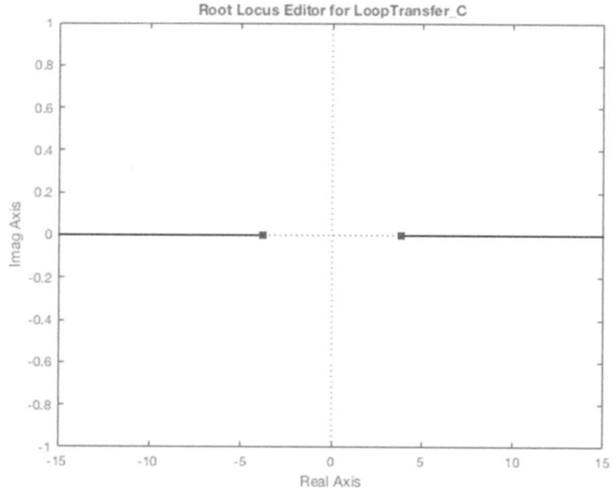

Among the proposed controllers, P, I, PI, PD, PID, only the PD and PID controllers make the system stable. However, even though the PD controller meets the overshoot and settlement time specifications, it does not meet the steady-state error rate. This is due to the fact that the direct loop $(C(s)G(s)H(s))$ has type 0. Therefore, the type must be increased. This way you could choose a PD controller configuration with a pole at the origin (I) or a PID.

We proceed by choosing the type of PID controller, interacting with the design interface as shown in Fig. 6.69. The result is then adjusted by moving the poles taking into account the behaviour of the response to a unit step input (Figs. 6.70 and 6.71).

The synthesized controller turns out to be the following:

$$\mathtt{C} =$$

$$\frac{-6604.7(\mathtt{s}+2.71)(\mathtt{s}+1.979)}{\mathtt{s}}$$

This is:

$$C(s) = \frac{-6604.7(\mathrm{s}+2.71)(\mathrm{s}+1.979)}{s}$$

Which corresponds with the following parameters values:

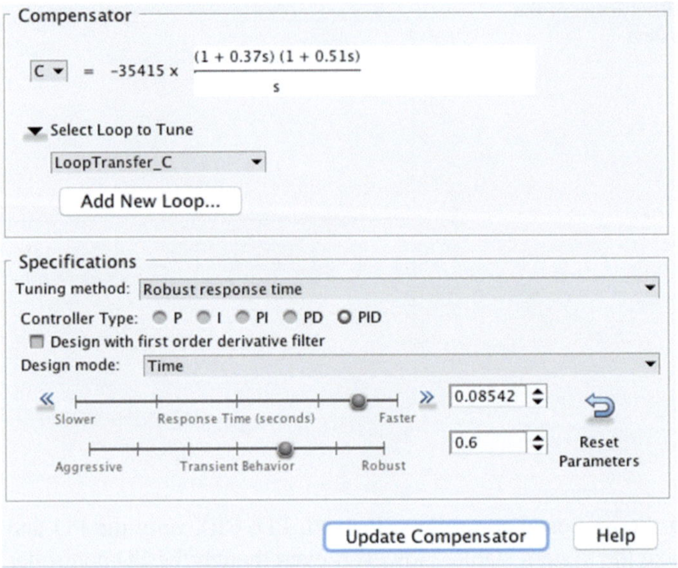

Fig. 6.69 PID controller optimization window

Fig. 6.70 Root locus of the controlled system

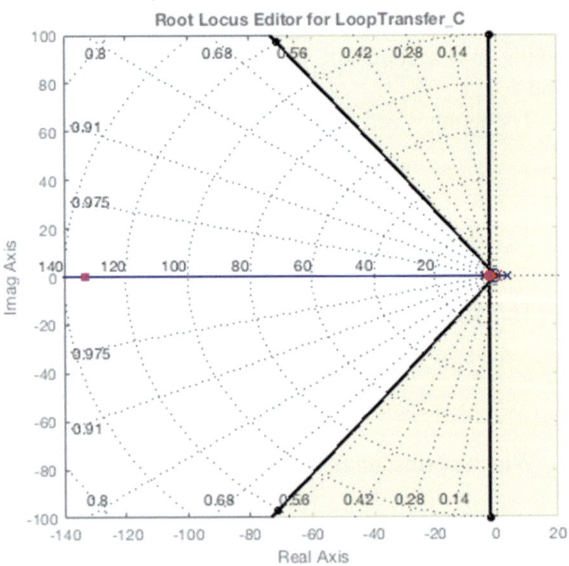

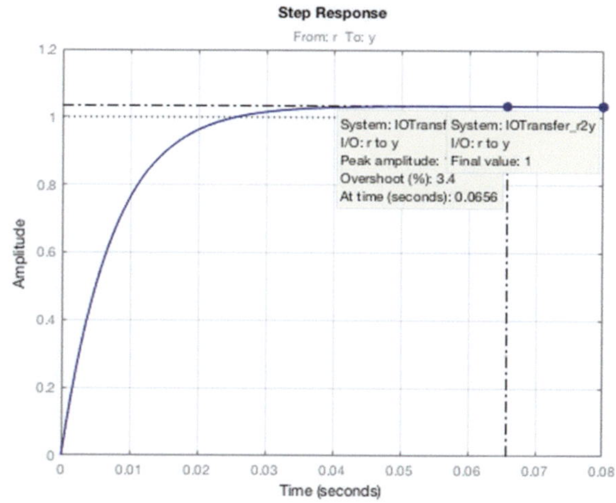

Fig. 6.71 Response to reference input unit step

```
pid(C)

ans =

Kp + Ki* 1/s + Kd * s

withKp = - 3.1e + 04, Ki = - 3.54e + 04, Kd = - 6.6e + 03

Name : C
Continuous - timePIDcontrollerinparallelform.
```

Thus, the tuned controller is

$$C(s) = -31{,}000 - \frac{35{,}400}{s} - 6600s$$

The behaviour of the output signal as a function of a disturbance input of the unitary step type can be obtained in several ways.

First, considering both null, the reference input, $\theta_d(t)$, and noise, $n(t)$, the influence of the disturbance signal on the output can be determined from the following expression

$$\Theta(\mathrm{s}) = [D(s) + C(s)E(s)]$$

Considering,

$$E(s) = -\Theta(\mathrm{s})H(s)$$

The relation between the output and the disturbance is obtained as

$$\Theta(\mathrm{s}) = \frac{G(s)}{1 + C(s)H(s)G(s)} D(s)$$

Substituting, $G(s)$, $H(s)$ y $C(s)$,

$$\Theta(\mathrm{s}) = \frac{\frac{-0.0209}{s^2 - 14.56}}{1 + \frac{-6604.7(\mathrm{S}+2.71)(\mathrm{S}+1.979)}{s} 1 \frac{-0.0209}{s^2 - 14.56}} D(s)$$

The influence on the inclination, $\theta(t)$, after a transient, can be assessed when sufficient time has elapsed, that is

$$\theta_{ss} = \lim_{t \to \infty} \theta(t)$$

Considering the final value theorem and a disturbance of unit step type, $D(s) = 1/s$, results in

$$\theta_{ss} = \lim_{s \to 0} s\Theta(s)$$

$$\theta_{ss} = \lim_{s \to 0} \left(\frac{s \frac{-0.0209}{s^2 - 14.56} \frac{1}{s}}{1 + \frac{-6604.7(\mathrm{S}+2.71)(\mathrm{S}+1.979)}{s} 1 \frac{-0.0209}{s^2 - 14.56}} \right)$$

$$\theta_{ss} = \frac{\frac{-0.0209}{0 - 14.56}}{1 + \frac{-6604.7(\mathrm{S}+2.71)(\mathrm{S}+1.979)}{0} 1 \frac{-0.0209}{0 - 14.56}}$$

$$\theta_{ss} = 0^\circ$$

Secondly, thanks to the `sisotool` utility and the *New Plot* → *New Step* option (Fig. 6.72); it is possible to monitor the effect of the output in the presence of a unit step-type disturbance. The result is shown in Fig. 6.73.

Using the same procedure as the followed one in the previous section, it is possible to get the response to a noise input of unit impulse type (Fig. 6.74). In this case, it used the option *New Plot* → *New Impulse*.

In both cases of inputs (disturbance as step and noise as impulse), it can be observed how robust the controller is, cancelling the effects due to both inputs.

Problem 6.2

The application of an electric current to a skeletal muscle in order to stimulate it has the purpose of activating it. The caused effect can correct disorders caused by damage to the central nervous system. This technique, known as functional electrical stimulation (FES), can be used to correct, for example, the symptom of the drop foot. Figure 6.75 shows a system formed by an actuator (band below the knee)

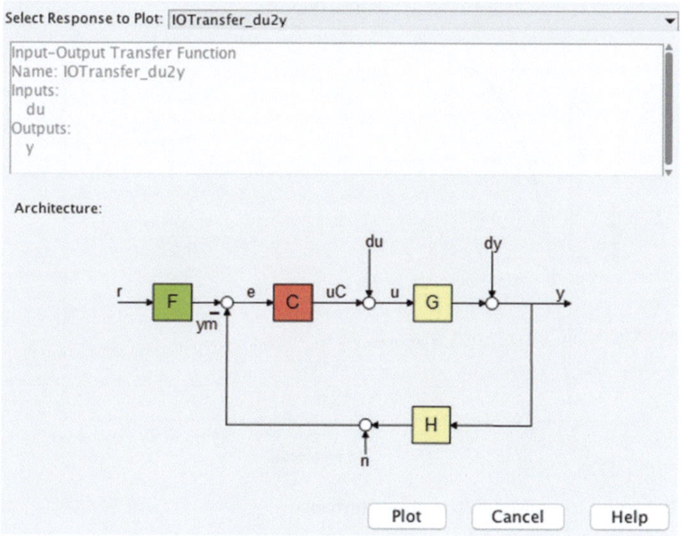

Fig. 6.72 Interface of settings for responses to be shown

Fig. 6.73 Response to unit step-type disturbance

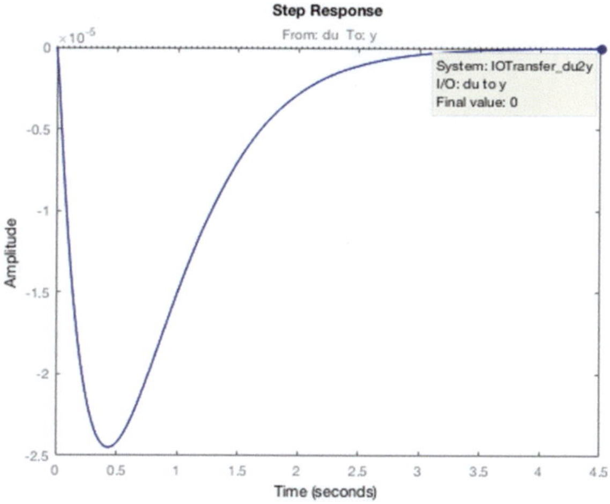

and a switch (band located in the foot). The operation of the system is based on the activation of a control of stimulation of the common peroneal nerve when the foot stops being in contact with the ground (switch on). In this situation, the FES control allows the contraction of the extensor digitorum longus (EDL), correcting the effect of the fallen foot. The contraction of this muscle is measured with a sensor located on the band placed on the foot, as shown in the figure. When the sole of the foot is in contact with the ground, the control system is deactivated (switch off).

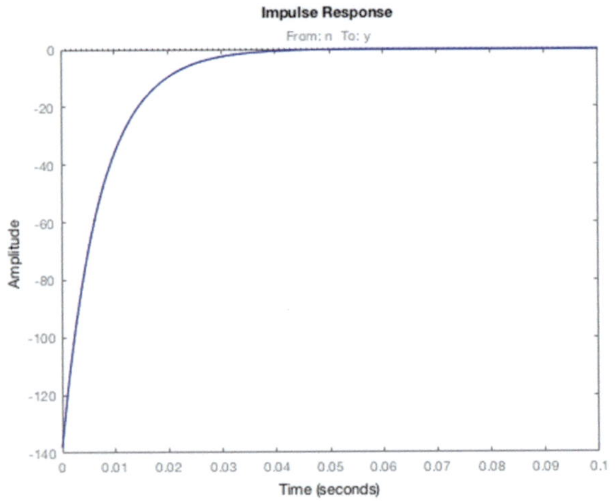

Fig. 6.74 Response to unit-impulse-type disturbance

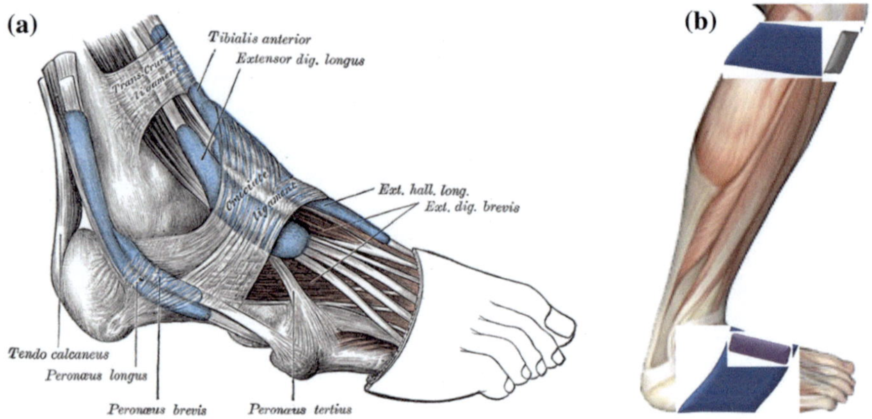

Fig. 6.75 **a** Identification of the muscle extensor digitorum longus, **b** FES system

The electrical stimulation application, FES, constitutes a feedback-loop control system, as shown in Fig. 6.76.

In this system, the muscle has been modelled as a linear system which has a relation between the tension it performs (muscle–tendon force, $f_{mt}(t)$) and the length of the same, $x(t)$. This force can be stimulated by a stimulation system, FES. The system is designed with the purpose to obtain a desired length of the muscle, $x_d(t)$. This value is compared in real time with the effect obtained, providing an error, $e(t)$, which must be corrected.

The Hill model can be used to describe the behaviour of the muscle as a mechanical system (Fig. 6.77).

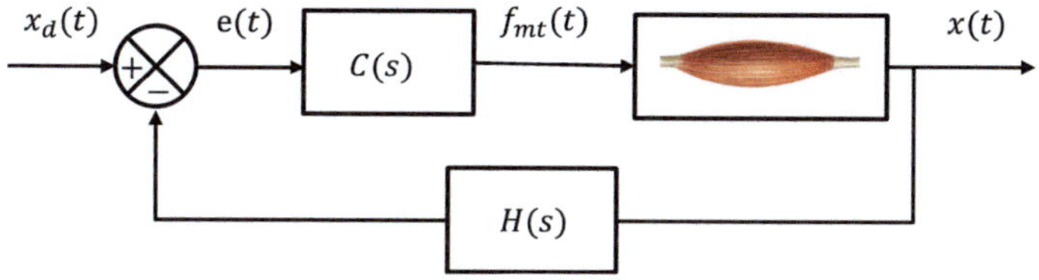

Fig. 6.76 System control for FES

Fig. 6.77 Muscle lineal model

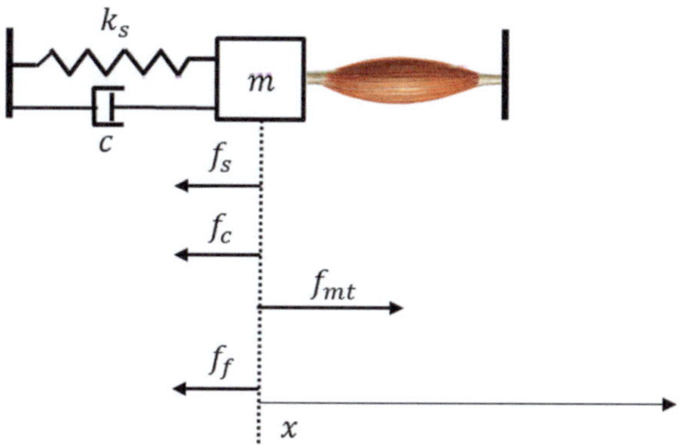

Different forces that appear in the behaviour of the muscle have been considered. An interpretation, according classical mechanics, has been performed.

Resistance to elongation or contraction can be described as a spring with a constant k_s (generating a force f_s), acting in parallel with a damper with damping factor c, which develops a force fc. In addition, we must consider the force of friction of Coulomb, $f_f(t)$, and the force of the muscle–tendon assembly, f_{mt}.

The resulting model can be described with the following equation, obtained from the application of Newton's laws

$$\sum_i f_i = f_{mt} - f_f - f_s - f_c = m\ddot{x}$$

Being:

$$f_f = \mu m g \dot{x}$$

$$f_s = k_s x$$

$$f_c = c\dot{x}$$

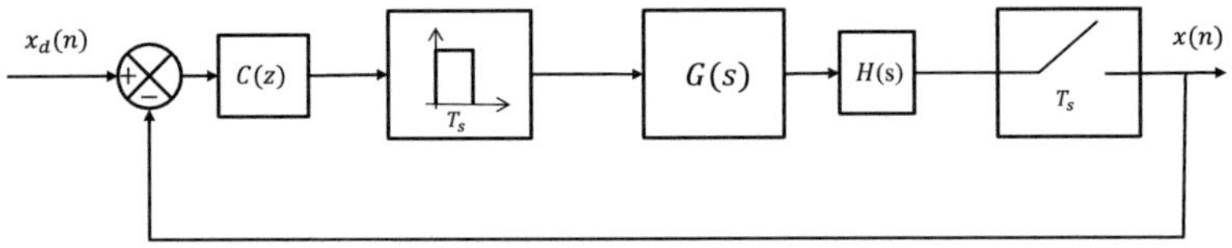

Fig. 6.78 Discrete control system for FES

The system is to be regulated in order to obtain the desired distance. In order to get this goal, a strain gauge is used as sensor, which introduces a delay T in the measurement.

In this scenario, the following questions are considered:

(a) Is it possible to use the Ziegler–Nichols tuning methods for controller design?
(b) Design a PID controller such that the performance to a unit step input would have an overshoot below 15% and a peak time below $2s$. Furthermore, a null steady-state error is required.
(c) Assume that a digital controller is used, implementing the architecture shown in the following Fig. 6.78, where it has been considered a zero-order hold and a sampling time T_s.
Determine:

i. The equivalent discrete transfer function, $G_c(z) = \frac{X(z)}{X_d(z)}$.
ii. The map of poles and zeros of $G(z)$. Is the digitally controlled system stable?
iii. Design the controller that meets the requirements that where specified in the previous section.
iv. Obtain the output sequence if the following the sequence is applied to the input:

$$\{x_d\}_k = \{0\ 0.5\ 1\ 0.5\ 0.1\ 0\ 0\ 0\ 0\ 0\ 0\ 0\ 0\ 0\ 0\ 0\ 0\ 0\ 0\ 0\ 0\}$$

Plot the samples of $\{x\}_k$ and $\{x_d\}_k$ on the same graph.
v. Detail how would be the implementation on a computer of the designed controller, using a difference equation.

Data: $m = 100\,\text{g}$, $k_s = 20\,\frac{\text{N}}{\text{m}}$, $c = 4\,\frac{\text{Ns}}{\text{m}}$, $g = 9.81\,\frac{\text{m}}{\text{s}^2}$, $\mu = 0.1$, $T = 0.7\,\text{s}$ and $T_s = 0.2\,\text{s}$.

Solution 6.2

(a) According to the described model of the muscle, the open-loop transfer function can be obtained according to the following equations, where the derivative property of the Laplace transform has been considered

$$\mathcal{L}\left[\frac{\mathrm{d}f(t)}{\mathrm{d}t}\right] = sF(s) - f(0^-)$$

$$\mathcal{L}\left[\frac{\mathrm{d}^2f(t)}{\mathrm{d}t^2}\right] = s^2F(s) - sf(0^-) - f(0^-)$$

Thus,

$$ms^2X(s) + (c + \mu mg)sX(s) + k_sX = F_{mt}(s)$$

$$G(s) = \frac{X(s)}{F_{mt}(s)} = \frac{1}{ms^2 + (c + \mu mg)s + k_s}$$

Substituting the values, we obtain $G(s)$ as

$$G(s) = \frac{1}{100s^2 + 102.1s + 20}$$

The constituted transfer function represents a second-order system and 0 type.

Due to fact that the open-loop transfer function represents a second-order system, the first method of Ziegler–Nichols could not be applied.

Consequently, the performance of the closed-loop system with a proportional control is analysed. A specific value of the gain that makes the system critically damped is looked for. In order to obtain this value, the root locus of the system is analysed, considering as direct loop $G(s)H(s)$. The following MATLAB code shows how this value is obtained, where the Padé approximation of order 1 has been used:

```
s=tf('s');
mu=0.1;
c=4;
ks=20;
m=100;
g=9.81;
T=0.7;
G=1/(m*s^2+(c+mu*m*g)*s+ks)
H=exp(-T*s);
Hpade=pade(H,1);
rltool(G*Hpade)
```

Figure 6.79 shows the value of the gain that places the poles on the imaginary axis. This value corresponds to $C(s) = K_{\text{cr}} = 170.99$.

The period, P_{cr}, corresponding with this K_{cr} can be obtained from the time response of the feedback system facing a unit step input

$$\texttt{step(feedback(Kcr} * \texttt{G, Hpade), 20)}$$

The result is shown in Fig. 6.80.

Measuring the time period experimentally found, $P_{\text{cr}} = 5.2\,\text{s}$.

Fig. 6.79 Root locus of the system

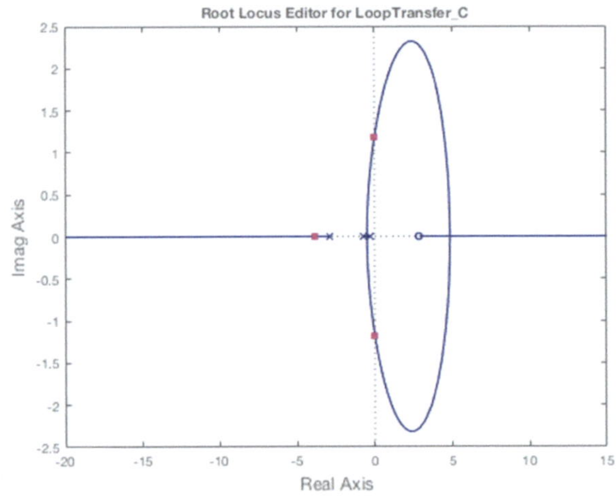

Fig. 6.80 Unit step response of the closed-loop system

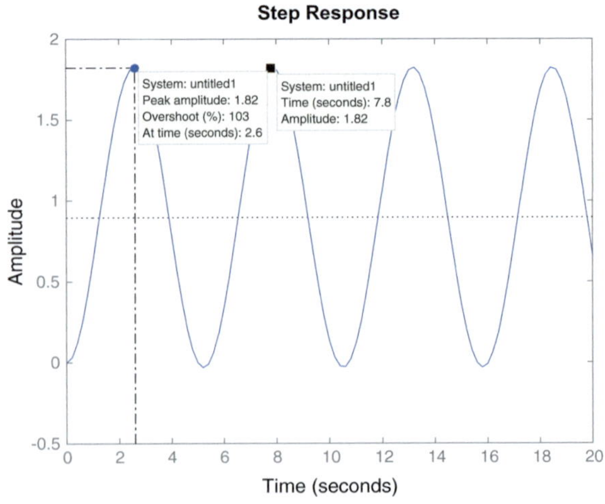

According to the second Ziegler–Nichols method, the PID parameters can be obtained from the following formulas

$$K_p = 0.6 \quad K_{cr} = 102.594$$

$$T_i = 0.5P_{cr} \quad K_i = \frac{K_p}{T_i} = 39.46$$

$$T_d = 0.125P_{cr} \quad K_d = K_pT_d = 66.69$$

This way, the tuned controller is the following

$$C(s) = \frac{66.69s^2 + 102.6s + 39.46}{s}$$

The system response with this controller is shown in Fig. 6.81.

This tuning of the PID, according to the Ziegler–Nichols method, could have been obtained with MATLAB, through its tuning tool (*Tuning Methods* → *Automated Tuning* → *PID Tuning* Fig. 6.82), accessible through graphical interfaces such as `sisotool`, `rltool` or SIMULINK.

(b) The PID tuned in the previous section does not meet the required constraints, so a modification must be made. In addition, the PID design is not implementable because it is non-causal. Therefore, we proceed to perform an optimization based on an algorithm of automatic tuning with derivative filter. As a result, the following controller is offered as

$$C_o = \frac{4373.9(s+0.5723)(s+0.4011)}{s(s+45.84)}$$

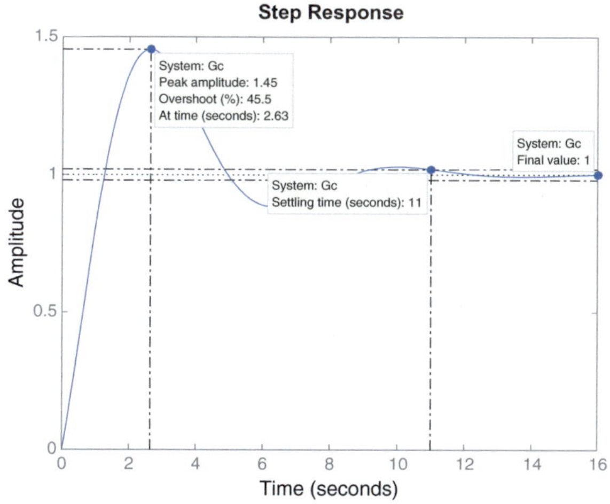

Fig. 6.81 Response of the feedback system with PID controller tuned by Ziegler–Nichols method, when a unit step input is applied

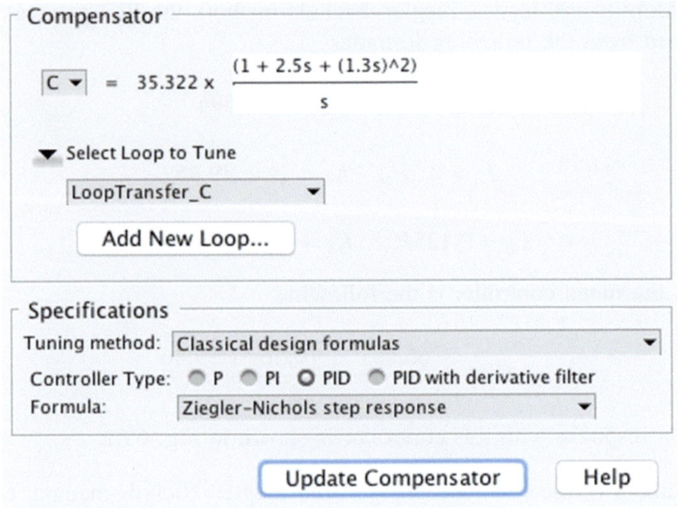

Fig. 6.82 Tuning a PID controller in MATLAB

That meets the requirements. Figure 6.83 shows a comparison of the results obtained with both controllers.

(c) Digitalization of the direct-loop system $(G(s)H(s))$ with a zero-order hold and a T_s sampling time can be performed using the following MATLAB code

```
Ts = 0.2;
Gz = c2d(G * Hpade, Ts)
pzmap(Gz)
```

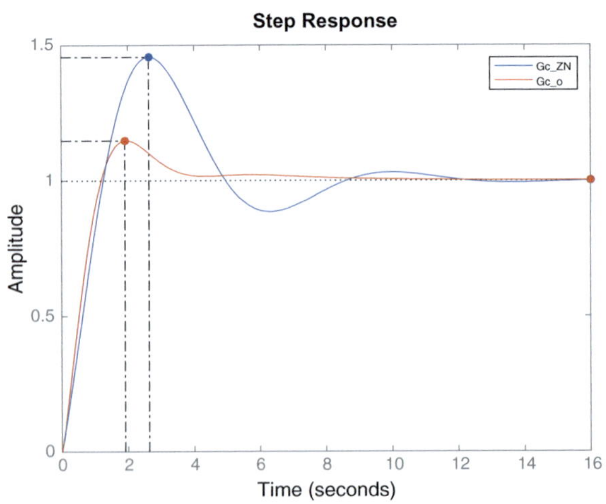

Fig. 6.83 Comparison of the obtained results for unit step input using a tuning by Ziegler–Nichols method and an automatic optimization

Yielding,

$$G(z) = \frac{-0.0001239z^2 + 0.0001398z + 0.0001414}{z^3 - 2.373z^2 + 1.836z - 0.4604}$$

As can be seen in Fig. 6.84, where the map of poles and zeros of the discretized system is shown, all poles are inside the unit circle, so the system is stable.

Since the system has been discretized, it is necessary to use the MATLAB PID controller tuning optimization tool again.

Proceeding in a similar way, i.e., launching sisotool and adjusting the location of the poles according to the desired performance (Fig. 6.85), a discrete PID controller can be obtained by

$$C(z) = \frac{1185.7(z - 0.9672)(z - 0.8933)}{(z - 1)(z + 0.8345)}$$

Using the following MATLAB code:

```
z=tf('z');
Cz=1185.7*(z-0.9672)*(z-0.8933)/((z-1)*(z+0.8345))
Gcz=feedback(C*G,1);
[b,a]=tfdata(Gcz,'v');
xdk=[0 0.5 1 0.5 0.1 0 0 0 0 0 0 0 0 0 0 0 0 0 0];
xk=filter(b,a,xdk)
stem(xdk,'b')
hold on
stem(xk,'r')
```

it is possible to obtained the wanted output (Fig. 6.86).

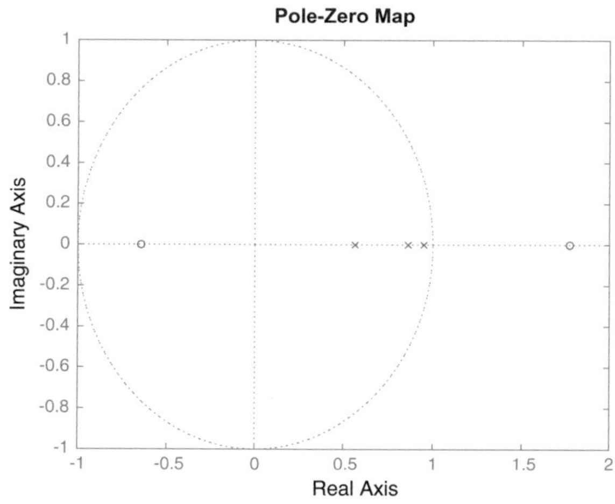

Fig. 6.84 Map of poles and zeros of $G(z)$

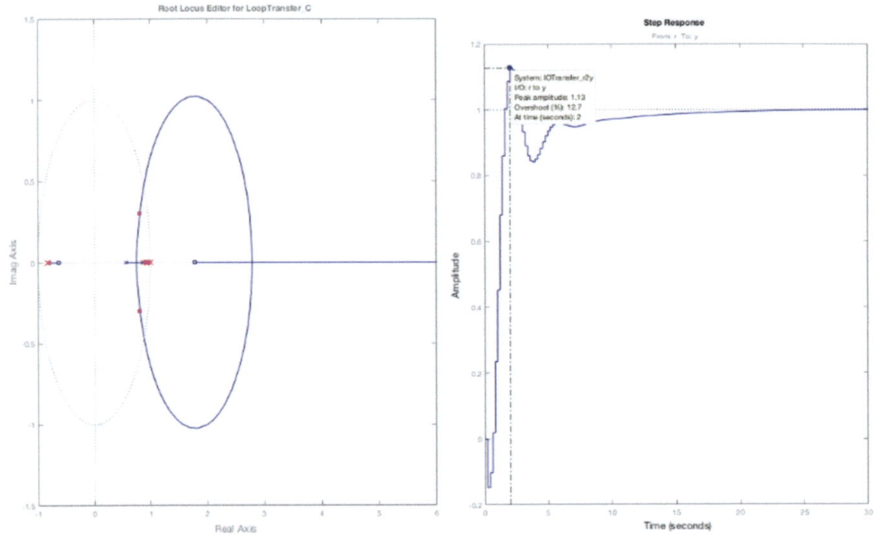

Fig. 6.85 Root locus of the discrete system

Fig. 6.86 Evolution of the signals $\{x\}_k$ and $\{x_d\}_k$

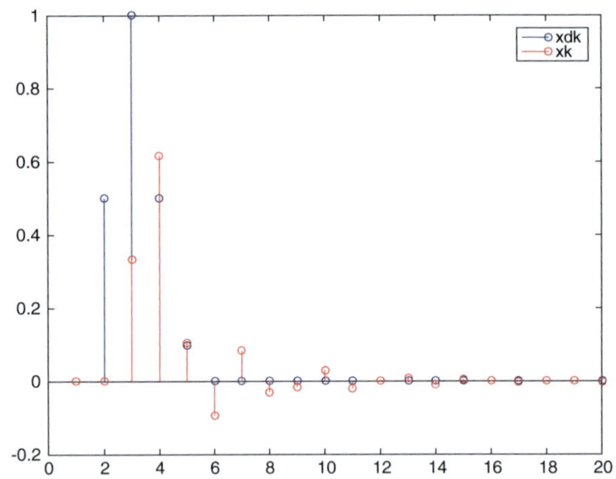

The implementation of the controller, $C(z)$, expressed through the difference equation, can be done starting from its expression in the domain z and performing an anti-transformation. This process is detailed below:

$$C(z) = \frac{F_{mt}(z)}{E(z)} = \frac{1185.7(z - 0.9672)(z - 0.8933)}{(z - 1)(z + 0.8345)}$$

$$\frac{F_{mt}(z)}{E(z)} = \frac{1186z^2 - 2206z + 1024}{z^2 - 0.1655z - 0.8345}$$

$$F_{mt}(z)\left(z^2 - 0.1655z - 0.8345\right) = E(z)\left(1186z^2 - 2206z + 1024\right)$$

$$\begin{aligned} z^2F_{mt}(z) - 0.1655zF_{mt}(z) - 0.8345F_{mt}(z) \\ = 1186z^2E(z) - 2206zE(z) + 1024E(z) \end{aligned}$$

Multiplying by z^{-2}

$$\begin{aligned} F_{mt}(z) - 0.1655z^{-1}F_{mt}(z) - 0.8345z^{-2}F_{mt}(z) \\ = 1186E(z) - 2206z^{-1}E(z) + 1024z^{-2}E(z) \end{aligned}$$

Applying the Z transform property of time shifting

$$\mathcal{Z}[x(n - n_0)] = z^{-n_0}X(z)$$

$$\begin{aligned} f_{mt}(n) - 0.1655f_{mt}(n - 1) - 0.8345f_{mt}(n - 2) \\ = 1186e(n) - 2206e(n - 1) + 1024e(n - 2) \end{aligned}$$

yielding

$$\begin{aligned} f_{mt}(n) = 0.1655f_{mt}(n - 1) + 0.8345f_{mt}(n - 2) \\ + 1186e(n) - 2206e(n - 1) + 1024e(n - 2) \end{aligned}$$

Problem 6.3

A brain–computer interface (BCI), is a communication system between a human brain and an external device. The exogenous BCI systems depend on the electrophysiological activity evoked by external stimuli and do not need an intensive training stage.

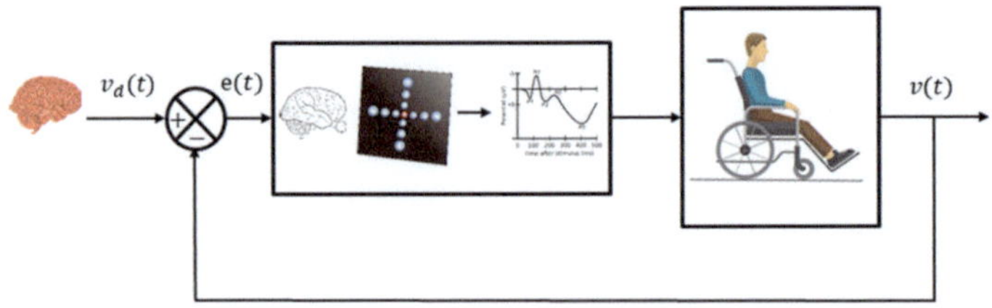

Fig. 6.87 Control system representation for a BCI to command a wheelchair

The P300 potential is signal with a peak amplitude that appears in the electroencephalography (EEG) approximately 300 ms after an unusual auditory or visual stimulus has been produced. Usually, the user is exposed to a series of stimuli. Among these stimuli, there are a few that are related to the user's intention. In this situation, the stimuli of interest, being infrequent and mixed with other much more common stimuli, cause the appearance of a P300 potential in the user's brain activity. This potential is observed mainly in the central and parietal areas of the cerebral cortex and can be registered by a simple sensor that develops a non-invasive measurement. These features could be used to control different devices such as wheelchair as control system for patients with physical paralysis. Figure 6.87 represents this system, where the user wants to move the wheelchair with a desired speed $v_d(t)$, and the system produces a real velocity $v(t)$.

The wheelchair speed is produced by a DC motor that can be modelled with a first-order system. The interpretation of the desired speed is obtained thanks to the registration of the P300 signal. This signal is model as a delay signal: Hence, the open-loop system could be described according the following equation

$$G(s) = \mathrm{e}^{-\theta s} \frac{K}{\tau s + 1}$$

where θ corresponds to the reaction delay to the stimuli, typically 300 ms. K and τ model the first-order model of the DC motor. These parameters could be measured experimentally according a step response. In our case, $K = 0.5$ and $\tau = 0.1$ s.

In order to control the system, a PID controller has to be implemented. Tune in this controller using SIMC and AMIGO methods.

Solution 6.3

Since the plant fits a first-order system, the first procedure of the SIMC method could be applied. According to Eq. (6.15)

$$K_c = \tfrac{1}{K} \cdot \tfrac{\tau}{\tau_c + \theta}$$
$$T_i = \min\{\tau, 4(\tau_c + \theta)\}$$

where τ_c, the unique parameter to be tuned is set to θ.

Using MATLAB, the procedure could be easily implemented:

```
s=tf('s');
theta=0.3;  % P300 signal delay
tau=0.1;
K=0.5;
num=K;
den=[tau 1];
G=tf(num,den,'InputDelay',theta)

% SIMC PI

tau_c=theta;
Kc_SIMC=1/K*tau/(tau_c+theta);
Ti_SIMC=min(tau,4*(tau_c+theta));

PI_SIMC= Kc_SIMC*(Ti_SIMC*s+1)/(Ti_SIMC*s);


step(G)
hold on

Gc=feedback(G,1)
step(Gc)

Gc_SIMC=feedback(PI_SIMC*G,1);
step(Gc_SIMC)
```

Hence, the tuned controller is given by

$$PI_{SIMC} = 0.3333\left(1 + \frac{1}{0.1s}\right)$$

The AMIGO tuning method could be applied considering that in this case $\tau < \theta$. Then, the parameters K_c and T_i can be obtained using the following formulae

$$K_c = \frac{0.1\tau}{K_p\theta} + \frac{0.15}{K_p}$$
$$T_i = 0.3\theta + 0.5\tau$$

Considering $K_p = K$, and using the following MATLAB code, the PI controller could be obtained.

```
% AMIGO
% In this case, tau < theta
Kc_AMIGO = 0.1*tau/(K*theta)+0.15/K;
Ti_AMIGO = 0.3*theta+0.5*tau;

PI_AMIGO= Kc_AMIGO*(Ti_AMIGO*s+1)/(Ti_AMIGO*s);

Gc_AMIGO=feedback(PI_AMIGO*G,1);
step(Gc_AMIGO)
```

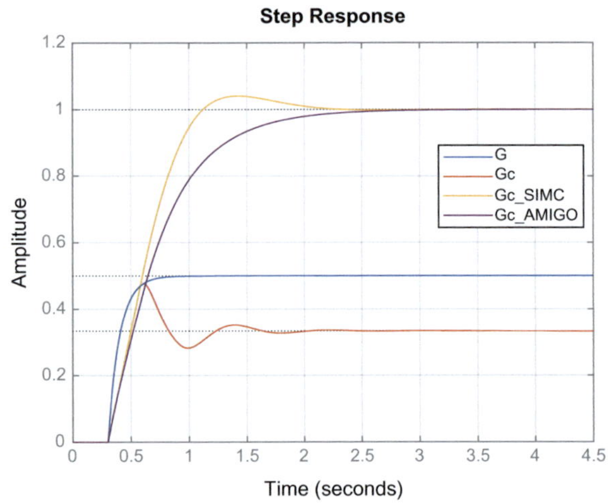

Fig. 6.88 Step response of the controlled system using different tuning methods

As a result we have

$$\mathrm{PI}_{\mathrm{AMIGO}} = 0.3667\left(1 + \frac{1}{0.14s}\right)$$

The comparison according the unitary step response is shown in Fig. 6.88.

As it can be seen, the response of AMIGO is slower than SIMC, which has an overdamped response. Finally, both responses have null steady-state error, due to the effect of the integrator.

MIX
Papier aus verantwortungsvollen Quellen
Paper from responsible sources
FSC® C105338

If you have any concerns about our products,
you can contact us on
ProductSafety@springernature.com

In case Publisher is established outside the EU,
the EU authorized representative is:
Springer Nature Customer Service Center GmbH
Europaplatz 3, 69115 Heidelberg, Germany

Printed by Libri Plureos GmbH
in Hamburg, Germany